Advances on Applications of Bioactive Natural Compounds

Advances on Applications of Bioactive Natural Compounds

Editors

Ana M. L. Seca
Eugenia Gallardo

MDPI • Basel • Beijing • Wuhan • Barcelona • Belgrade • Manchester • Tokyo • Cluj • Tianjin

Editors
Ana M. L. Seca
cE3c-GBA, FCT- Universidade dos Açores
Portugal

Eugenia Gallardo
Universidade da Beira Interior (CICS-UBI)
Portugal

Editorial Office
MDPI
St. Alban-Anlage 66
4052 Basel, Switzerland

This is a reprint of articles from the Special Issue published online in the open access journal *Applied Sciences* (ISSN 2076-3417) (available at: https://www.mdpi.com/journal/applsci/special_issues/Bioactive_Natural_Compounds).

For citation purposes, cite each article independently as indicated on the article page online and as indicated below:

LastName, A.A.; LastName, B.B.; LastName, C.C. Article Title. *Journal Name* **Year**, *Volume Number*, Page Range.

ISBN 978-3-0365-3975-1 (Hbk)
ISBN 978-3-0365-3976-8 (PDF)

Cover image courtesy of Ana M. L. Seca

Contents

About the Editors

Ana M. L. Seca has been an Assistant Professor at the University of Azores since 2000, and she conduts research a the Center for Ecology, Evolution and Environmental Changes—cE3c (ABG) as full member and at LAQV-REQUIMTE as a collaborator. She has a degree in chemistry and an MSc. in the science and technology of paper and forest products, both of which were obtained at the University of Aveiro, Portugal, where she also received her PhD in chemistry in 2000. She has published 58 SCI papers and 8 book chapters. Her current research interests cover the isolation and identification of secondary metabolites with potential pharmacological applications and the synthesis of pharmaceutically relevant natural compound analogues.

Eugenia Gallardo has been an Assistant Professor of Toxicology at the Faculty of Health Sciences of the University of Beira Interior since 2009 and conducts her research at the Health Sciences Research Centre at the University of Beira Interior. She has a degree in pharmaceutical sciences (2000), a Master's of advanced studies in toxicology (2002), and a European PhD in toxicology (2006) that she obtained from the University of Santiago de Compostela (Spain). Since 2011, she has been the Head of the Laboratory of Fármaco-Toxicologia at UBImedical (Portugal). She has published 100 SCI papers and 16 book chapters and has guest-edited 1 book. Her main areas of interest are bioanalysis, toxicology, and drug monitoring, namely the development of analytical methods for the analysis of environmental substances, pesticides, drugs, and drugs of abuse with a special focus on biofluids as well as other bioactive compounds in non-biological specimens (plants and extracts).

Editorial

Secondary Metabolites and Their Applications

Eugenia Gallardo [1,2,*] and Ana M. L. Seca [3,4,*]

1   Centro de Investigação em Ciências da Saúde (CICS-UBI), Universidade da Beira Interior, Av. Infante D. Henrique, 6200-506 Covilhã, Portugal

2   Laboratório de Fármaco-Toxicologia, UBIMedical, Universidade da Beira Interior, Estrada Municipal 506, 6200-284 Covilhã, Portugal

3   cE3c—Centre for Ecology, Evolution and Environmental Changes/Azorean Biodiversity Group & Faculty of Sciences and Technology, University of Azores, Rua Mãe de Deus, 9500-321 Ponta Delgada, Portugal

4   LAQV-REQUIMTE, Department of Chemistry, Campus de Santiago, University of Aveiro, 3810-193 Aveiro, Portugal

*   Correspondence: egallardo@fcsaude.ubi.pt (E.G.); ana.ml.seca@uac.pt (A.M.L.S.); Tel.: +351-275-329-002 (E.G.); +351-296-650-174 (A.M.L.S.)

Citation: Gallardo, E.; Seca, A.M.L. Secondary Metabolites and Their Applications. *Appl. Sci.* **2022**, *12*, 2317. https://doi.org/10.3390/app12052317

Received: 15 February 2022
Accepted: 17 February 2022
Published: 23 February 2022

Publisher's Note: MDPI stays neutral with regard to jurisdictional claims in published maps and institutional affiliations.

1. Introduction

The identification of secondary metabolites present in both terrestrial and marine species continues to be a fundamental and privileged path for the emergence of new and fundamental natural products available on the market with very different applications. For example, aplidine is a new natural anticancer agent, and it was approved in Australia in 2018 to treat multiple myeloma [1] and was isolated from the first time from tunicate *Aplidium albicans* Milne Edwards. (+)-Nootkatone is a natural sesquiterpene compound, found in very small amounts in several species such as *Chamaecyparis nootkatensis* (D. Don) Sudworth, *Chrysopogon zizanioides* (L.) Roberty and *Citrus paradisi* Macfad., which exhibits highly appreciated organoleptic properties and is, therefore, highly demanded as a flavoring agent or adjuvant in the food, pharmaceutical and perfumery industries [2,3]. This same natural secondary metabolite is applied as insecticide and acaricide, and it was very recently authorized by United Sates Environmental Protection Agency to be include on formulations to control the spread of mosquitoes that transmit infectious diseases as dengue and zika [4]. And who does not know the application of ascorbic acid, found for example in acerola and lemon fruits, as an antioxidant agent [5], widely used by the food industry?

The successful application of secondary metabolites in diverse requests often involves the use of, more or less, complex mixtures of metabolites, extracted by different methods and from different natural sources, but whose chemical composition and active principles are perfectly established. For example, there are the defined botanical mixtures, called natural product botanicals, which are recognized by the FDA as medicinal entities, successfully used in clinical therapy for the treatment of various diseases. This is the case of the mixture of the secondary metabolites of *Solanum sodomaeum* L., solamargine, solasonine, and mono- and di-glycosides derivatives of solasodine, approved by the European Medicine Agency (trade name Curaderm) for the treatment of basal cell carcinoma of the skin [6].

One area in which the application of secondary metabolites mixtures rather than of pure compounds is common is in the formulation of biopesticides. For example, the insecticide Grandevo® is a mixture of metabolites produced in the fermentation of *Chromobacterium subtsugae*, which includes pigments from the violacein family and proteins that are repellent and antifeeding [7].

Despite the great successes already achieved regarding the secondary metabolites' identification and the development of new applications for these metabolites, this is an area of research that should not slow down. Research must continue identifying and isolating secondary metabolites in unexplored natural sources; new methodologies for extracting secondary metabolites should be tested and optimized to develop greener and more efficient processes; pure secondary metabolites and chemically characterized mixtures must be tested in different biological activities in order to enhance new applications responding to the growing needs of humanity.

The Special Issue of *Applied Sciences*, "Advances in Applications of Natural Bioactive Compounds", aims to contribute to the desired continuous advance in this scientific field, bringing together publications focused on the most recent advances in the identification of secondary metabolites from terrestrial and marine sources, in new extractive methodologies, and in proposals for applications that add value to natural resources and contributes to a more sustainable development.

2. Contributions

This Special Issue brings together seven research articles and four review articles presenting the latest advances in the study of secondary metabolites from different natural sources and their potential applications. The scientific results are presented that led to the identification of secondary metabolites in samples of terrestrial and marine origin, obtained by classic and innovative extractive methodologies, as well as studies to evaluate their effect in different areas such as health, biopesticides, etc.

Invasive alien species are recognized for triggering various impacts, such as habitat modification, alteration of community structure, and affecting ecosystem processes. The characterization of secondary metabolites in invasive species could be a way of valuing and increasing the search for available biomass promoting its harvest and contribute to the conservation of biodiversity and sustainability of the ecosystem. This is one of the objectives of the studies carried out by Paula et al. [8] and Pinto et al. [9], in which the secondary metabolite composition of two invasive species, one terrestrial (*Acacia dealbata* L.) and one marine (seaweed *Asparagopsis armata* Harvey), respectively, are presented. The results show that the hydroethanolic extract of non-mature flowers of *Acacia dealbata* exhibits the greatest potential as an antioxidant and acetylcholinesterase inhibitor, these activities being correlated with the higher content of chalcones present [8]. These authors suggest the extract as a source of compounds capable of improving brain functions, while collecting the flower at an early stage of flowering contributes to interrupting the plant's reproduction, which will help to control its invasive potential. Pinto et al. [9] address new perspectives for the valuation of *Asparagopsis armata*, an invasive alga in Europe, and have determined the profile of the secondary metabolites of this species using GC-MS and UHPLC-MS techniques. The authors established the species as a source of halogenated compounds and fatty acids, with 1,4-dibromobuten-1-ol and palmitic acid being the two most abundant compounds identified in the lipophilic fraction, while in the polar fraction, brominated phenolic compounds derived from cinnamic acid are predominant. This work also shows that changing the extraction method modifies the metabolomic profile of the obtained extract. The use of ethanol under microwave energy extracted almost exclusively one compound, the 2,3-diphenylpropanoic acid derivative with one hydroxyl group and two bromine atoms in each aromatic ring [9].

The Martí-Quijal group [10] also uses samples of marine origin, in this case the cyanobacterium *Arthrospira platensis* Gomont, and optimizes the extraction with pulsed electric fields to propose a more selective, sustainable, and efficient methodology to extract chlorophylls, carotenoids, and phenolic compounds, which have applications as nutritional and antioxidants agents.

In recent years, natural phytotoxic compounds derived from plants have been extensively investigated for herbicidal properties and have been shown to suppress weeds. These compounds can then be used as promising models for standard bioherbicides. In this sense, Rob et al. [11] have demonstrated in their manuscript for the first time the presence of methyl phloretate in *Garcinia xanthochymus* Hook.f. ex T.Anderson, as well as its phytotoxic properties. Vanillic acid was also identified. Both compounds have significant growth inhibitory activities in watercress and timothy, demonstrating as such the potential of the leaves of *Garcinia xanthochymus* in weed management.

The leaves of *Afzelia xylocarpa* (Kurz) Craib. were investigated as a natural source of herbicides. Krumsri et al. [12] have identified vanillic acid and *trans* ferulic acid as the major bioactive compounds, both being responsible for the phytotoxic activity of *Afzelia xylocarpa*. The concentration of those compounds and differences in the sensitivity of the plant models are key to their phytotoxic action. Additionally, a mixture of both compounds at low concentrations had synergistic effects concerning phytotoxic activity.

Shedding light on the mechanisms of action that justify the observed secondary metabolite bioeffects is an essential point where more research is needed before it can advance towards clinical therapy. The secondary metabolite thymoquinone can be extracted from the seed oil of *Nigella sativa* L. (black seed oil) and exhibits several biological activities [13], among which are relevant antitumor properties, and has already been the subject of clinical studies (ClinicalTrials.gov Identifier: NCT03208790). However, its mechanism of action is not yet fully understood. Alhosin et al. [14] make a very significant contribution to clarifying this point. They show that thymoquinone, unlike doxorubicin, specifically targets the oncogene UHRF1. Probably through the downregulation of HAUSP, thymoquinone induces the rapid ubiquitination of UHRF1, an essential event for the induction of cancer cell apoptosis without, however, affecting the Bcl-2 oncogene [14].

The total synthesis of natural secondary metabolites contributes not only to confirm the proposed chemical structure of the natural compound but also to satisfy the demand for more compounds for different applications without degrading the environment. Thus, the work by Kim et al. [15] is a great contribution to the development of new applications for the secondary metabolite Aaptoline B, a pyrroloquinoline alkaloid extracted from the marine sponge *Aaptos aaptos* Schmidt and considered a privileged scaffold in drug discovery. The authors suggest a short but comprehensive and efficient synthetic process for Aaptolin B and demonstrate, using *Caenorhabditis elegans* as a model of the Parkinson's disease-like neuronal environment, that this compound exhibits a neuroprotective effect [15]. Thus, the door is open for investigation into a new application for this secondary metabolite.

Laboratory synthesis is not the only strategy applied to obtain new and greater amounts of bioactive secondary metabolites. *Streptomyces* is a group of Gram-positive bacteria capable of biosynthesizing a wide range of natural compounds from different families, with different commercial applications and high economic value [16]. Doxorubicin, an antitumor antibiotic, and avermectin, used to treat parasitic worms and insect pests, are just two well-known examples. However, the potential of these microorganisms to produce secondary metabolites is much greater than achieved so far. The problem is, under laboratory conditions, many of the genes responsible for this biosynthesis remain silent. Heterologous biosynthesis has been an applied approach to overcome this limitation. Pham et al. [17] publish in this Special Issue a review of the literature on the latest advances in heterologous biosynthesis to produce natural compounds from *Streptomyces*, which include the overexpression/deletion of regulatory genes, promoter replacement strategies, and engineering of ribosomes.

On the other hand, many natural secondary metabolites, semi-synthesized or synthesized in the laboratory or by a biotechnological approach, exhibit excellent pharmacodynamics but very poor pharmacokinetic properties. The development of nano delivery systems has been a successful way to overcome the pharmacokinetic limitations of many natural compounds, allowing them to broaden their medicinal application [18]. Sindhu et al. [19] present in this Special Issue a review of the literature on the development of nano formulations involving natural compounds with application in the treatment of cardiovascular diseases, highlighting those with the most significant results for the prevention and treatment of this type of disease, that is, cases involved in clinical studies.

As referred to before, invasive species are considered one of the most significant global threats to biodiversity. However, these species can also be quite helpful to humans, providing a complex number of services, which makes it difficult to assess their positive and negative effects. Caramelo et al. [20] review the knowledge of the chemical composition and the discovery of the biological properties of one such invasive alien species—*Ailanthus altissima* (Mill.) Swingle—and its potential use as medicine and additive by the pharmacy and food industries. Methods for the extraction and detection to know the chemical composition of extracts have been discussed in detail by the authors.

Finally, Silva-Beltrán et al. [21] evaluate the chemical profiles of propolis produced in Brazil and Mexico. Both countries' geographic and ecosystem differences provide propolis with different chemical profiles. However, data in both countries indicate that more research is still needed to determine the optimal characteristics of propolis and its components, such as period of intake or type of extract to use before it can be administered to humans as medicine The authors have also review the use of propolis as environmental bioindicator, concluding that more studies are necessary.

In the guest editors' opinion, this Special Issue of *Applied Sciences* brings together great contributes to the desired continuous advance in secondary metabolites and the development of new applications for these metabolites. In fact, the published papers are focused on the most recent advances in the identification of secondary metabolites from terrestrial and marine sources obtained by classical and greener extractive methodologies. They are also focused on proposals for new applications that enhance the natural resources and their secondary metabolites, contributing to more sustainable development.

Author Contributions: A.M.L.S. and E.G. conceived, designed, and wrote the editorial. All authors have read and agreed to the published version of the manuscript.

Funding: This study was financed by Portuguese National Funds, through FCT—Fundação para a Ciência e Tecnologia, the European Union, QREN, FEDER, and COMPETE, through funding the cE3c center (UIDB/00329/2020), the LAQV-REQUIMTE (UIDB/50006/2020) and the CICS-UBI (UIDB/00709/2020 and UIDP/00709/2020).

Institutional Review Board Statement: Not applicable.

Informed Consent Statement: Not applicable.

Data Availability Statement: Not applicable.

Acknowledgments: Thanks are due to all the authors and peer reviewers for their valuable contributions to this Special Issue.

Conflicts of Interest: The authors declare no conflict of interest.

References

1. Newman, D.J.; Cragg, G.M. Natural Products as Sources of New Drugs over the Nearly Four Decades from 01/1981 to 09/2019. *J. Nat. Prod.* **2020**, *83*, 770–803. [CrossRef] [PubMed]
2. Fraatz, M.A.; Berger, R.G.; Zorn, H. Nootkatone—A biotechnological challenge. *Appl. Microbiol. Biotechnol.* **2009**, *83*, 35–41. [CrossRef] [PubMed]
3. Fan, J.; Liu, Z.; Xu, S.; Yan, X.; Cheng, W.; Yang, R.; Guo, Y. Non-food bioactive product (+)-nootkatone: Chemistry and biological activities. *Ind. Crops Prod.* **2022**, *177*, 114490. [CrossRef]
4. United States Environmental Protection Agency. Available online: https://www3.epa.gov/pesticides/chem_search/ppls/0918 73-00001-20200807.pdf (accessed on 8 February 2022).
5. Ribeiro, J.S.; Santos, M.J.M.C.; Silva, L.K.R.; Pereira, L.C.L.; Santos, I.A.; Lannes, S.C.S.; Silva, M.V. Natural antioxidants used in meat products: A brief review. *Meat Sci.* **2019**, *148*, 181–188. [CrossRef] [PubMed]
6. Chase, T.; Cham, K.E.; Cham, B.E. Curaderm, the Long-Awaited Breakthrough for Basal Cell Carcinoma. *Int. J. Clin. Med.* **2020**, *11*, 579–604. [CrossRef]
7. Marrone, P.G. Pesticidal natural products–status and future potential. *Pest Manag. Sci.* **2019**, *75*, 2325–2340. [CrossRef] [PubMed]
8. Paula, V.; Pedro, S.I.; Campos, M.G.; Delgado, T.; Estevinho, L.M.; Anjos, O. Special Bioactivities of Phenolics from *Acacia dealbata* L. with Potential for Dementia, Diabetes and Antimicrobial Treatments. *Appl. Sci.* **2022**, *12*, 1022. [CrossRef]
9. Pinto, D.C.G.A.; Lesenfants, M.L.; Rosa, G.P.; Barreto, M.C.; Silva, A.M.S.; Seca, A.M.L. GC- and UHPLC-MS Profiles as a Tool to Valorize the Red Alga *Asparagopsis armata*. *Appl. Sci.* **2022**, *12*, 892. [CrossRef]
10. Martí-Quijal, F.J.; Ramon-Mascarell, F.; Pallarés, N.; Ferrer, E.; Berrada, H.; Phimolsiripol, Y.; Barba, F.J. Extraction of Antioxidant Compounds and Pigments from Spirulina (*Arthrospira platensis*) Assisted by Pulsed Electric Fields and the Binary Mixture of Organic Solvents and Water. *Appl. Sci.* **2021**, *11*, 7629. [CrossRef]
11. Rob, M.M.; Hossen, K.; Khatun, M.R.; Iwasaki, K.; Iwasaki, A.; Suenaga, K.; Kato-Noguchi, H. Identification and Application of Bioactive Compounds from *Garcinia xanthochymus* Hook. for Weed Management. *Appl. Sci.* **2021**, *11*, 2264. [CrossRef]

12. Krumsri, R.; Ozaki, K.; Teruya, T.; Kato-Noguchi, H. Isolation and Identification of Two Potent Phytotoxic Substances from *Afzelia xylocarpa* for Controlling Weeds. *Appl. Sci.* **2021**, *11*, 3542. [CrossRef]
13. Malik, S.; Singh, A.; Negi, P.; Kapoor, V.K. Thymoquinone: A small molecule from nature with high therapeutic potential. *Drug Discov. Today* **2021**, *26*, 2716–2725. [CrossRef]
14. Alhosin, M.; Abdullah, O.; Kayali, A.; Omran, Z. A Fast Ubiquitination of UHRF1 Oncogene Is a Unique Feature and a Common Mechanism of Thymoquinone in Cancer Cells. *Appl. Sci.* **2021**, *11*, 7633. [CrossRef]
15. Kim, S.; Yang, W.; Han, Y.-T.; Cha, D.-S. Synthesis of Proposed Structure of Aaptoline B via Transition Metal-Catalyzed Cycloisomerization and Evaluation of Its Neuroprotective Properties in *C. elegans*. *Appl. Sci.* **2021**, *11*, 9125. [CrossRef]
16. Khushboo, K.P.; Dubey, K.K.; Usmani, Z.; Sharma, M.; Gupta, V.K. Biotechnological and industrial applications of *Streptomyces* metabolites. *Biofuels Bioprod. Bioref.* **2022**, *16*, 244–264. [CrossRef]
17. Pham, V.T.T.; Nguyen, C.T.; Dhakal, D.; Nguyen, H.T.; Kim, T.-S.; Sohng, J.K. Recent Advances in the Heterologous Biosynthesis of Natural Products from *Streptomyces*. *Appl. Sci.* **2021**, *11*, 1851. [CrossRef]
18. Saw, P.E.; Lee, S.; Jon, S. Naturally Occurring Bioactive Compound-Derived Nanoparticles for Biomedical Applications. *Adv. Ther.* **2019**, *2*, 1800146. [CrossRef]
19. Sindhu, R.K.; Goyal, A.; Algın Yapar, E.; Cavalu, S. Bioactive Compounds and Nanodelivery Perspectives for Treatment of Cardiovascular Diseases. *Appl. Sci.* **2021**, *11*, 11031. [CrossRef]
20. Caramelo, D.; Pedro, S.I.; Marques, H.; Simão, A.Y.; Rosado, T.; Barroca, C.; Gominho, J.; Anjos, O.; Gallardo, E. Insights into the Bioactivities and Chemical Analysis of *Ailanthus altissima* (Mill.) Swingle. *Appl. Sci.* **2021**, *11*, 11331. [CrossRef]
21. Silva-Beltrán, N.P.; Umsza-Guez, M.A.; Ramos Rodrigues, D.M.; Gálvez-Ruiz, J.C.; de Paula Castro, T.L.; Balderrama-Carmona, A.P. Comparison of the Biological Potential and Chemical Composition of Brazilian and Mexican Propolis. *Appl. Sci.* **2021**, *11*, 11417. [CrossRef]

Article

Special Bioactivities of Phenolics from *Acacia dealbata* L. with Potential for Dementia, Diabetes and Antimicrobial Treatments

Vanessa Paula [1,†], Soraia I. Pedro [2,3,4,†], Maria G. Campos [5,6,*], Teresa Delgado [2], Letícia M. Estevinho [1] and Ofélia Anjos [2,4,7,*]

1. Centro de Investigação de Montanha, Instituto Politécnico de Bragança, Campus de Santa Apolónia, 5300-253 Bragança, Portugal; vanessapaula@ipb.pt (V.P.); leticia@ipb.pt (L.M.E.)
2. Centro de Biotecnologia de Plantas da Beira Interior, 6001-909 Castelo Branco, Portugal; soraia_p1@hotmail.com (S.I.P.); teresadelgado86@hotmail.com (T.D.)
3. Centro de Investigação em Ciências da Saúde (CICS-UBI), Universidade da Beira Interior, Av. Infante D. Henrique, 6200-506 Covilhã, Portugal
4. Centro de Estudos Florestais, Instituto Superior de Agronomia, Universidade de Lisboa, Tapada da Ajuda, 1349-017 Lisboa, Portugal
5. Observatory of Drug-Herb Interactions, Faculty of Pharmacy, Health Sciences Campus, University of Coimbra, Azinhaga de Santa Comba, 3000-548 Coimbra, Portugal
6. CQ-Centre of Chemistry—Coimbra, Department of Chemistry, Faculty of Sciences and Technology, University of Coimbra, Rua Larga, 3004-535 Coimbra, Portugal
7. Instituto Politécnico de Castelo Branco, 6001-909 Castelo Branco, Portugal
* Correspondence: mgcampos@ff.uc.pt (M.G.C.); ofelia@ipcb.pt (O.A.)
† These authors contributed equally to this work.

Citation: Paula, V.; Pedro, S.I.; Campos, M.G.; Delgado, T.; Estevinho, L.M.; Anjos, O. Special Bioactivities of Phenolics from *Acacia dealbata* L. with Potential for Dementia, Diabetes and Antimicrobial Treatments. *Appl. Sci.* 2022, 12, 1022. https://doi.org/10.3390/app12031022

Academic Editor: Ana M. L. Seca

Received: 21 December 2021
Accepted: 14 January 2022
Published: 19 January 2022

Publisher's Note: MDPI stays neutral with regard to jurisdictional claims in published maps and institutional affiliations.

Abstract: Some diseases still need better therapeutic approaches, including the prevention of development. Natural resources are investigated with this purpose; among them, we decided to use an invasive plant as a main strategy. This will help in two ways: screening new compounds in flowers prevents the plant from causing widespread damage by controlling the dissemination and also obtains crude material for further applications. In the present study, flower extracts from *Acacia dealbata* Link harvested in Portugal were studied during three stages of flowering. Phenolic compounds were evaluated using HPLC/DAD and the total phenolics as the total flavonoids content was determined. The bioactivities screened were antioxidant potential, inhibitory activities of some enzymes (acetylcholinesterase, lipase and α-glucosidase) and, to complete the screening, the inhibition of microbial growth was determined against Gram-negative and Gram-positive bacteria, as well as for yeasts. The data obtained suggested that the hydroethanolic extracts gave good results for all these biological activities and varied according to the maturation status of the flowers, with the early stage being the most active, which can be related to the chalcones content. This new approach will lead to the possible control of the invasive plant and also future perspective research for therapeutic purposes.

Keywords: invasive species; flowers; antioxidant; microbial activity; bioactivity; enzyme inhibitory potential; acetylcholinesterase; lipase; α-glucosidase

1. Introduction

In drug discovery, screening for potential bio-active structures is often started with natural product extracts. The present work reflects the preliminary study carried out with extracts of an invasive plant, *Acacia dealbata* L., which may be used in a win–win relationship, that is, as a raw material to supply bioactive molecules, while the harvest itself will contribute to controlling its invasiveness by decreasing the proliferation of this plant.

Among the possible bioactivities to explore that are believed to be linked to the compounds expected to be in the Acacia extracts we chose, the acetylcholinesterase inhibitors (AChEIs) can be explored further for the treatment of dementia symptoms [1] or even to delay the disease. The current extracts used to slow the progression of many of these pathologies still have modest outcomes.

Acacia species are aggressive invaders that affect ecosystem integrity worldwide [2]. *Acacia dealbata* Link is considered to be one of the most aggressive in Portugal. These plants are distributed throughout all Portuguese provinces [3–5] and may invade farmland and autochthonous forests, establishing monocultures, modifying the ecosystem structure and impacting the economy. Their flowers play a key role in the colonizing capacity, limiting the growth of other species due to the allopathic interference caused by vegetal material decomposition in the soil, which maintains the toxicity of soil percolates for a long time [6]. Extracts from the flowers are already explored as hydrogels for personal care products, cosmetics or pharmaceuticals, as well as perfumes based on their antiradical and anti-proliferative potential [7].

Along with this research, other possibilities were explored based on previous studies on ethanol extracts of leaves and flowers of Acacia species that suggested the presence of secondary metabolites, such as coumarins, cyanogenic glycosides, tannins, alkaloids and steroids [8]. Notwithstanding, this genus is recognized as a source of phenolic compounds [9–11], with flavonoids being the predominant chemical group [12]. In the study performed by Farid et al. [13] on the leaves, flowers and pods of *A. seyal*, *A. nilotica* and *A. laeta*, the antioxidant activity was promising. Furthermore, according to Ziani et al. [14], *A. tortilis* extracts exhibited various biological activities and their phenolic compounds were studied for the development of cytotoxic and anti-inflammatory drugs.

Antimicrobial activity was also identified in several Acacia species against *Escherichia coli*, *Staphylococcus aureus* and *Salmonella typhi* bacteria and the fungal strain *Candida albicans* and *Aspergillus niger* [15–17].

According to the aforementioned studies, the bark and pods contain secondary metabolites, such as flavonoids, alkaloids, saponins and tannins, which may contribute to the bioactivities. Among the most relevant potential applications in health, the inhibitory activities of the enzymes α-glucosidase, acetylcholinesterase and lipase will have an important impact on the development of new drugs. For instance, lipases, which are present in pancreatic secretions and are responsible for fat digestion, are able to break down triglycerides into free fatty acids and glycerol. Acetylcholinesterase inhibitors are relevant for Alzheimer's disease treatment strategies.

For the possible applications cited above, especially regarding the antioxidant potential, the need to screen new molecules for the development of, for example, new drugs for dementia-related diseases, and the need to use a crude material that will leave a low environmental impact, are all good reasons to explore *A. dealbata* flowers. This reflection leads to the goal of the present work, which consisted of obtaining more robust data for possible future research using selected assays to evaluate the inhibitory activities of the enzymes acetylcholinesterase, α-glucosidase and lipase. To complete the screening, the inhibition of microbial growth was determined against Gram-negative and Gram-positive bacteria, as well as for the yeasts *Candida albicans* ATCC 10231TM and amphotericin-B-resistant *Candida albicans* ESA100.

In addition, phenolic compounds and flavonoids were screened using HPLC/DAD to characterize the extracts.

2. Materials and Methods

2.1. Plant Samples

The flowers of *A. dealbata* Link used in this study were harvested during three flowering stages (early flower—EF, mid flower—MF and late flower/ripening—LF), as shown in Figure 1, in 2018. After each harvest, the flowers were manually separated from the thick stems and leaves, and then lyophilized in dark conditions and milled. The fraction lower than 60 mesh was frozen at −20 °C until the extraction procedures. To study the characteristics and biological properties of the flowers, samples from two locations (A and

B) in the central region of Portugal were analyzed and compared. These two regions were selected because there are large stands of this invasive species.

Figure 1. Different stages of *A. dealbata* flower used in this study (EF—early flower, MF—mid flower and LF—late flower/ripening). Photo from Ofélia Anjos.

2.2. Extraction Conditions

The extraction procedure was performed according to Paula et al. [18]. Briefly, 10 g of flower samples were extracted with 100 mL of solvent with 50/50 ethanol/water following 15 h of stirring at 120 rpm at room temperature. Afterward, the flower samples were centrifuged at 5000 rpm for 10 min and the supernatant was collected and stored at $-20\ ^\circ$C until further analysis. MilliQ water was used for the extraction and 96% ethanol was purchased from Merck. Therefore, in this study, the extracts were prepared via mixing. All extractions were performed in duplicate and all subsequent measurements and analyses were performed in triplicate. The final concentration of the extract was 100 mg/mL, which involved a calculation based on the differences between the dried weight before and after the extraction of plant material. Then, all the dilutions were corrected so that these presented the same concentration, namely, 100 mg/mL.

2.3. HPLC/DAD Conditions

Chromatographic analyses were carried out using HPLC/DAD [19]. Briefly, 50 μL of each extract prepared were submitted to analysis in a Gilson 170 and Waters Spherisorb ODS2 (5 mm) column (4.6 × 250 mm) using water acidified with *o*-phosphoric acid to achieve pH = 2.4 (A) and an acetonitrile (B) gradient with a 0.8 mL/min flow rate and a column temperature of 24 $^\circ$C. All samples were analyzed in triplicate. Standard chromatograms were plotted at λ_{max} values of 260 and 340 nm. Spectral data for all peaks were accumulated in the range 220–500 nm using DAD (Gilson 170; World Headquarters, Middleton, WI, USA), and the software Unipoint was used to analyze each of them. The suitability of the method was previously evaluated by Campos [19]. All the extracts were submitted to a results analysis of the main phenolic constituents. Structural determination of the phenolic compounds was performed according to the theoretical rules presented in Campos and Markham [20].

2.4. Total Phenol (TPC) and Flavonoid (TFC) Contents

The total phenolic compounds were determined using the Folin–Ciocalteu method [21]. A volume of 0.5 mL of the sample or standard was diluted with 2.5 mL of Folin–Ciocalteu (10% *v/v*) reagent, and afterward, 2 mL of sodium carbonate (Na_2CO_3, 75 g/L) was added. The absorbance was measured at a wavelength of 760 nm after incubation for 2 h in the dark. The calibration curve was prepared using standard solutions of gallic acid (AG) with concentrations between 50 and 350 mg/L.

The flavonoid content was estimated according to Woisky et al. [22]. Different extract concentrations (2.5 mL) were diluted with 2.5 mL of ethanol solution of $AlCl_3$ (2%). After 1 h of incubation in the dark at room temperature, the absorbance of the mixture was read at 420 nm.

2.5. Antioxidant Activity

The original ferric reducing antioxidant power (FRAP) assay (without incubation) was monitored according to a method reported by Berker et al. [23]. A sample extract concentration (0.1 mL) was diluted with 0.3 mL distilled water and 3 mL FRAP reagent. The mixture was shaken and the absorbance was read at 595 nm after 6 min in the dark.

Previous to this assay, a DPPH screening was performed following the methodology described in Alencar et al. [24].

2.6. Acetylcholinesterase Inhibition Assay

The acetylcholinesterase inhibition was assessed spectrophotometrically as described by Sukumaran et al. [25]. The acetylcholinesterase activity was quantified using 5,5′-dithiobis-(2-nitrobenzoic acid) (15 mM, DTNB, Sigma Chemical, St. Louis, MO, USA); *Electrophorus electricus* acetylcholinesterase (electric eel Type-VI-S, Sigma Chemical, St. Louis, MO, USA) was used as the substrate. The hydrolysis of acetylcholine iodide was monitored through the formation of the anion 2-nitro-5-thiobenzoate (yellow) at 412 nm for 15 min. The percentage of inhibition was determined by comparing the rates of reaction of the samples with those of the blank (ethanol in phosphate buffer 0.2 M, pH = 8) using the expression $((E - S)/E) \times 100$, where E is the enzymatic activity without the extract and S is the activity of the sample. IC50 values were estimated from the graphic of inhibition percentage versus sample concentration. Eserine was used as a control.

2.7. Lipase Inhibitory effect

The inhibition of pancreatic lipase was assessed as described by Habib et al. [26]. In brief, 25 µL of the sample or the lipase inhibitor (Orlistat, Sigma Chemical, St. Louis, MO, USA) were mixed with 25 µL of lipase solution (16.7 U/mL in Tris-HCl, pH 8.0). For a 30 min incubation at 25 °C, the 4-methylumbelliferone product was released and the reaction was stopped by adding sodium citrate (0.1 mL of 0.1 M, pH 4.2). The measurements were performed with a fluorescence microplate reader (at 460 nm emission and 355 nm excitation). The results are expressed as percentages of α-lipase inhibition.

2.8. Inhibition Assay for α-Glucosidase Activity

The α-glucosidase activity was measured as reported by Tadera et al. [27]. The α-glucosidase-mediated transformation of the substrate pNPG into α-D-glucose and p-nitrophenol at 405 nm was monitored. In brief, the enzyme (0.05 U/mL, Sigma Chemical, St. Louis, MO, USA) was dissolved in 100 mM phosphate buffer (pH 6.8) in a 96-well plate. It was pre-incubated at 37 °C for 5 min with the samples (10–60 µg/mL); p-nitrophenyl-α-D-glucopyranoside (600 µM, Sigma Chemical, St. Louis, MO, USA) was then added and the mixture was incubated at 37 °C for 30 min. The enzymatic reaction was monitored spectrophotometrically by measuring the absorbance at 405 nm. The obtained values corresponded to the slope measured between 5 and 20 min. Acarbose (0–200 µM, Sigma Chemical, St. Louis, MO, USA) was used as a positive control.

2.9. Antimicrobial Potential

The in vitro antimicrobial activity was evaluated through a broth microdilution assay in 96 multi-well microtiter plates, as recommended by the Clinical and Laboratory Standards Institute (CLSI) Guidelines. The minimum inhibitory concentration (MIC) of the samples, defined as the lowest concentration of extract necessary to inhibit microbial growth, was determined against Gram-negative bacteria (*Escherichia coli* ATCC 29998[TM] and *Escherichia coli* ESA37 cephalosporins-resistant), Gram-positive bacteria (*Staphylococcus aureus* ATCC 43300TM and *Staphylococcus aureus* ATCC ESA111) and the yeasts *Candida albicans* ATCC 10231[TM] and amphotericin B-resistant *Candida albicans* ESA100. Gentamicin and amphotericin B were used as controls. The cell mass was diluted with 0.85% NaCl solution to reach 0.5 on the "MacFarland scale", which was checked at 580 nm for bacteria and 640 nm for yeast. Muller–Hinton broth was used for bacteria or Yeasts Peptone Dextrose broth for yeast. Samples were diluted with broth to reach a concentration ranging from 5 to 30 mg/mL. The cell mass (10^6 colony-forming units (CFU)/mL) was added to all wells and incubated at 37 °C for 24 h and 25 °C for 48 h for bacteria and yeast, respectively. The antimicrobial activity was then determined by adding 20 µL of 2,3,5-triphenyl-2H-tetrazoliumchloride (TTC, 5 µg/mL) solution. In each assay, a negative control (containing only broth), a positive control (inoculated broth) and a DMSO control (DMSO with inoculated broth) were analyzed. The results were expressed in micrograms per milliliter.

2.10. FT-Raman Spectral Acquisition

The spectra of the plant extracts were obtained using the methodology proposed by [28] using an FT-Raman spectrometer (BRUKER, MultiRAM, Bruker Portugal Unipessoal, Lisbon, Portugal), equipped with a 180° high-throughput collecting lens, an ultra-high sensitivity liquid nitrogen cooled Ge Diode detector and an integrated 1064 nm diode-pumped Nd:YAG laser with a maximum output power of 500 mW. The spectra were collected with 50 scans per spectrum at a spectral resolution of 8 cm^{-1} using a scanner velocity of 5 kHz in the wavenumber range from 4000 to 70 cm^{-1}. The measurements were performed in duplicate in a 5 mm optic space quartz cell with the opposite face mirrored. The spectra were collected at constant room temperature (22 °C).

2.11. Statistical Analysis

The analysis of variance (ANOVA) was performed to identify the significance of the results and to determine the percentage of the variance of each factor using STATISTICA 7 (StatSoft Inc., Tulsa, OK, USA) software. Whenever the *p*-value was below 0.05, Tukey's test was applied. Mean values and standard deviations were calculated and are presented in the figures as error bars.

Principal component analysis (PCA) was performed to identify the relationship between the factor (site and flower maturation) and the different variables measured.

For ANOVA and PCA, the StatSoft software was used; for spectral data analyses OPUS OPUS®, version: 7.5.18 (Bruker Optik, Ettlingen, Germany) and UnscramblerX 10.5 (CAMO, Oslo, Norway) were applied.

3. Results and Discussion

3.1. Characterization of the Extract

3.1.1. HPLC/DAD

From the HPLC/DAD chromatograms, six compounds were identified (Figure 2) and the corresponding ultraviolet absorption spectra of the most concentrated are provided in Figure 3. The minor compounds were related to the nucleus of flavonols (compounds **1–4**) and the major were related to chalcones (compounds **5** and **6**).

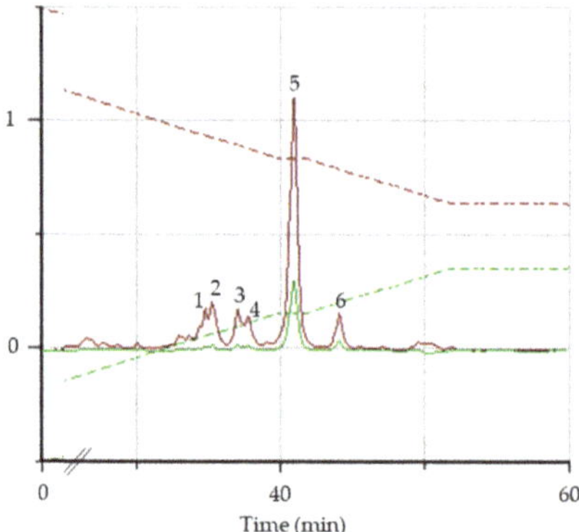

Figure 2. Example of an HPLC/DAD profile of the early maturation stage flowers from region A obtained with the most active *Acacia dealbata* L. hydroethanolic extract (50% (*v/v*)). Compounds **1–4** were flavonols and **5–6** were chalcones.

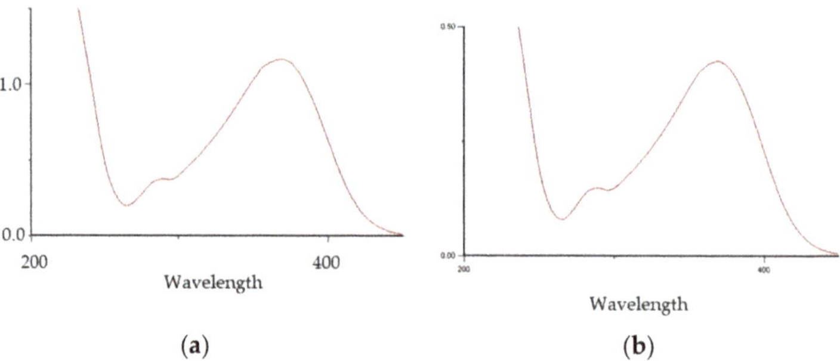

Figure 3. Ultraviolet absorption spectra obtained using HPLC/DAD analysis of the flavonoid content. Compounds (**a,b**) were chalcone derivatives.

These latter components also belong to the flavonoid group of polyphenols; however, the C-ring remained open between the oxygen in position 1 and the carbon in position 2 (for a better understanding, please check the structure in Figure 4). These two compounds were chalcone derivatives that were previously identified by Campos [19].

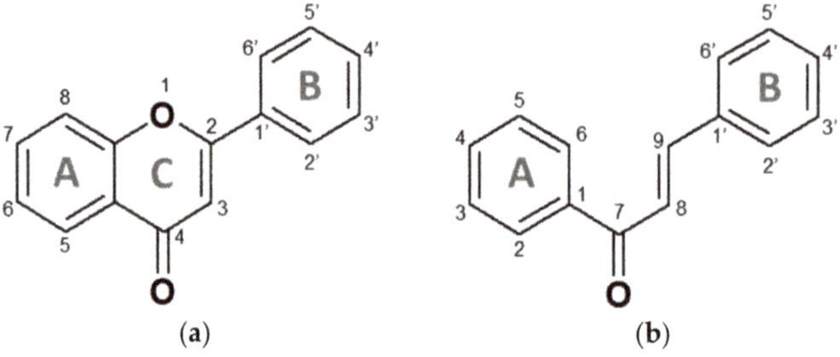

Figure 4. Basic structures of flavone (**a**) and chalcone (**b**).

Chalcones are polyphenols are structured as α,β-unsaturated aromatic ketones and can be found in various plants. They have been targeted in various studies that provide data about their potential in various therapeutic purposes, attributing bioactivities to them, such as helpfulness regarding oncology, infections and even in neurodegenerative diseases, including Alzheimer's [29]. For the latter, the antioxidant and anti-inflammatory effects are also very important, and the data obtained with the present extract was found to be promising for some of these applications, as shown below.

3.1.2. Total Phenol (TPC) and Flavonoid (TFC) Contents

Nevertheless, phenolic acid derivatives were not found in significate amounts in the hydroethanolic extracts, where its full quantification was performed using the fast, cheap and conventional Folin–Ciocalteu assay; chalcones can also be quantified with this methodology [30]. These results are shown in Figure 5, as well as the total content for flavonoids. The high TPC and TFC for both region extracts corresponded to the hydroethanolic solutions (50%) of the EF. However, even with lower differences, the highest concentration corresponded to EF from plants in region A.

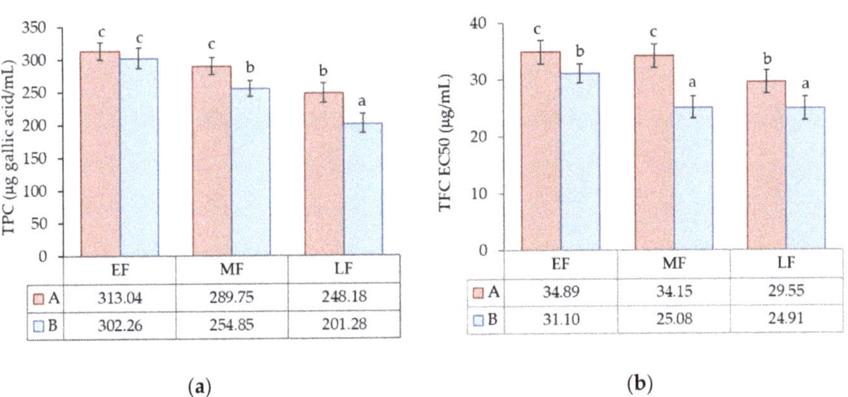

	EF	MF	LF
A	313.04	289.75	248.18
B	302.26	254.85	201.28

(a)

	EF	MF	LF
A	34.89	34.15	29.55
B	31.10	25.08	24.91

(b)

Figure 5. Total phenol and flavonoid contents ((a) TPC and (b) TFC) of *A. dealbata* flower extracts with different maturation stages from two central regions of Portugal. A—region A, B—region B; EF—early flower, MF—mid flower, LF—late flower. Different lowercase letters (a, b, c) denote significant differences between values (Tukey's test; $p < 0.05$).

The amount of TPC in flowers from the two different regions ranged from 201.28 to 313.04 µg gallic acid/mL for region B and region A, respectively, among the various statuses of maturation. The results obtained are in agreement with other authors, who obtained, for example, a concentration of 206.4 mg gallic acid/g sample in the Acacia flowers [31]. Wu et al. [32] obtained higher results for TPC with an ethanol extraction for *A. confusa* (544.4 mg/g of flower). In the aforementioned study, it was also found that the TPC was higher when ethanol was used as extract solvent. They also obtained high values for the antioxidant bioactivity due to this content in phenolic compounds.

The flowers harvested in region A presented a higher value of total phenol and flavonoid content, where this difference represented 24.4% of the total variation for the TPC and 37.4% for the TFC (Table 1). The maturation status was also of key importance where the higher values were obtained for the early flowering status (55.5% of the total variation for the TPC and 27.3% of the total variance for the TFC). This data was in accordance with the function of chalcones as precursors of flavonoids.

Table 1. Percentage of variance obtained in a two-way ANOVA of the total phenol and flavonoid contents of the flowers extracts.

	Site (S)	Maturation Status (M)	S × M	Residual
TPC (µg gallic acid/mL)	24.4 ***	55.5 ***	11.7 **	8.4
TFC EC50 (µg/mL)	37.4 ***	27.3 ***	n.s.	35.3

MS—maturation status, EF—early flower, MF—mid flower, LF—late flower; n.s. for $p > 0.05$, * $0.01 < p < 0.05$, ** $0.001 < p < 0.01$, *** $p < 0.001$.

One of the goals in this work was to understand whether the early flowering status had better potential as a source of antioxidant compounds, or at least equal to the bioactivity of the compounds in the maturated flowers in the ripening stage, in order to collect them earlier and not give the possibility for the pollen to disperse. In this case, the results showed better activity than was expected, certainly due to the highest amounts of chalcones in this phase of the vegetative cycle of Acacia. These values were in line with the data obtained for the antioxidant activities. Studies from various authors with other *Acacia* spp. also showed that the TPC can vary according to the plant organ, solvent used in the extraction procedure and season [33,34], which were in accordance with the data here discussed for *A. dealbata* extracts.

3.2. Bioactivities

3.2.1. Antioxidant Activity

Among many other bioactivities associated with polyphenolic compounds, the antioxidant activity is one of the most reported [35,36] and can be explored for various purposes. The concerns about the increase in neurodegenerative diseases have increased the interest in some of these most promising molecules [29]. In the present work, our main goal was to verify its previous potential for the evaluation of the freshness of the matrix, for further standardization of the extracts and its potential for future research regarding anti-inflammatory drugs or to help in the prevention of degenerative diseases in accordance with its inhibitory activity for the enzyme acetylcholinesterase.

Various methodologies were chosen to evaluate this potential, as presented below. One of the most used to obtain preliminary data is the DPPH scavenging assay, which determines the capacity of a compound or group of compounds to stabilize the 2,2-diphenyl-1-picrylhydrazyl (DPPH·) radical. The amount of extract required to scavenge 50% of DPPH solution (and for FRAP reagent) is defined as the IC50 (mg/mL). The antiradical free scavenging radical activity is measured through the stable radical DPPH, which shows the ability of substances present in the extracts to sequester free radicals from the existing medium. The extracts under evaluation showed free radical scavenging activity with EC50 values ranging from 6.79 to 9.93 mg/mL (Figure 6), which are good activities. The hydroethanolic extracts prepared with the EF collected in both locations showed similar results between the analyzed samples. Nevertheless, the data collected presented a decrease in this bioactivity with increasing maturation status. The results could be correlated to the content of phenols and flavonoids quantified in the samples. These first results suggested that *A. dealbata* could be a good source of natural antioxidants, which may be used to stabilize free radicals, which induce oxidative stress.

The other analysis performed to validate this data was a FRAP (ferric reducing antioxidant power) assay, which measures the ability of antioxidants to reduce the Fe(III)–TPTZ complex to the intensely blue Fe(II)–TPTZ complex in an acidic medium. The antioxidant activity can be measured spectrophotometrically by the change in color of the solution to blue resulting from the ferric to ferrous ion reduction, i.e., Fe(II) to Fe(III), while coordinated by 2,4,6-tris(2-pyridyl)-s-triazine (TPTZ). The data on the ferric reducing potential of the analyzed extracts also indicated better values for the EF extracts in both locations. All other extracts recorded moderate FRAP activity.

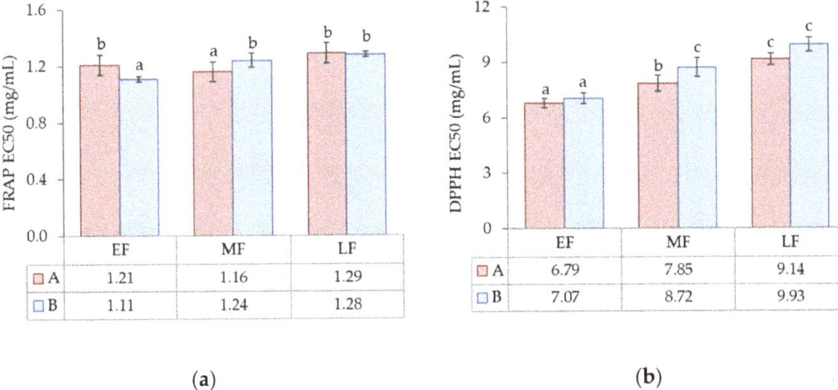

(**a**)   (**b**)

Figure 6. Antioxidant activity ((**a**) FRAP EC50 and (**b**) DPPH EC50) of *A. dealbata* flowers extracts with different maturation stages with a mixture of 50/50% (*v/v*) ethanol/water from two central regions of Portugal. A—region A, B—region B; EF—early flower, MF—mid flower, LF—late flower. Different lowercase letters (a, b, c) denote significant differences between values (Tukey's test; $p < 0.05$).

The data acquired with the extracts under study are shown in Figure 6.

Table 2 depicts that the material region harvested was not significant for the FRAP assay; however, the maturation status represented nearly 83% of the total variation in DPPH and nearly 54% for FRAP with a high significance.

Table 2. Percentage of variance obtained in a two-way ANOVA (factor 1: site (S) and factor 2: maturation status (M)) of the antioxidant activity parameters for the flower's extracts.

	Site (S)	Maturation Status (M)	S × M	Residual
FRAP EC50 (mg/mL)	n.s.	54.3 ***	n.s.	45.7
DPPH EC50 (mg/mL)	10.0 **	83.3 ***	n.s.	6.7

MS—maturation status, EF—early flower, MF—mid flower, LF—late flower; n.s. for $p > 0.05$, * $0.01 < p < 0.05$, ** $0.001 < p < 0.01$, *** $p < 0.001$.

3.2.2. Enzyme Inhibitory Activity (Acetylcholinesterase, Lipase and α-Glucosidase Enzymes)

The enzyme inhibitory activity of *A. dealbata* flower extracts with different maturation stages was evaluated due to the potential application in drug discovery in different possible applications. The data shown here correlates with the three enzymes studied. First, the most relevant of them, which can be directly associated with neurodegenerative diseases, is the effect of acetylcholinesterase. This enzyme is able to metabolize acetylcholine, and its inhibition will increase this neurotransmitter in the synaptic cleft. The activity is expected to increase cerebral functions [37], which can contribute to delaying the development of neurodegenerative diseases.

All the flower extracts showed the ability to inactivate acetylcholinesterase, lipase and α-glucosidase enzymes (Figure 7). Despite all the concentrations of the extracts being able to inhibit the acetylcholinesterase enzyme, the most effective, with values between 44.1 and 49.16 µg/mL, was correlated with the EF maturation stage. Even using different *Acacia mearnsii* material (bark rich in tannin compounds), Ogawa and Yazaki [38] also verified similar results for lipase and glucosidase inhibition.

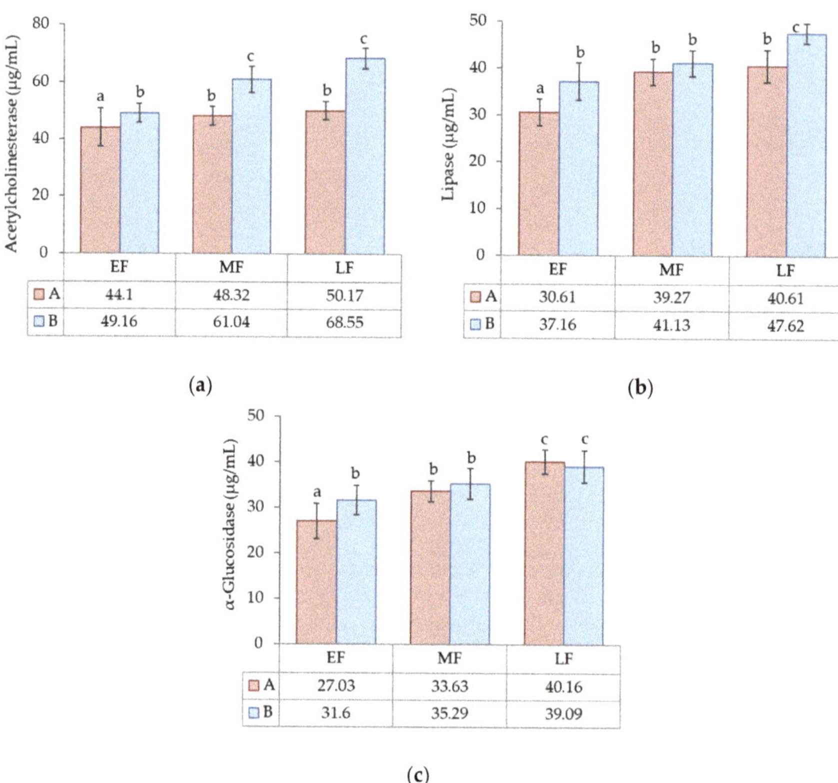

(a)

(b)

(c)

Figure 7. Enzyme inhibitory activity of *A. dealbata* flowers' hydroethanolic extracts with different maturation stages for (**a**) acetylcholinesterase, (**b**) lipase and (**c**) α-glucosidase activities. A—region A, B—region B; EF—early flower, MF—mid flower, LF—late flower. Different lowercase letters (a, b, c) denote significant differences between values (Tukey's test; $p < 0.05$).

The flowers harvested in region A had a higher value of inhibition capacity for all the enzymes and this difference represented 51.9% of the total variation for acetylcholinesterase inhibition, 25.7% for lipase and without effect for 13.6% for α-glucosidase (Table 3). The maturation status (Table 3), once again, was also of key importance, with a higher variability attributed to these factor (27.7% of the total variation for acetylcholinesterase, 52.6% of the total variation for lipase and 68.5% of the total variation for α-glucosidase).

Table 3. Variance percentages obtained in a two-way ANOVA of enzyme inhibitory activity of *A. dealbata* flowers.

	Site (S)	Maturation Status (M)	S × M	Residual
ACET (µg/mL)	51.9 ***	27.7 ***	7.6 **	12.9
Lipase (µg/mL)	25.7 ***	52.6 ***	n.s.	21.7
GLUC (µg/mL)	n.s.	68.5 *	n.s.	31.5

GLUC—α-glucosidase, ACET—acetylcholinesterase; MS—maturation status, EF—early flower, MF—mid flower, LF—late flower; n.s. for $p > 0.05$, * $0.01 < p < 0.05$, ** $0.001 < p < 0.01$, *** $p < 0.001$.

The importance associated with these three enzymes is the added value in the bioactivity, for instance, for acetylcholinesterase inhibition.

3.2.3. Antimicrobial Bioactivity

This bioactivity can be explored alone or even in parallel to others. For instance, if the target patients with neurodegenerative diseases need to be medicated with corticosteroids, which induce an immunodepression effect that makes them more vulnerable to infections, a molecule that accumulates acetylcholinesterase inhibition and activity against the most common bacteria and yeasts will improve the therapy [39]. Therefore, to complete the bioactivities for these extracts, we chose this last bioactivity, which will contribute better performance to further studies of the compounds in the extracts.

The extracts were screened for the antimicrobial activity for various microorganisms *E. coli* ATCC 29998 TM, *Staphylococus aureus* ATCC 43300TM, *Candida albicans* ATCC 10231 TM, *E. coli* ESA37, *S. aureus* ATCC ESA111 and amphotericin-B-resistant *Candida albicans* ESA 100 were tested.

The minimum inhibitory concentration of the *A. dealbata* flower extracts with different maturation stages is shown in Table 4.

Table 4. Antimicrobial bioactivity activity of the *A. dealbata* flower extracts with different maturation stages.

		MS	EcoliA (µg/mL)	Saur (µg/mL)	Calbi (µg/mL)	EcoliESA (µg/mL)	SaurESA (µg/mL)	AB_r (µg/mL)
Ethanol	A	EF	4.57 ± 0.29 [a]	0.79 ± 0.11 [a]	8.37 ± 0.50 [a]	9.30 ± 0.34 [a]	2.63 ± 0.24 [a]	15.00 ± 1.08 [a]
		MF	4.96 ± 0.35 [c]	0.89 ± 0.09 [a]	8.90 ± 0.55 [a]	10.22 ± 0.61 [a,b]	2.85 ± 0.1 [a,b]	19.16 ± 1.57 [b]
		LF	5.67 ± 0.21 [b]	1.21 ± 0.12 [b]	10.34 ± 0.50 [b]	11.19 ± 0.79 [a,b]	3.02 ± 0.22 [a,b]	19.89 ± 3.48 [b]
	B	EF	6.00 ± 0.26 [b]	0.87 ± 0.08 [a]	10.53 ± 0.58 [b]	9.72 ± 1.15 [a,b]	3.21 ± 0.24 [b]	19.43 ± 1.43 [b]
		MF	7.28 ± 0.43 [d]	1.32 ± 0.16 [b]	11.48 ± 0.74 [b]	11.47 ± 1.51 [b]	3.83 ± 0.25 [c]	22.59 ± 1.68 [b,c]
		LF	8.46 ± 0.29 [e]	1.57 ± 0.21 [c]	13.40 ± 0.71 [c]	13.77 ± 1.00 [c]	4.05 ± 0.24 [c]	25.61 ± 0.87 [c]
			Gentamicin (µg/mL)	Gentamicin (µg/mL)	Anfotericin B (µg/mL)	Gentamicin (µg/mL)	Gentamicin (µg/mL)	Anfotericin B (µg/mL)
			6.16 ± 0.08	1.11 ± 0.04	10.50 ± 0.005	10.95 ± 0.21	3.27 ± 0.08	20.20 ± 0.22

A—region A, B—region B; EF—early flower, MF—mid flower, LF—late flower; EcoliA—*E. coli* ATCC 29998 TM, Saur—*S. aureus* ATCC 43300TM, Calbi—*C. albicans* ATCC 10231 TM, EcoliESA—*E. coli* ESA37, SaurESA—*S. aureus* ATCC ESA111, AB_r—amphotericin-B-resistant *Candida albicans* ESA 100. Different lowercase letters (a, b, c, d, e) in the same column denote significant differences between values (Tukey's test; $p < 0.05$).

The results shown in Table 5 report the higher effects of the region and maturation status on the antimicrobial bioactivity. Once again, the extracts from the EF harvested in region A gave the best activity. It seems that maybe a possible change occurred due to the different climatic conditions, soils or other factors. The effect of the harvest place should be explored more in future research designed with this purpose.

Table 5. Variance percentages obtained in a two-way ANOVA of the antimicrobial bioactivity of *A. dealbata* flowers.

	Site (S)	Maturation Status (M)	S × M	Residual
EcoliA (µg/mL)	68.3 ***	22.6 ***	6.3 ***	2.8
Saur (µg/mL)	27.9 ***	50.5 ***	9.2 **	12.4
Calbi (µg/mL)	64.2 ***	28.8 ***	n.s.	7.0
EcoliESA (µg/mL)	21.3 ***	47.8 ***	9.6 *	21.2
SaurESA (µg/mL)	69.1 ***	17.7 ***	4.1 *	9.1
AB_r (µg/mL)	47.3 ***	36.0 ***	n.s.	16.8

EF—early flower, MF—mid flower, LF—late flower; EcoliA—*E. coli* ATCC 29998 TM, Saur—*S. aureus* ATCC 43300TM, Calbi—*C. albicans* ATCC 10231 TM, EcoliESA—*E. coli* ESA37, SaurESA—*S. aureus* ATCC ESA111, AB_r—amphotericin-B-resistant *Candida albicans* ESA 100; n.s. for $p > 0.05$, * $0.01 < p < 0.05$, ** $0.001 < p < 0.01$, *** $p < 0.001$.

To summarize, in Figure 8, it is possible to see that all the data collected clearly highlighted that for all the bioactivities explored in these extracts, the EF status of maturation was the most significant. The principal component analysis explained 79.2% of the total variability observed for the different parameters analyzed concerning the two factors (site and maturation status). It is possible to note that regions A and B showed different behaviors concerning the studied properties, highlighting an influence of the harvest place regarding the properties of the plants given to the extracts. PC1 (which explained 64.3% of the total variation) clearly separated the three maturation stages, with EF showing the better results, which could be an advantage because collecting the flower in an early stage inhibits the possibility of the reproduction of the plant, which prevents further reproduction and invasion of other lands.

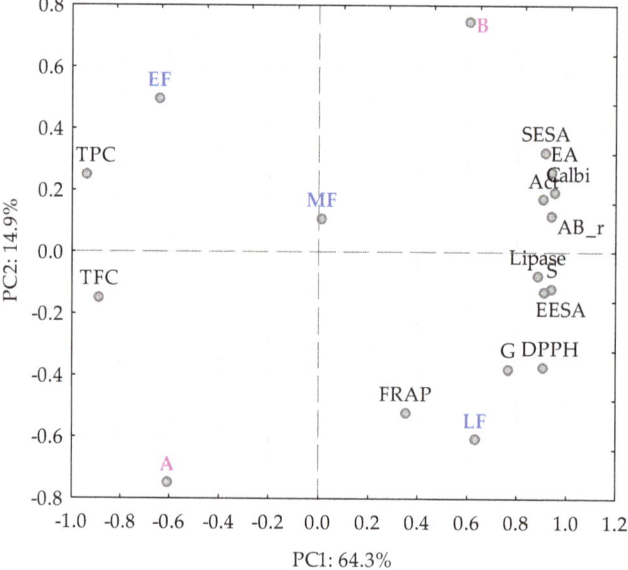

Figure 8. Principal component analysis of ethanol:water (50:50) extracts from the *A. dealbata* flowers in different maturation stages from two central regions of Portugal. A—region A, B—region B; EF—early flower, MF—mid flower, LF—late flower; TPC—total phenol content, TFC—total flavonoid content; GLUC—α-glucosidase, ACET—acetylcholinesterase; EcoliA—*E. coli* ATCC 29998 TM, Saur—*S. aureus* ATCC 43300TM, Calbi—*C. albicans* ATCC 10231 TM, EcoliESA—*E. coli* ESA37, SaurESA—*S. aureus* ATCC ESA111, AB_r—amphotericin-B-resistant *Candida albicans* ESA 100.

3.3. FT-Raman Spectral Acquisition

The spectra obtained with the hydroethanolic flowers extract (50:50) shown in Figure 9 display the strong influence of the compounds in this matrix [40]. The intense bands around 2975, 2931 and 2883 cm^{-1} were assigned to C–H stretching of some compounds present in the plant extracts, but also of the CH_2 and CH_3 stretching of the ethanol [18], and were influenced by the other substances from the plant extracts. The peak at 1450 cm^{-1} corresponded to the C–OH bending of ethanol, but also originated from the combination of bending vibration of CH_2 and the vibration of the COO– group in the flavanol and organic acids [41,42].

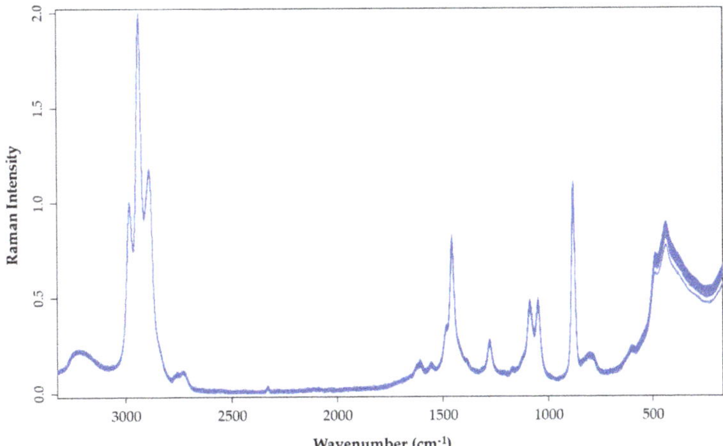

Figure 9. Raman spectrum, baseline corrected spectrum for Acacia flowers samples.

The band at 1270 cm^{-1} was found to be due to the deformation vibration of O–C–H, C–C–H and C–O–H.

The peaks at 1090 and 1050 cm^{-1} were from the in-plane movement of the carbon ring and the substituents, which are regions linked to the characterization of alcohols, ethers, esters, carboxylic acids and anhydrides [43].

The bands around 705 cm^{-1} corresponded to the C–O and C–C–O stretching and O–C–O bending, and the bands in the 880 cm^{-1} region were assigned to the C–H, C–O–H and CH$_2$ bending.

Concerning the results of the PCA (mean centered) made with the spectral data (Figure 10) the separation between regions as observed for the analytical data is clear, as well as the separation between the maturation stages of the flowers.

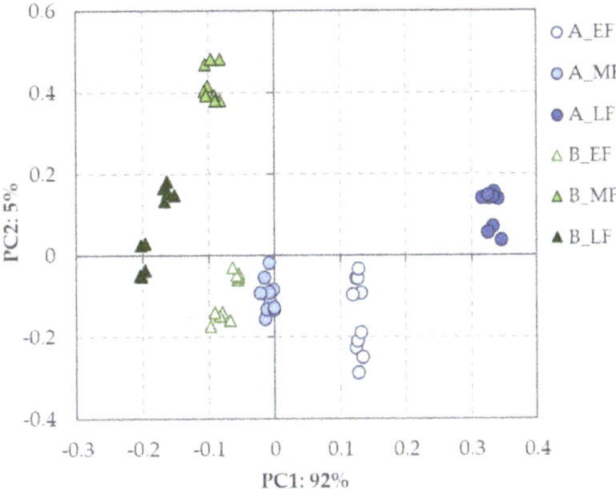

Figure 10. Score plot of the first two principal components after the PCA performed on the Raman spectra recorded from Acacia flowers samples using the first derivative Savitzky–Golay spectra transform with 15 smoothing points.

These results showed that Raman spectroscopy is a promising technique for discriminating between flower raw materials to identify possible differences in the biological activities of their extracts.

To our knowledge, this is the first time that these extracts have been studied and these bioactivities were explored together to provide a possible new drug strategy in neurodegenerative treatments, among other possibilities, and to help to contribute to the control of an invasive plant.

4. Conclusions

One of the challenges to be faced soon is the green chemistry and development of new drugs with low environmental impacts. The compounds of *A. dealbata* in the EF stage were found to have the necessary bifocality that fulfilled these two requirements. The antioxidant activity and the acetylcholinesterase inhibition of the compounds in the extracts enrolled in this work are presented for the first time as promising for further potential research for some therapeutic purposes, for instance, neurodegenerative diseases. These compounds can also be redesigned in new structures for new drugs, for instance, to enhance memory and other brain functions by influencing chemical activity in the brain. From the data presented in this work, all the assays revealed the best potential for the non-mature flowers of *A. dealbata*. This allows for the possibility to harvest them at this preliminary status and stop the reproduction of the plant, which will help to control its invasive potential.

In this study, it was also shown that Raman spectroscopy is a promising technique for identifying the differences between flower extracts in order to better select raw materials based on their biological activities.

Author Contributions: Conceptualization, O.A., M.G.C. and L.M.E.; methodology, V.P., M.G.C., S.I.P., T.D., L.M.E. and O.A.; software, O.A.; validation, V.P., M.G.C., S.I.P., T.D., L.M.E. and O.A.; formal analysis, V.P., M.G.C., S.I.P., T.D., L.M.E. and O.A.; investigation, V.P., M.G.C., S.I.P., L.M.E. and O.A.; writing—original draft preparation, V.P., M.G.C., S.I.P., L.M.E. and O.A.; writing—M.G.C., L.M.E. and O.A.; supervision O.A., M.G.C. and L.M.E.; project administration, O.A., M.G.C. and L.M.E.; funding acquisition, M.G.C., L.M.E. and O.A. All authors have read and agreed to the published version of the manuscript.

Funding: This research was funded by project PCIF/GVB/0145/2018 (Acacia4fireprev). This work is also funded by National Funds through the FCT—Foundation for Science and Technology under the Projects UIDB/00239/2020 (CEF), (UI0204): UIDB/00313/2020 and UIDB/00690/2020 (CIMO).

Institutional Review Board Statement: Not applicable.

Informed Consent Statement: Not applicable.

Data Availability Statement: We choose to exclude this statement because the study did not report any data.

Conflicts of Interest: The authors declare that they have no conflict of interest.

References

1. Mohammad, D.; Chan, P.; Bradley, J.; Lanctôt, K.; Herrmann, N. Acetylcholinesterase inhibitors for treating dementia symptoms—A safety evaluation. *Expert Opin. Drug Saf.* **2017**, *16*, 1009–1019. [CrossRef]
2. Lorenzo, P.; Pereira, C.S.; Rodríguez-Echeverría, S. Differential impact on soil microbes of allelopathic compounds released by the invasive *Acacia dealbata* Link. *Soil Biol. Biochem.* **2013**, *57*, 156–163. [CrossRef]
3. Marchante, E.; Freitas, H.; Marchante, H. *Guia Prático para a Identificação de Plantas Invasoras de Portugal Continenta*; Imprensa da Universidade de Coimbra: Coimbra, Portugal, 2008; ISBN 9789892603988.
4. Lorenzo, P.; González, L.; Reigosa, M.J. The genus Acacia as invader: The characteristic case of *Acacia dealbata* Link in Europe. *Ann. For. Sci.* **2010**, *67*, 101. [CrossRef]
5. Domingues De Almeida, J.; Freitas, H. Exotic naturalized flora of continental Portugal—A reassessment. *Bot. Complut.* **2006**, *30*, 117–130. [CrossRef]
6. Souza-Alonso, P.; González, L.; Cavaleiro, C. Ambient has Become Strained. Identification of *Acacia dealbata* Link Volatiles Interfering with Germination and Early Growth of Native Species. *J. Chem. Ecol.* **2014**, *40*, 1051–1061. [CrossRef] [PubMed]
7. Casas, M.P.; Conde, E.; Ribeiro, D.; Fernandes, E.; Domínguez, H.; Torres, M.D. Bioactive properties of *Acacia dealbata* flowers extracts. *Waste Biomass Valorization* **2020**, *11*, 2549–2557. [CrossRef]
8. Pinto, F.; Silva, F.; Barbosa, A. Evaluation of Haemolytic Activity of Leaves from *Acacia podalyriifolia*. *Eur. J. Med. Plants* **2016**, *17*, 1–5. [CrossRef]

9.   Silva, B.M.; Andrade, P.B.; Seabra, R.M.; Ferreira, M.A. Determination of selected phenolic compounds in quince jams by solid-phase extraction and HPLC. *J. Liq. Chromatogr. Relat. Technol.* **2001**, *24*, 2861–2872. [CrossRef]

10. Lin, H.Y.; Chang, T.C.; Chang, S.T. A review of antioxidant and pharmacological properties of phenolic compounds in *Acacia confusa*. *J. Tradit. Complement. Med.* **2018**, *8*, 443–450. [CrossRef]

11. Li, X.-C.; Liu, C.; Yang, L.-X.; Chen, R.-Y. Phenolic compounds from the aqueous extract of *Acacia catechu*. *J. Asian Nat. Prod. Res.* **2011**, *13*, 826–830. [CrossRef]

12. De Andrade, C.A.; Miguel, M.D.; Miguel, O.G.; Ferrnato, M.d.L.; Peitz, C.; Cunico, M.; Dias, J.d.F.G.; Balestrin, L.; Kerber, V.A. Efeitos Alelopáticos das Flores da *Acacia podalyriaefolia* A. Cunn. *Visão Acadêmica* **2003**, *4*, 93–98. [CrossRef]

13. Abdel-Farid, I.B.; Sheded, M.G.; Mohamed, E.A. Metabolomic profiling and antioxidant activity of some Acacia species. *Saudi J. Biol. Sci.* **2014**, *21*, 400–408. [CrossRef]

14. Ziani, B.E.C.; Carocho, M.; Abreu, R.M.V.; Bachari, K.; Alves, M.J.; Calhelha, R.C.; Talhi, O.; Barros, L.; Ferreira, I.C.F.R. Phenolic profiling, biological activities and in silico studies of *Acacia tortilis* (Forssk.) Hayne ssp. raddiana extracts. *Food Biosci.* **2020**, *36*, 100616. [CrossRef]

15. Saini, M.L.; Saini, R.; Roy, S.; Kumar, A. Comparative pharmacognostical and antimicrobial studies of Acacia species (Mimosaceae). *J. Med. Plants Res.* **2008**, *2*, 378–386.

16. Banso, A. Phytochemical and antibacterial investigation of bark extracts of *Acacia nilotica*. *J. Med. Plants Res.* **2009**, *3*, 82–85.

17. Mahesh, B.; Satish, S. Antimicrobial Activity of Some Important Medicinal Plant Against Plant and Human Pathogens. *World J. Agric. Sci.* **2008**, *4*, 839–843.

18. Paula, V.B.; Delgado, T.; Campos, M.G.; Anjos, O.; Estevinho, L.M. Enzyme inhibitory potential of ligustrum lucidum aiton berries. *Molecules* **2019**, *24*, 1283. [CrossRef] [PubMed]

19. Campos, M.; Markham, K.R.; Mitchell, K.A.; da Cunha, A.P. An approach to the characterization of bee pollens via their flavonoid/phenolic profiles. *Phytochem. Anal.* **1997**, *8*, 181–185. [CrossRef]

20. Campos, M.d.G.; Markham, K.R. *Structure Information from HPLC and On-Line Measured Absorption Spectra: Flavones, Flavonols and Phenolic Acids*; Imprensa da Universidade de Coimbra: Coimbra, Portugal, 2007; ISBN 9789892604800.

21. Singleton, V.L.; Orthofer, R.; Lamuela-Raventós, R.M. [14] Analysis of total phenols and other oxidation substrates and antioxidants by means of folin-ciocalteu reagent. *Methods Enzymol.* **1999**, *299*, 152–178.

22. Woisky, R.G.; Salatino, A. Analysis of propolis: Some parameters and procedures for chemical quality control. *J. Apic. Res.* **1998**, *37*, 99–105. [CrossRef]

23. Berker, K.I.; Güçlü, K.; Tor, İ.; Apak, R. Comparative evaluation of Fe(III) reducing power-based antioxidant capacity assays in the presence of phenanthroline, batho-phenanthroline, tripyridyltriazine (FRAP), and ferricyanide reagents. *Talanta* **2007**, *72*, 1157–1165. [CrossRef] [PubMed]

24. Alencar, S.M.; Oldoni, T.L.C.; Castro, M.L.; Cabral, I.S.R.; Costa-Neto, C.M.; Cury, J.A.; Rosalen, P.L.; Ikegaki, M. Chemical composition and biological activity of a new type of Brazilian propolis: Red propolis. *J. Ethnopharmacol.* **2007**, *113*, 278–283. [CrossRef] [PubMed]

25. Sukumaran, S.; Chee, C.; Viswanathan, G.; Buckle, M.; Othman, R.; Abd Rahman, N.; Chung, L. Synthesis, Biological Evaluation and Molecular Modelling of 2'-Hydroxychalcones as Acetylcholinesterase Inhibitors. *Molecules* **2016**, *21*, 955. [CrossRef]

26. Habib, H.M.; Kheadr, E.; Ibrahim, W.H. Inhibitory effects of honey from arid land on some enzymes and protein damage. *Food Chem.* **2021**, *364*, 130415. [CrossRef]

27. Tadera, K.; Minami, Y.; Takamatsu, K.; Matsuoka, T. Inhibition of alpha-glucosidase and alpha-amylase by flavonoids. *J. Nutr. Sci. Vitaminol.* **2006**, *52*, 149–153. [CrossRef]

28. Anjos, O.; Caldeira, I.; Pedro, S.I.; Canas, S. FT-Raman methodology applied to identify different ageing stages of wine spirits. *LWT* **2020**, *134*, 110179. [CrossRef]

29. Thapa, P.; Upadhyay, S.P.; Suo, W.Z.; Singh, V.; Gurung, P.; Lee, E.S.; Sharma, R.; Sharma, M. Chalcone and its analogs: Therapeutic and diagnostic applications in Alzheimer's disease. *Bioorg. Chem.* **2021**, *108*, 104681. [CrossRef]

30. Isla, M.I.; Salas, A.; Danert, F.C.; Zampini, I.C.; Ordoñez, R.M. Analytical methodology optimization to estimate the content of non-flavonoid phenolic compounds in Argentine propolis extracts. *Pharm. Biol.* **2014**, *52*, 835–840. [CrossRef]

31. De Andrade, C.A.; Costa, C.K.; Bora, K.; Miguel, M.D.; Miguel, O.G.; Kerber, V.A. Determinação do conteúdo fenólico e avaliação da atividade antioxidante de *Acacia podalyriifolia* A. Cunn. ex G. Don, Leguminosae-mimosoideae. *Rev. Bras. Farmacogn.* **2007**, *17*, 231–235. [CrossRef]

32. Wu, J.-H.; Tung, Y.-T.; Wang, S.-Y.; Shyur, L.-F.; Kuo, Y.-H.; Chang, S.-T. Phenolic Antioxidants from the Heartwood of *Acacia confusa*. *J. Agric. Food Chem.* **2005**, *53*, 5917–5921. [CrossRef] [PubMed]

33. Faujdar, S.; Sharma, S.; Sati, B.; Pathak, A.K.; Paliwal, S.K. Comparative analysis of analgesic and anti-inflammatory activity of bark and leaves of *Acacia ferruginea* DC. *Beni-Suef Univ. J. Basic Appl. Sci.* **2016**, *5*, 70–78. [CrossRef]

34. Gabr, S.; Nikles, S.; Pferschy Wenzig, E.M.; Ardjomand-Woelkart, K.; Hathout, R.M.; El-Ahmady, S.; Motaal, A.A.; Singab, A.; Bauer, R. Characterization and optimization of phenolics extracts from Acacia species in relevance to their anti-inflammatory activity. *Biochem. Syst. Ecol.* **2018**, *78*, 21–30. [CrossRef]

35. Mazur, W.M.; Duke, J.A.; Wähälä, K.; Rasku, S.; Adlercreutz, H. Isoflavonoids and Lignans in Legumes: Nutritional and Health Aspects in Humans 11The method development and synthesis of the standards and deuterium-labelled compounds was supported by National Institutes of Health Grants No. 1 R01 CA56289-01 and No. 2 R0. *J. Nutr. Biochem.* **1998**, *9*, 193–200. [CrossRef]

36. Hertog, M.G.; Feskens, E.J.; Kromhout, D.; Hertog, M.G.; Hollman, P.C.; Hertog, M.G.; Katan, M. Dietary antioxidant flavonoids and risk of coronary heart disease: The Zutphen Elderly Study. *Lancet* **1993**, *342*, 1007–1011. [CrossRef]

37. Merino, E.N.; Sendin, M.A.C.; Osorio, J.A.V. Enfermedad de Alzheimer. *Med. Programa Form. Médica Contin. Acreditado* **2015**, *11*, 4306–4315. [CrossRef]

38. Ogawa, S.; Yazaki, Y. Tannins from *Acacia mearnsii* De Wild. Bark: Tannin Determination and Biological Activities. *Molecules* **2018**, *23*, 837. [CrossRef]

39. Pierpaoli, W. Neurodegenerative Diseases: A Common Etiology and a Common Therapy. *Ann. N. Y. Acad. Sci.* **2005**, *1057*, 319–326. [CrossRef]

40. Jin, Z.; Chu, Q.; Xu, W.; Cai, H.; Ji, W.; Wang, G.; Lin, B.; Zhang, X. All-Fiber Raman Biosensor by Combining Reflection and Transmission Mode. *IEEE Photonics Technol. Lett.* **2018**, *30*, 387–390. [CrossRef]

41. Paradkar, M.M.; Sakhamuri, S.; Irudayaraj, J. Comparison of FTIR, FT-Raman, and NIR Spectroscopy in a Maple Syrup Adulteration Study. *J. Food Sci.* **2002**, *67*, 2009–2015. [CrossRef]

42. Nickless, E.M.; Holroyd, S.E.; Stephens, J.M.; Gordon, K.C.; Wargent, J.J. Analytical FT-Raman spectroscopy to chemotype Leptospermum scoparium and generate predictive models for screening for dihydroxyacetone levels in floral nectar. *J. Raman Spectrosc.* **2014**, *45*, 890–894. [CrossRef]

43. Larkin, P. *Infrared and Raman Spectroscopy: Principles and Spectral Interpretation*; Elsevier Inc.: Amsterdam, The Netherlands, 2011; Volume 9, ISBN 9780123869845.

applied
sciences

MDPI

Article

GC- and UHPLC-MS Profiles as a Tool to Valorize the Red Alga *Asparagopsis armata* †

Diana C. G. A. Pinto [1,*], Marie L. Lesenfants [2], Gonçalo P. Rosa [1,2,3], Maria Carmo Barreto [2,3], Artur M. S. Silva [1] and Ana M. L. Seca [1,2,3]

[1] LAQV-REQUIMTE & Department of Chemistry, Campus de Santiago, University of Aveiro, 3810-193 Aveiro, Portugal; goncalo.p.rosa@uac.pt (G.P.R.); artur.silva@ua.pt (A.M.S.S.); ana.ml.seca@uac.pt (A.M.L.S.)
[2] Faculty of Sciences and Technology, University of Azores, Rua Mãe de Deus, 9501-321 Ponta Delgada, Portugal; lesenfantsmarie@hotmail.com (M.L.L.); maria.cr.barreto@uac.pt (M.C.B.)
[3] cE3c-Centre for Ecology Evolution and Environmental Changes, Azorean Biodiversity Group, Rua Mãe de Deus, 9501-321 Ponta Delgada, Portugal
*   Correspondence: diana@ua.pt; Tel.: +351-234-401407
†   To the memory of Professor Ana I. Neto.

Abstract: *Asparagopsis armata* Harvey is a red alga native from the southern hemisphere and then introduced in the Mediterranean Sea and the Atlantic Ocean, including the Azores Archipelago, where it is considered an invasive alga. Some studies show that the extracts exhibit antimicrobial and antifouling activities, and it is incorporated in some commercialized cosmetic products. (e.g., Ysaline®). However, knowledge of this species chemical composition is scarce. The GC-MS and UHPLC-MS profiles of both the nonpolar and polar extracts were established to contribute to this problem solution. According to the results, *A. armata* is rich in a great structural variety of halogenated lipophilic and aromatic compounds, some of them identified here for the first time. In the lipophilic extract, 25 compounds are identified, being the halogenated compounds and fatty acids, the two major compound families, corresponding to 54.8% and 35.7% of identified compounds (224 and 147 mg/100 g of dry algae, respectively). The 1,4-dibromobuten-1-ol and the palmitic acid are the two most abundant identified compounds (155 and 83.4 mg/100 g of dry algae, respectively). The polar extract demonstrated the richness of this species in brominated phenolics, from which the cinnamic acid derivatives are predominant. The results obtained herein open new perspectives for valuing the *A. armata* as a source of halogenated compounds and fatty acids, consequently improving its biotechnological and economic potential. Promoting this seaweed and the consequent increase in its demand will contribute to biodiversity conservation and ecosystem sustainability.

Keywords: *Asparagopsis armata*; Rhodophyta; GC-MS; UHPLC-MS; dibrominated compounds; 1,4-dibromobuten-1-ol; palmitic acid; brominated phenolics; red seaweed; invasive seaweed

Citation: Pinto, D.C.G.A.; Lesenfants, M.L.; Rosa, G.P.; Barreto, M.C.; Silva, A.M.S.; Seca, A.M.L. GC- and UHPLC-MS Profiles as a Tool to Valorize the Red Alga *Asparagopsis armata*. *Appl. Sci.* **2022**, *12*, 892. https://doi.org/10.3390/app12020892

Academic Editor: Magdalena Biesaga

Received: 26 December 2021
Accepted: 13 January 2022
Published: 16 January 2022

Publisher's Note: MDPI stays neutral with regard to jurisdictional claims in published maps and institutional affiliations.

1. Introduction

Marine organisms live in complex habitats and are often at the mercy of too variable environmental conditions. To protect themselves from predators, competitors, and many other threats, sea creatures were forced to develop protective mechanisms that include producing a vast number of bioactive compounds, many of which are unique in chemical structure and biological function [1,2]. Seaweeds are among the most explored marine organisms due to their richness in bioactive compounds [3]. However, there are still several species for which the chemical profile is unknown, and *Asparagopsis armata* is one example.

Asparagopsis species belong to the phylum Rhodophyta (red seaweeds), order Bonnemaisoniales, genus *Asparagopsis* with three taxonomically accepted species: *A. armata* Harvey, *A. svedelii* Taylor, and *A. taxiformis* Trevisan [4]. *Asparagopsis taxiformis* has been used as food by Hawaiians [5] while *A. armata* is commercially farmed in northern Europe to extract bioactive molecules [6]. *Asparagopsis armata* is native to Australia and is now widely distributed from the North Atlantic to the coast of Senegal and in the Mediterranean basin [7]. This seaweed seems to exhibit an ecological advantage due to its toxic metabolites that affect the competing algae fixation, forming monospecific coverages, reducing the habitat's species richness, and causing an indirect economic impact by negatively affecting fishing and aquaculture [8]. Despite its negative impact, *A. armata* is commercially farmed in northern Europe to extract bioactive molecules such as sulphated polysaccharides with iodine and bromine groups [6], that are applied as a natural preservative in cosmetics and products to anti-acne treatment (e.g., Invincity®). In Azores, this invasive seaweed presents an enormous rate of biomass production for which there is no use. Uses such as food or technological application that increase their commercial value and thus create a dynamic that mitigates the negative effects mentioned above are essential.

Although several studies have been published on the biological activity of *A. armata* extracts showing their potential as a source of natural compounds with antiviral, antimicrobial, and antifouling activity [9–11], there are few published works dedicated to the characterization of secondary metabolites existing in the extracts of this species, apart from sterols and volatile halogenated compounds [12–14]. Herein we present and discuss the composition of the polar and lipophilic extracts of *A. armata*, obtained by greener and more efficient methodologies and using GC-MS and UHPLC-MS as tools for separation and identification of the metabolites present.

2. Materials and Methods

2.1. Chemicals

The following chemicals are used as standards for UHPLC-MS analysis: 2-bromopropionic acid, 2-bromo-5-methoxybenzoic acid and methyl 2-bromo-5-methoxybenzoate were purchased from Alfa Aesar (Kandel, Germany). 2-Bromobenzaldehyde and 3-bromobenzaldehyde were purchased from Sigma-Aldrich GmbH (Darmstadt, Germany). 2′-Bromoacetophenone was obtained from TCI (Oxford, UK). Caffeic, *p*-coumaric, and chlorogenic acids were supplied by Acros Organics (Geel, Belgium).

UHPLC mobile phase eluents: formic acid and acetonitrile HPLC-grade solvents were purchased from Panreac (Barcelona, Spain), and ultrapure water was obtained by a Direct-Q® water purification system (Merck Life Science, Darmstadt, Germany). All other chemicals were of analytical grade.

Solvents used for GC–MS analysis were purchased from Panreac and Acros Organics and were of analytical grade. Chemicals such as *N,O*-bis(trimethylsilyl)trifluoroacetamide (BSTFA) (99%) and trimethylsilyl chloride (TMSCl) (99%), purchased from Sigma-Aldrich, were used as derivatization agents. Octadecane (99%), eicosane (99%), palmitic acid (99%), stearic acid (>97%), (Z,Z) 9,12-octadecadienoic acid (99%), 1-bromononane (98%), 1,8-dibromooctane (98%), (3β) cholest-5-en-3-ol (99%), 1-octadecanol (99%), and 1-mono palmitin (99%) were purchased from Sigma-Aldrich and used as standards to qualitative and quantitative GC-MS analysis.

2.2. Plant Material and Extracts Preparation

Asparagopsis armata Harvey (on gamethophyte phase) was collected in June 2017, in the intertidal zone of S. Miguel Island coast, in Ponta Delgada, Azores. Prof. Ana Isabel Neto, expert on seaweeds from University of Azores, was responsible for the taxonomic identification of the harvested alga. A voucher specimen (SMG-2017-06) was deposited in Ruy Telles Palhinha Herbarium (AZB).

Asparagopsis armata sample was cleaned from salt and epiphytes, drained between sheets of absorbent paper, and cut into pieces approximately 1 cm long. Fresh algae after draining contains 92.9% (±0.97%) of water.

Dried *A. armata* (231.06 g) were obtained from fresh material by air drying, protected from light, and then dried in an oven with forced ventilation at 40 °C, until constant weight.

The lipophilic extract (extract A) was prepared from 229.2 g of dried seaweed by maceration (1:10 algae:dichloromethane) at room temperature, under stirring and in the dark, for 3 cycles of 3 days each, with solvent renewal at the end of each cycle. The solvent from the combined extraction steps was first evaporated with a rotary vacuum evaporator at 40 °C and then dryed under vacuum at room temperature (yield 1.25%).

From fresh *A. armata*, collected from the same place and date, several polar extracts were prepared, using different polar solvents (ethanol, ethanol:water (50%), and water) and different methods of extraction (maceration, ultrasounds, and microwave) in order to optimize the extraction procedure according to the Table 1.

Table 1. Experimental conditions used in the preparation of the *A. armata* polar extracts.

Polar Extracts	Extraction Method	Time/Temperature	Solvent
1	Maceration	1 h/boiling	H_2O [1]
2	Maceration	24 h/room temp.	H_2O [1]
3	Maceration	24 h/room temp.	Ethanol [1]
4	Maceration	24 h/room temp.	Ethanol/H_2O (1:1) [1]
5	Microwave	10 min/50 °C	Ethanol/H_2O (1:1) [2]
6	Ultrasounds	10 min/room temp.	Ethanol/H_2O (1:1) [2]
7	Microwave	10 min/50 °C	Ethanol [2]
8	Ultrasounds	10 min/room temp.	Ethanol [2]

[1] The proportion algae:solvent was 1 g:4 mL. [2] The proportion algae:solvent was 1 g:10 mL.

The dried polar extracts were obtained after evaporation and/or lyophilization. The yields obtained are shown in the Table 2.

Table 2. Extraction yield of the polar *A. armata* extracts prepared.

Polar Extracts Prepared	Extraction Yield (g Extract/100 g Fresh Alga)
1	1.7 ± 0.1
2	1.7 ± 0.2
3	1.8 ± 0.1
4	1.6 ± 0.1
5	1.6 ± 0.4
6	1.3 ± 0.1
7	1.9 ± 0.3
8	2.0 ± 0.2

2.3. Biologic Activities

The following paragraphs present the main experimental details of each of the bioactivity assays such as range of concentrations tested, and the positive controls used. A more detailed description of each of the methods employed can be found in Zárate et al. [15].

2.3.1. DPPH Scavenging Activity

Antioxidant activity was assayed by the DPPH (1,1-diphenyl-2-picryl-hydrazyl) radical scavenging assay [16]. The extracts were tested at concentrations ranging between 0.244 µg/mL and 250 µg/mL in methanol. Trolox, at concentrations ranging from 0.100 to 100 µg/mL, was used as the positive control.

2.3.2. ABTS Scavenging Activity

A microplate adaptation of the method by Re et al. [17] was adopted to perform the ABTS radical scavenging assay. The extracts were tested in a range of concentrations of 0.244 to 250 µg/mL, and the positive control was Trolox (0.100–100 µg/mL).

2.3.3. Ferric Chelating Activity

The Fe^{2+} chelating ability of the extracts was assayed by the ferrous iron-ferrozine complex method [18]. The positive control used was EDTA (0.195–100 µg/mL). Samples were assayed at serial concentrations between 0.244 and 250 µg/mL.

2.3.4. Anticholinesterasic Activity

A modification of the method described by Ellman et al. [19] and Arruda et al. [20] was used for measuring the extracts' ability to inhibit the activity of acetylcholinesterase (AChE) and butyrylcholinesterase (BuChE). The algal extracts were tested at concentrations ranging from 0.293 to 150 µg/mL and compared to the positive control donepezil (0.010–5.0 µg/mL).

2.3.5. Inhibition of Tyrosinase Activity

The extracts were assayed by adapting the tyrosinase inhibition method described by Shimizu et al. [21] and modified by Manosroi et al. [22]. Serial concentrations of the extracts (0.488 µg/mL to 250 µg/mL dissolved in 100 mM phosphate buffer, pH 6.8 were tested. Kojic acid at 0.293–150 µg/mL was used as positive control.

2.3.6. Inhibition of Collagenase Activity

The anti-collagenase activity was assayed through an adaptation of the method of Mandl et al. [23]. The extracts and the reference compound EDTA were tested with concentrations ranging between 15.6–250 µg/mL.

2.3.7. Inhibition of Elastase Activity

A modified version of the described by Ndlovu et al. [24] was used to measure the inhibition of elastase activity by the extracts. The concentration of extract used in the assay ranged from 0.488 µg/mL to 250 µg/mL. N-Methoxysuccinyl-Ala-Ala-Pro-chloromethyl ketone at 0.019–20 µg/mL was used as a positive control.

2.4. UHPLC-DAD-ESI-MS/MS^n Characterization of A. armata Aqueous Extract

A methanolic solution of *A. armata* aqueous extract (prepared under the experimental conditions described in Table 1 for extract **2** and then lyophilized) was analyzed using an UHPLC-ESI-DAD/MS apparatus. The column used was a thermo scientific hypersil gold column (1000 mm × 20 mm) with a part. size of 1.9 µm at 30 °C. The mobile phase was composed of (A) 0.1% formic acid aqueous solution (*v/v*) and (B) acetonitrile, and the flow rate was 0.2 mL/min. The solvent gradient started with 5% of solvent A over 14 min followed by 40% of solvent A for 2 min, 100% over 7 min and finally 5% over 10 min. The injection volume was 2 µL, and the chromatographic profiles were documented at 280 nm. The mass spectrometer used was an LTQ XL linear ion trap 2D equipped with an orthogonal electrospray ion source (ESI). The equipment was operated in negative-ion mode with electrospray ionization source of 5.00 kV and ESI capillarity temperature of 275 °C. The full scan covered a mass range of 50–2000 *m/z*. The identification of individual phenolic compounds by UHPLC-MS was achieved by comparing with data available on the literature and interpretation of the MS fragmentation.

2.5. GC-MS Analysis of A. armata Dichloromethan Extract

Before GC-MS analysis, four replicates of dried dichloromethane extract of *A. armata* (extract A, nearly 20 mg each), was dissolved in 1 mL of dichloromethane, added the internal standard (octadecane) and was silylated using 250 µL of pyridine, 250 µL of BSTFA, and 50 µL of TMSCl in a screw glass tube at 70 °C during 30 min. This derivatization procedure is well known [25,26] and allows the hydroxyl groups that may exist in the extract constituents to be transformed into silyloxyl groups. Thus, compounds with -OH substituents will be detected by GC-MS analysis as silylated derivatives which are more volatile, stable, and less polar [25,26].

Each replicate was analyzed by GC–MS twice using a QP2010 Ultra Shimadzu apparatus with Xcalibur software, equipped with a DB-5- J &W capillary column (30 m × 0.25 mm i.d. and a film thickness of 0.25 μm). The injection was in split mode. The temperature of the column was 70 °C during 2 min, at 3.5°/min until 120 °C, at 9°/min until 170 °C, 170 °C during 10 min, at 9°/min until 300 °C, which was maintained for 6 min. Injector temperature was at 320 °C, and the transfer-line temperature was at 200 °C. The mass spectrometer was operated in the EI mode with energy of 70 eV, and data were collected at a rate of 1 scan s^{-1} over a range of m/z 50–1000. The carrier gas was helium (purity of 99.995%) at 1.2 mL/min.

The compounds were identified, from total ion chromatogram, by comparing the retention times and by comparing their mass spectra with (i) those existing in the MS databases of Wiley229, NIST14, and Shimadzu Pesticide Library; (ii) those published in the literature and by analyzing the fragmentation pattern; (iii) those obtained in the same experimental conditions to standard compounds.

Quantitative analysis was carried out using internal standard method (octadecane, 28.68 min), by intrapolation on the calibration curves obtained with at least 4 distinct but known concentrations of representative standards of each family of organic compounds identified in the dichloromethane extract (namely, palmitic acid, stearic acid, (Z,Z) 9,12-octadecadienoic acid, 1-octadecanol, monopalmitin, cholesterol, sitosterol, phenyl acetic acid, 1,8-dibromooctane, 1-bromononane) relative to octadecane. The correlation coefficient (r^2) of each calibration curve was greater than 0.99. All standards with hydroxyl substituents were subjected to the silylation procedure described above.

3. Results and Discussion

3.1. *Asparagopsis armata Polar Extracts Yield and Chemical Diversity*

Extracts from fresh *A. armata* were prepared using polar solvents (water, ethanol) and different energy sources according to the data present in Table 1. The yields of extractions are presented in Table 2. Preparing polar extracts from fresh instead of dry seaweed has the additional advantage of circumventing the difficulties of drying seaweed and grinding after drying. The alga has a high-water content (92.9%), and the drying process is prolonged. However, no signs of chemical degradation were evident (no changes in color or observed microorganism development). After drying, the seaweed becomes extremely hard, requiring very robust equipment for its grinding.

The different methods and solvents tested reveal identical performance in terms of extraction yield.

Considering the yields obtained and the more sustainable experimental conditions (economically and environmentally "greener"), the profiles of the most interesting extracts were qualitatively analyzed by UHPLC-MS, with the most relevant results shown in Figure 1.

The chemical profiles of the ethanol extracts (extracts 7 and 8) are very similar, being both ultrasound and microwave the most selective extraction method. The use of ethanol under microwave energy extracted almost exclusively one compound. Although it was not our aim to identify this compound, its isotopic pattern and m/z in the mass spectra suggests that it is a 2,3-diphenylpropanoic acid derivative, having two hydroxyl groups and four bromine atoms distributed in the aromatic rings ([M-H]$^-$ m/z 568/570/572/574/576). Due to the fragment with m/z 319/321/323, we suggest that each aromatic ring possesses one hydroxyl group and two bromine atoms.

The chemical profile of the water extracts (extracts **1** and **2**) showed greater chemical diversity. Thus, the extraction method must be chosen according to the intended application to the extracted compounds that are different according to the method used. If the extraction time is the most relevant, it should choose ultrasound extraction since it is cheaper than microwave extraction, which requires a more expensive apparatus. Using water extraction by maceration at room temperature will be cheaper and allow chemically richer extracts.

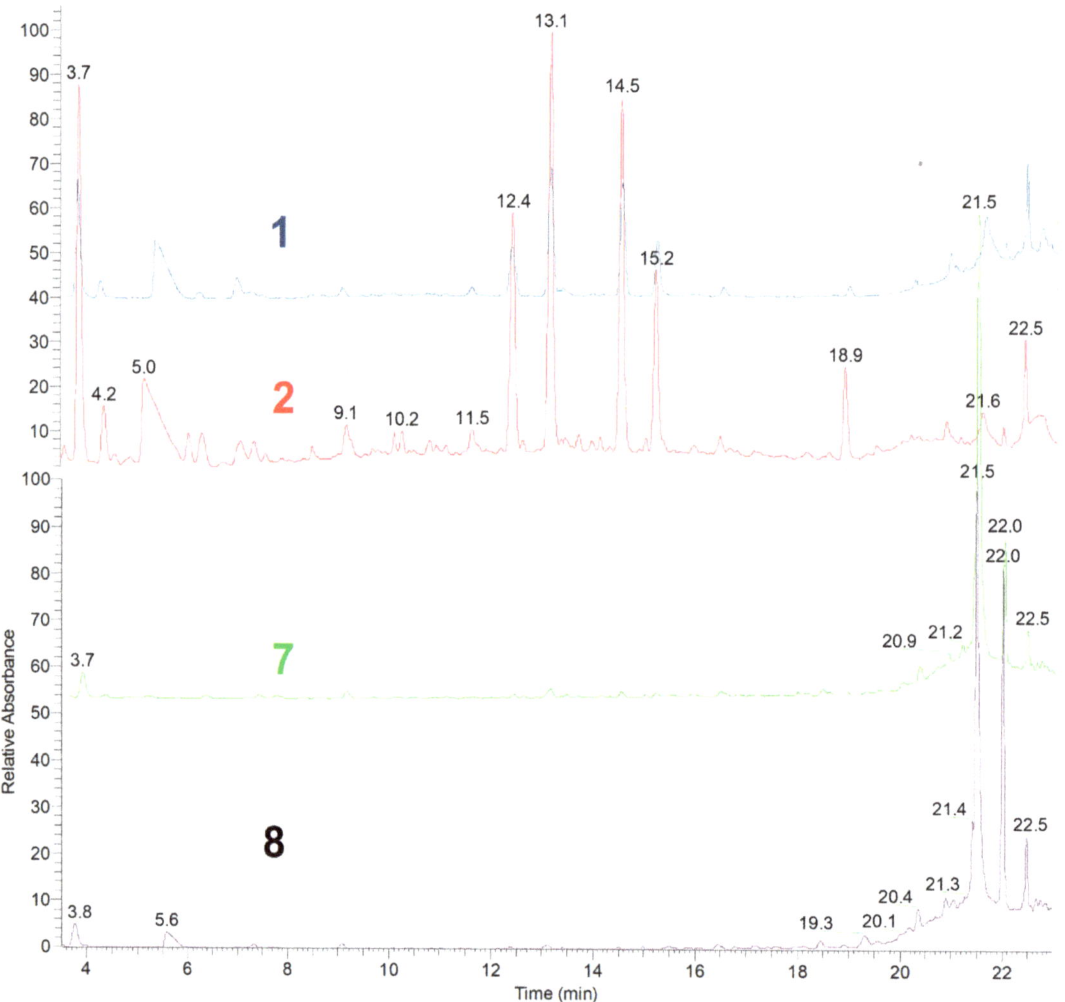

Figure 1. UHPLC chromatograms of *greener A. armata* polar extract with higher yield, recorded at 280 nm.

As this work intends to study the chemical profile of the polar extract of *A. armata*, the characterization of the chemical composition of the aqueous extract obtained by maceration at room temperature (extract **2**) constitutes a significant scientific contribution to knowledge about new polar secondary metabolites in this alga.

3.2. Biological Activities of Asparagopsis armata Extracts

Considering the potential application of *A. armata* extracts in the cosmetic industry, the dichloromethane extract (extract A) and the aqueous extract (extract **2**), which are in our perspective the most interesting ones, were evaluated for their potential as anti-aging agents. In fact, if we consider some green chemistry aspects, such as extraction yields and energy consumption, these extracts are the best option. Moreover, their chemical profile characterization was achieved.

In this regard, the antioxidant, antiacetylcholinesterase, and antibutyrylcolinesterase activities of extract A were evaluated. Regarding the antioxidant activity, this extract demonstrated an interesting ability to scavenge both the DPPH and ABTS radicals (23.6 ± 2.6%

and 31.4 ± 2.8%, respectively), if we compare with the positive control values (Trolox-inhibition of DPPH 89.7 ± 0.5% and inhibition of ABTS 92.4 ± 0.2%). However, these values were obtained at concentrations of 250 µg/mL, and it was not possible to calculate EC_{50} value for extract A. Concerning the cholinesterases inhibition, the extract A showed, at a concentration of 150 µg/mL, inhibition values of 17.1 ± 2.3% and 18.9 ± 2.0%, respectively for acetylcholinesterase and butyrylcholinesterase. Again, it is imperative to highlight that it was impossible to obtain the EC_{50} and the positive control value for donepezil at a concentration of 2 µg/mL was 95.2 ± 0.4%.

Regarding the extract **2** biological assays, we evaluated the ferric chelating ability, the antioxidant, anticollagenase, and antielastase activities. All the activities were evaluated using 250 µg/mL extract concentration, and the EC_{50} could not be obtained. The extract ability to scavenge the DPPH radical and to chelate iron(III) can be considered low and moderate, with percentage inhibition values of 1.47 ± 0.42% and 7.75 ± 0.44%, respectively for DPPH (trolox 89.71 ± 0.50%, concentration of 250 µg/mL) and ferric chelating (EDTA 93.73 ± 0.24%, concentration of 100 µg/mL) assays. In the case of anticollagenase, the same extract showed an inhibition percentage of 7.42 ± 1.79%, which was much lower than the value for positive control EDTA (96.31 ± 1.20%). Finally, the extract inhibited elastase by 43.20 ± 2.58% at a 250 µg/mL concentration, an interesting value, although not comparable to the positive control, *N*-methoxy succinil-Ala-Ala-Pro-Val-chloromethyl ketone which at a 20 µg/mL inhibited the enzyme by 95.51 ± 3.96%.

Although the results of the bioassays were not exceptional, it is possible to conclude that *A. armata* produce bioactive metabolites, and we can infer that their isolation will improve the activity.

3.3. UHPLC-MS Characterization of A. armata Aqueous Extract

Given the above discussed, the aqueous extract obtained at room temperature (extract **2**) was analyzed by UHPLC-MS in the negative mode and showed the presence of several phenolic compounds (Figure 2 and Table 3).

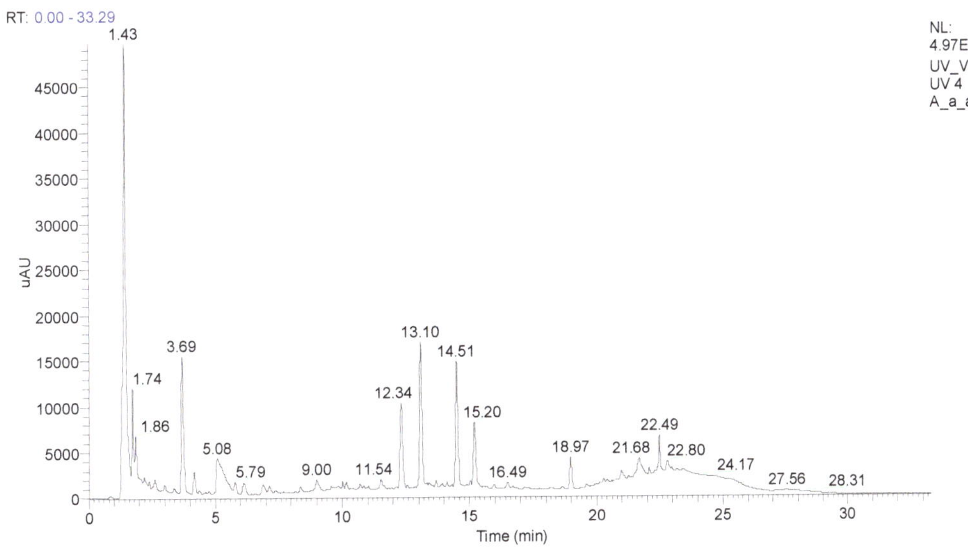

Figure 2. UHPLC chromatogram of *A. armata* aqueous extract, recorded at 280 nm.

The most noticeable aspect of the UHPLC-MS profile is the richness in compounds having bromine atoms (Table 3 and Figure 3), aspect that it is not completely new in view of the publications concerning *A. armata* [11,27] and other red algae [28–30].

Table 3. Tentatively identified compounds on *A. armata* aqueous extract by UHPLC-MS.

Rt (min)	UV-Vis (nm)	[M-H]$^-$ (*m/z*)	MS^2 (*m/z*)	Possible Compound
1.43	201, 324	333/335 [a,b]	302/304, 254, 195, 125	Methyl 3-(2-bromo-3,4,5-trihydroxyphenyl)acrylate
1.74	193, 262, 324	353	191, 179	Caffeic acid derivative
1.86	192, 249, 327	341	179, 135	Caffeic acid derivative
3.69	195, 232, 285	524/526/528 [c,e]	473/475/477, 429/431/433, 286/288, 270/272, 255/257, 218/220, 139, 123	Methyl 3-(5-bromo-3-(2-bromo-3,4,5-trihydroxyphenoxy)-2,4-dihydroxyphenyl)acrylate
5.08	193, 214	586/588/590/592/594 [d]	572/574/576/578/580, 556/558/560/562/564, 322/324/326, 306/308/310, 280/282/284, 264/266/268	Methyl 2,3-dibromo-5-(2,5-dibromo-3,4-dihydroxyphenoxy)-4-hydroxybenzoate
12.34	195, 232	454/456/458/460/462 [d]	272/274/276, 227/228/230	Unknown
13.10	243, 274sh	320/322 [a,e]	274/276, 241, 136	3-(3-Bromo-4,5-dimethoxyphenyl)acrylic acid
14.51	243, 274sh	550/552/554 [c,e]	505/507/509, 470/472/474, 320/322, 274/276, 242, 191, 150, 136	2-(2-Bromo-3,4-dimethoxy-5-methylphenyl)-3-(3-bromo-4,5-dimethoxyphenyl)propanoic acid
15.20	257, 282sh	369	163	*p*-Coumaric acid derivative
18.97	249, 286sh	323	163	*p*-Coumaric acid derivative

Rt = retention time, [M-H]$^-$ = pseudomolecular and MS^2 = fragment ions, [a] ion cluster for one bromine, [b] adduct with formic acid, [c] ion cluster for two bromine, [d] ion cluster for four bromine, [e] adduct with chloride.

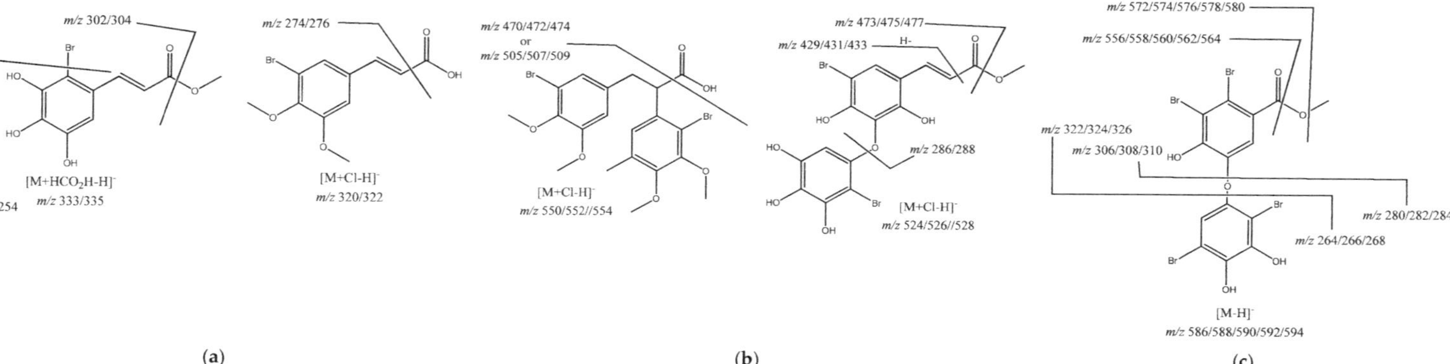

Figure 3. (**a**) Proposed structures for monobromide derivatives; (**b**) proposed structures for dibromide derivatives; (**c**) proposed structure for tetrabromide derivative.

Among the identified compounds are four cinnamic acid derivatives, specifically from caffeic and *p*-coumaric acids (Table 3), prevalent compounds produced by natural resources, including seaweeds and other marine organisms [31,32]. The fragments observed in MS^2, *m/z* 179, and *m/z* 163 are an essential clue to establish that the compounds are respectively caffeic and *p*-coumaric acid derivatives (Table 3).

The other metabolites present in the extract are brominated derivatives, and the compound with a retention time of 12.34 min could not be identified, but the isotopic pattern suggests that it has four bromine atoms (Table 3). A structure was proposed for the remaining five metabolites (Figure 3) based on the pseudomolecular ion, isotopic pattern, and fragment ions observed. However, we should highlight that we did not establish the aromatic rings substitution pattern, and in the proposed structures, the substituents order was made considering the most common ones described in the literature [28,30].

Compounds eluted at 1.43 min and 13.10 min were identified as cinnamic acid derivatives bearing one bromine atom (Figure 3a) due to their pseudomolecular ion isotopic pattern (Table 3). Moreover, brominated polyphenols and polymethoxybenzenes are commonly found in seaweeds [30]. The main fragments ions suggest that the compound eluted at 1.43 min is the methyl 3-(2-bromo-3,4,5-trihydroxyphenyl)acrylate or one isomer, being the fragment ion *m/z* 254 obtained through the loss of the bromine atom and the fragment ion *m/z* 195 obtained through an extra loss of CO_2CH_3. In the case of compound eluted at 13.10 min 3-(3-bromo-4,5-dimethoxyphenyl)acrylic acid, it can be observed the typical loss of CO_2 and also the loss of the bromine atom to give the fragment ion *m/z* 241 (Figure 3a).

Next, we identified two compounds, methyl 3-(5-bromo-3-(2-bromo-3,4,5- trihydroxyphenoxy)-2,4-dihydroxyphenyl)acrylate eluted at 3.69 min and 2-(2-bromo-3,4-dimethoxy-5-methylphenyl)-3-(3-bromo-4,5-dimethoxyphenyl)propanoic acid eluted at 14.51 min, both having an ion cluster for two bromine atoms. The main fragment ions confirm the proposed structures (Figure 3b); however, we should again stress that the bromine atoms' position in the aromatic rings was not demonstrated. The fragments just suggest that they are one in each ring.

Finally, we have methyl 2,3-dibromo-5-(2,5-dibromo-3,4-dihydroxyphenoxy)-4-hydroxybenzoate (Figure 3c), eluted at 5.08 min, showing an ion cluster for four bromine atoms (Table 3) and presenting fragment ions consistent with the proposed structure. Once more, it is important to highlight that some fragment ions establish that the bromine atoms are equally distributed in the aromatic rings, but their position can be interchanged.

3.4. GC-MS of Lipophilic Extract

To prepare the lipophilic extract, in a preliminary study, hexane and dichloromethane solvents were used, and the results (unpublished results) showed that dichloromethane extraction yield was superior to that obtained with hexane under the same experimental conditions. Furthermore, dichloromethane is a widely used non-polar and unspecific solvent capable of dissolving a wide variety of less polar metabolites. Thus, to obtain broader knowledge, the dichloromethane extract (extract A) was selected for analysis by GC-MS to determine the chemical composition of the less polar fraction of *A. armata*. The results of qualitative and quantitative analysis are summarized in Table 4.

Table 4. Tentative identified compounds on dichloromethane extract (extract A) from dried *A. armata* by GC-MS.

Rt (min)	Compound Identified *	Medium Content (SD)		
		mg/100 mg Extract	mg/100 g Dried Alga	mg/kg Fresh Alga
	Fatty Acids			
24.1	Lauric acid (C12:0) [b]	0.489 (0.024)	6.11 (0.30)	4.33 (0.21)
30.6	Myristic acid (C14:0) [b]	1.88 (0.07)	23.5 (0.9)	16.6 (0.6)
34.6	Pentadecanoic acid (C15:0) [b]	0.404 (0.012)	5.05 (0.14)	3.58 (0.10)
36.5	Palmitoleic acid (C16:1) [b]	0.767 (0.066)	9.59 (0.82)	6.80 (0.58)
36.8	Palmitic acid (C16:0) [a]	6.67 (0.29)	83.4 (3.6)	59.2 (2.6)
39.5	Oleic acid (C18:1) [a]	1.15 (0.13)	14.4 (1.6)	10.2 (1.1)
39.8	Stearic acid (C18:0) [a]	0.381 (0.017)	4.76 (0.21)	3.37 (0.15)
	Total	**11.7**	**147**	**104**
	Halogenated Compounds			
4.2	Chloro acetic acid [b]	0.497 (0.049)	6.21 (0.61)	4.40 (0.43)
6.1	Clorobutenol [c]	0.502 (0.012)	6.27 (0.15)	4.45 (0.11)
8.5	3-Bromoprop-2-enoic acid [c]	1.40 (0.20)	17.5 (2.51)	12.4 (1.78)
10.1	Bromopyruvic acid [c]	0.612 (0.061)	7.65 (0.77)	5.43 (0.54)
15.5	1,4-Dibromobuten-1-ol [c]	12.4 (1.0)	155 (12)	110 (9)
19.1	Bromotridecenol [c]	0.842 (0.048)	10.5 (0.6)	7.47 (0.42)
19.3	Isomer of bromotridecenol (19.1 min) [c]	0.807 (0.061)	10.1 (0.8)	7.16 (0.54)
23.7	Isomer of 1,4-dibromobuten-1-ol (15.5 min) [c]	0.940 (0.019)	11.7 (0.2)	8.33 (0.17)
	Total	**18.0**	**224**	**159**
	Glycerol and derivatives			
15.2	Glycerol [a,b]	0.151 (0.020)	1.96 (0.20)	1.39 (0.14)
40.4	Glyceryl-glycoside [b]	0.163 (0.031)	2.04 (0.39)	1.44 (0.27)
41.5	1-Monomyristin [b,c]	0.356 (0.046)	4.44 (0.57)	3.15 (0.40)
43.3	1-Monopalmitin [a]	0.353 (0.013)	4.41 (0.16)	3.13 (0.11)
	Total	**1.02**	**12.9**	**9.11**
Sterols				
48.4	Cholesterol [a]	0.617 (0.076)	7.71 (0.95)	5.47 (0.68)
	Aromatic compounds			
16.1	Phenyl acetic acid [a]	0.683 (0.049)	8.54 (0.61)	6.06 (0.43)
18.6	4-Hydroxybenzaldehyde [b]	0.158 (0.009)	1.97 (0.11)	1.40 (0.08)
	Total	**0.841**	**10.5**	**7.46**
	Other compounds			
4.7	Ethyl amine [b]	0.058 (0.004)	0.727 (0.054)	0.516 (0.038)
8.1	Glycolic acid [b]	0.267 (0.037)	3.33 (0.47)	2.37 (0.33)
25.4	Heptadecane [b]	0.900 (0.043)	11.2 (0.5)	7.97 (0.38)
	Total	**1.23**	**15.3**	**10.9**

Rt—Retention time. SD—Standard deviation. * All the compounds possessing hydroxyl groups are identified as the corresponding TMS derivatives. Compounds were identified by: [a] comparison with pure standards; [b] comparison with the GC-MS spectral libraries with similarity index higher than 94% and confirmation of the peaks at *m/z* values corresponding to the characteristic fragmentation pattern of the family in question; [c] interpretation of pattern fragmentation of MS spectrum according to published literature.

The results in Table 4, graphically summarized in Figure 4, clearly show the predominance, in this lipophilic extract, of halogenated compounds (54.8%), being 40.6% dibromine mono-unsaturated alcohols and 14.2% mono-bromine and mono-chloride compounds.

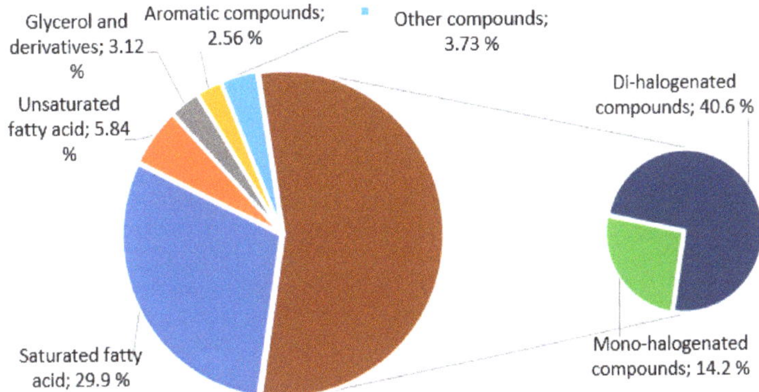

Figure 4. Graphical presentation of the percentage of each class of compounds dichloromethane extract of *A. armata*.

The most abundant compound in the extract (1,4-dibromobutenol, Rt 15.5 min) belongs to this class (68.8% of the total halogenated compounds and 37.8% of the total identified compounds). The presence of one or more halogen atoms, chlorine, or bromine, in the chemical structure of several of the identified compounds is assumed, with a high degree of certainty, from the characteristic isotopic pattern of each chemical element. The bromine exists as two isotopes in approximately equal amounts of atomic mass 79 and 81, while chlorine has 2 isotopes of atomic mass 35 and 37 in a ratio of approximately 3:1.

In the case of the most abundant compound, at Rt 15.5 min, the presence of 2 bromine atoms is deduced from the peak corresponding to molecular ion $[M]^{+\bullet}$ at *m/z* 300/302/304 (a very weak signal) in the ratio of 1:2:1. The mass spectrum of this compound also shows the base peak at *m/z* 75, corresponding to $[HOSi(CH_3)_2]^+$, and the second most intense peak at *m/z* 285/287/289 (also in a 1:2:1 ratio), corresponding to $[M-15]^+$ (loss of a methyl group from the TMS ether moiety). These signals support the hypothesis that the compound at Rt 15.5 min is a silylated primary alcohol [33], with two bromine atoms as substituents. Thus, it can be deduced that the compound in the extract will have a molecular mass of 230 corresponding to the chemical formula $C_4H_6OBr_2$, suggesting the presence of a double bond in the basic structure of dibromo butanol. The greatest diagnostic value is the fragments with *m/z* at 131/133 and 137/139. The presence of these peaks, with identical intensity in each pair, shows that the fragmentation of the molecule gives rise to fragments containing one bromine atom each, being a strong indication of bromine atoms on different carbons. The peak at *m/z* 131/133 will correspond to the fragment $[C_4H_4Br]^+$, resulting from loss of HBr by the molecular ion, following a δ-chain loss, indicating that the bromine atoms are found in the carbons 1 and 4. The peak at *m/z* 137/139 could result from loss of a methyl from the silyloxyl group, followed by bonding of the resulting siliconium ion to the terminal bromine atom, and a final rearrangement [33] to originate the fragment $[BrSi(CH_3)_2]^+$.

Comparing the mass spectrum of 1,4-dibromobuten-1-ol (15.5 min) with that exhibited by the compound at 23.7 min, it is concluded that both are dibrominated mono-unsaturated alcohols with molecular formula $C_4H_6OBr_2$, which mean they are isomers. They have a very similar fragmentation pattern $[M-CH_3]^+$ at *m/z* 285/287/289, $[BrSi(CH_3)_2]^+$ at *m/z* 137/139, and the base peak at *m/z* 73 ($[Si(CH_3)_3]^+$, being the last one ubiquitous to the spectra of TMS derivatives. The most significant differences are the absence of signals at *m/z* 131/133 and the presence of the peak at *m/z* 197/199/201 corresponding to the fragment $[C_3H_3Br_2]^+$, resulting from an α-cleavage with retention of the 2 bromine atoms and the double bond on the cation. These fragments suggest that the difference between the compounds at 15.5 and 23.7 min is the position of the bromo substituents. These two compounds are the only di-halogenated compounds identified in the extract and correspond to 166.2 mg/100 g of dry algae (40.6% of identified compounds).

The peaks at 19.1 and 19.3 min also correspond to two unsaturated alcohols silylated, but in this case mono-brominated, as deduced from the isotopic pattern analysis of fragments $[M-15]^+$ at *m/z* 333/335 (1:1); $[C_2H_6SiBr]^+$ at *m/z* 137/139 (1:1). The peak at *m/z* 179 on the mass spectra of both isomers, corresponding to the non-brominated fragment $[C_{13}H_{23}]^+$ ($[M-90-Br]^+$), suggests the same bromine substitution pattern on both isomers. The molecular formula of the two compounds on the extract is $C_{13}H_{25}BrO$, named bromotridecenol. The difference between them should be the position of the double bond, difficult to determine by MS.

The second most abundant halogenated compound (7.8%) corresponds to the peak at 8.5 min, identified as 3-bromoprop-2-enoic acid TMS derivative. The mass spectrum shows the signals at *m/z* 207/209 (~1:1), corresponding to the fragment $C_5H_8BrO_2Si^+$ ($[M-15]^+$) and 137/139 (~1:1) corresponding to $[(CH_3)_2SiBr]^+$, both mono-bromine silylated fragments typical from halocarboxylic acids pattern fragmentation [33]. The *m/z* value of these two fragments differs by 2 units from those observed for 3-bromopropanoic acid, evidencing the existence of a double bond at C2-C3. Moreover, the three signals observed in the spectrum of 3-bromoprop-2-enoic acid TMS derivative at *m/z* 143 $[M-Br]^+$, 75 (base peak, $[HOSi(CH_3)_2]^+$) and 53 (loss of trimethylsilanol and HBr resulting the fragments $[C_3H_3O]^+$), all corresponding to non-brominate fragments, confirm the proposed identification of compound at 8.5 min. This compound was previously identified on the *A. armata* as ethyl ester [14].

The presence of halogenated alcohols and carboxylic acids in *Asparagopsis* species was already reported in the literature [11,14,34]. The results of several studies show that these are toxic metabolites that, in addition to exhibiting antimicrobial activity, confer a competitive advantage over predators and other species [35,36]. Moreover, to our best knowledge, this is the first time that 1,4-dibromobuten-1-ol and its isomer have been reported and quantified in the *Asparagopsis* genus.

The second most abundant family of compounds was the fatty acids (35.8%), being the content of saturated acids 5.1 times higher than the content of unsaturated acids (Figure 4). This can be an asset since saturated fatty acids seem to have a beneficial effect on the modulation of the immune system on food-borne bacterial infection [37].

The results of Table 4 show that the second most abundant compound is palmitic acid (20.3%), being the most plentiful fatty acid (57.0%). In comparison, the second most abundant fatty acid is myristic acid (16.1%). These results are in line with those published by Pereira et al. [38] for the same species, where palmitic and myristic acids are, by far, the most abundant fatty acids. Interestingly, in line with this result, the monoacylglycerols of these two acids, monomyristin and monopalmitin, were identified and quantified in the extract (Table 4; Rt = 41.5 and Rt = 43.3 min, respectively), representing about 2.1% of the compounds detected.

As far as unsaturated fatty acids are concerned (Figure 4), the most abundant is oleic acid ((9Z) 9-octadecenoic acid). It is the third most abundant fatty acid (9.83%), representing only 3.50% of the identified compounds.

From the point of view of beneficial effects on health, the presence of oleic acid can be considered an asset, while the high content of palmitic acid can be seen as an obstacle to seaweed consumption. However, we draw readers' attention to the fact that data in the literature suggest that the potential adverse effect of saturated acids is controversial and should be reviewed [39,40]. Moreover, the results show that *A. armata* can be valued as a source of palmitic acid. In fact, this compound is described as modulating the immune response [37] and it is an interesting antifouling agent due to its action against the first bacterial colonies on submerged surfaces [41].

Neither polyunsaturated acid nor unsaturated acids with *trans* configuration were detected, contrary to what is reported by Pereira et al. [38], where *trans*-oleic acid is detected in greater amounts than *cis*-oleic acid, and small amounts of arachidonic and eicosapentaenoic (EPA) acids are also noticed. These discrepancies may be related to geographical, seasonal, and/or environmental factors, but above all, due to the significant differences between sample preparation procedures before GC-MS analysis.

Only one compound from the hydrocarbon family has been identified, the heptadecane (Table 4, Rt = 25.4 min, $[M]^{+\bullet}$ at *m/z* 240;). This alkane represents 2.7% of the total content of compounds, being the seventh most abundant. Although some hydrocarbons have been identified in the genus *Asparagopsis* [42], they have a chain of up to 5 carbons, while in this work, a long-chain alkane (17 carbons) is identified for the first time.

Cholesterol was the only sterol detected in the dichloromethane extract of *A. armata*, corresponding to 7.71 mg/100 g of dry seaweed (1.8% of identified compounds). This result aligns with the literature [12,13] that describes cholesterol as the most abundant sterol in this species. Lopes et al. [13] indicate a total cholesterol content of 28.92 mg/100 g of dry material, a value higher than the one reported here. This difference is justified because cholesterol exists mainly in non-free form as a constituent of the cell membrane, being released during the alkaline hydrolysis to which the sample was submitted in work by Lopes et al. [13]. It should be noted that the sterol content of *Asparagopsis* species is variable and depends on several factors (e.g., geographic, environmental, and seasonal) but mainly dependent on the life cycle phase [12]. Despite its "bad reputation" as a constituent of food products, cholesterol is a valuable natural product due to its chemical skeleton being easily transformable into new derivatives with diverse applications, such as drug delivery, production of liquid crystals and gelling agents, hormone precursor, and bioactive compounds [43].

4. Conclusions

The initial study on applying different methodologies to obtain polar extracts of *A. armata* showed that the choice of the extraction method would have to consider which compounds are intended to be extracted and the intended application for the obtained extract. The chemical profile of the water extract obtained by maceration (extract **2**) has greater chemical diversity and is economically cheaper.

Herein, for the first time, the aqueous and lipophilic *A. armata* extracts were tested as anti-aging agents with antioxidant and/or enzymatic inhibitors action. The polar extract was the most active, exhibiting the ability to inhibit $43.20 \pm 2.58\%$ of elastase activity at a 250 µg/mL concentration. On the other hand, although the lipophilic extract is a weak cholinesterase inhibitor, it is a dual inhibitor (it inhibits acetyl and butyryl-cholinesterase identically), which constitutes an advantage concerning its potential as an agent against Alzheimer's Disease.

Qualitative analysis by UHPLC-MS demonstrates that *A. armata* water extract is rich in phenolic compounds bearing bromine atoms; furthermore, the cinnamic acid scaffold is a prominent nucleus. The secondary metabolites herein described may explain the seaweed persistence and absence of predators. Furthermore, the anti- elastase activity mentioned above may be related to these phenolic compounds present in the water extract; nevertheless, more studies are essential.

The analysis of the most lipophilic fraction of *A. armata* (dichloromethane extract) by GC-MS allowed to identify and quantify 25 compounds, seven of them for the 1st time in this species, such as two dibrominated unsaturated alcohols, the 1,4-dibromobuten-1-ol and its isomer with the bromine atoms in different positions (a total of 166.2 mg/100 g of dry algae, 40.6% of identified compounds); two monobromine unsaturated alcohols, the bromotridecenol and its isomer with double bond on different position (a total of 20.6 mg/100 g of dry algae, 5.0% of identified compounds); one alkane, the heptadecane (a total of 11.2 mg/100 g of dry algae, 2.7% of identified compounds); and two monoacylglycerols, the monopalmitin and the monomyristin (a total of 8.85 mg/100 g of dry algae, 2.1% of identified compounds).

The halogenated compounds and fatty acids are the two major compound families in the extract, being the 1,4-dibromobuten-1-ol and the palmitic acid the two most abundant identified compounds (155 and 83.4 mg/100 g of dry algae, respectively) and the myristic acid the third most abundant compound (23.5 mg/100 g of dry algae). The unsaturated fatty acids family, mainly 9Z-octadecenoic acid (14.4 mg/100 g of dry algae), corresponds to nearly 6% of the identified compounds.

As it is already known that no completely satisfactory method allows the analysis of the entire range of halogenated compounds of the *Asparagopsis* species, as they have a wide range of solubilities and boiling points [44], the work presented here is innovative as it allows the identification of a variety of lipophilic and halogenated aromatic compounds structurally distinct from those identified so far in *A. armata*.

Considering what is already known about the ecotoxicological effect of several halogenated compounds identified in *Asparagopsis* species [11,45], which give them a competitive advantage, and the commercial value of several scaffold phytochemicals identified in this species, the results obtained herein open new perspectives for valuing the *A. armata* as a source of halogenated compounds and fatty acids with high biotechnological and economic potential.

Valuing invasive species as a source of value-added natural products is an approach that promotes biodiversity conservation and ecosystem sustainability. It increases the algae harvest, allowing the assisted recolonization of free space by native species, because it increases the demand for existing and available biomass.

Author Contributions: Conceptualization, D.C.G.A.P., M.C.B. and A.M.L.S.; formal analysis and investigation, D.C.G.A.P., M.L.L., G.P.R., M.C.B. and A.M.L.S.; writing—original draft preparation, D.C.G.A.P. and A.M.L.S.; writing—review and editing, D.C.G.A.P., G.P.R., M.C.B., A.M.S.S. and A.M.L.S. All authors have read and agreed to the published version of the manuscript.

Funding: This study was financed by ASPAZOR project (DRCT: ACORES-01-0145-FEDER-00060- AS-PAZOR); Portuguese National Funds, through FCT—Fundação para a Ciência e a Tecnologia, and as applicable co-financed by the FEDER within the PT2020 Partnership Agreement by funding the LAQV-REQUIMTE (UIDB/50006/2020) and the cE3c centre (FCT Unit funding (Ref. UID/BIA/00329/2013, 2015–2018) and UID/BIA/00329/2019).

Institutional Review Board Statement: Not applicable.

Informed Consent Statement: Not applicable.

Data Availability Statement: Not applicable.

Acknowledgments: Thanks are also due to the University of Aveiro and the University of Azores, to Mónica S.G.A. Válega and her support in UHPLC-MS analysis, and to Carlota Ferro for her support in biological activities.

Conflicts of Interest: The authors declare no conflict of interest.

References

1. Fusetani, N. Marine Natural Products. In *Natural Products in Chemical Biology*; Civjan, N., Ed.; John Wiley and Sons: New York, NY, USA, 2012; pp. 31–64.
2. Hamed, I.; Özogul, F.; Özogul, Y.; Regenstein, J.M. Marine bioactive compounds and their health benefits: A review. *Compr. Rev. Food Sci. Food Saf.* **2015**, *14*, 446–465. [CrossRef]
3. Rosa, G.P.; Sousa, P.; Tavares, W.R.; Pagès, A.K.; Seca, A.M.L.; Pinto, D.C.G.A. Seaweeds secondary metabolites with beneficial health effects: An overview of successes in in vivo studies and clinical trials. *Mar. Drugs* **2020**, *18*, 8. [CrossRef]
4. Guiry, G.M. AlgaeBase 2021. World-Wide Electronic Publication. National University of Ireland. Galway. Available online: http://www.algaebase.org (accessed on 1 October 2021).
5. Abbott, I.A. The uses of seaweed as food in Hawaii. *Econ. Bot.* **1978**, *32*, 409–412. [CrossRef]

6. Kraan, S.; Barrington, K.A. Commercial farming of *Asparagopsis armata* (Bonnemaisoniceae. Rhodophyta) in Ireland. Maintenance of an introduced species? *J. Appl. Phycol.* **2005**, *17*, 103–110. [CrossRef]

7. Chualáin, F.N.; Maggs, C.A.; Saunders, G.W.; Guiry, M.D. The invasive genus *Asparagopsis* (Bonnemaisoniaceae. Rhodophyta): Molecular systematics. morphology. and ecophysiology of Falkenbergia isolates. *J. Phycol.* **2004**, *40*, 1112–1126. [CrossRef]

8. Katsanevakis, S.; Wallentinus, I.; Zenetos, A.; Leppäkoski, E.; Çinar, M.E.; Oztürk, B.; Grabowski, M.; Golani, D.; Cardoso, A.C. Impacts of invasive alien marine species on ecosystem services and biodiversity: A pan-European review. *Aqua. Invas.* **2014**, *9*, 391–423. [CrossRef]

9. Pinteus, S.; Lemos, M.F.L.; Alves, C.; Silva, J.; Pedrosa, R. The marine invasive seaweeds *Asparagopsis armata* and *Sargassum muticum* as targets for greener antifouling solutions. *Sci. Total Environ.* **2021**, *750*, 141372. [CrossRef]

10. Pinteus, S.; Alves, C.; Monteiro, H.; Araújo, E.; Horta, A.; Pedrosa, R. *Asparagopsis armata* and *Sphaerococcus coronopifolius* as a natural source of antimicrobial compounds. *World J. Microbiol. Biotechnol.* **2015**, *31*, 445–451. [CrossRef]

11. Félix, R.; Dias, P.; Félix, C.; Cerqueira, T.; Andrade, P.B.; Valentão, P.; Lemos, M.F.L. The biotechnological potential of *Asparagopsis armata*: What is known of its chemical composition, bioactivities and current market? *Algal Res.* **2021**, *60*, 102534. [CrossRef]

12. Combaut, G.; Bruneau, Y.; Codomier, L.; Teste, J. Comparative sterols composition of the red alga *Asparagopsis armata* and its Tetrasporophyte *Falkenbergia rufolanosa*. *J. Nat. Prod.* **1979**, *42*, 150–151. [CrossRef]

13. Lopes, G.; Sousa, C.; Bernardo, J.; Andrade, P.B.; Valentão, P.; Ferreres, F.; Mouga, T. Sterol profiles in 18 macroalgae of the portuguese coast. *J. Phycol.* **2011**, *47*, 1210–1218. [CrossRef] [PubMed]

14. McConnell, O.; Fenical, W. Halogen chemistry of the red alga *Asparagopsis*. *Phytochemistry* **1977**, *16*, 367–374. [CrossRef]

15. Zárate, R.; Portillo, E.; Teixidó, S.; Carvalho, M.A.A.P.D.; Nunes, N.; Ferraz, S.; Seca, A.M.L.; Rosa, G.P.; Barreto, M.C. Pharmacological and cosmeceutical potential of seaweed beach-casts of macaronesia. *Appl. Sci.* **2020**, *10*, 5831. [CrossRef]

16. Blois, M.S. Antioxidant determinations by the use of a stable free radical. *Nature* **1958**, *181*, 1199–1200. [CrossRef]

17. Re, R.; Pellegrini, N.; Proteggente, A.; Pannala, A.; Yang, M.; Rice-Evans, C. Antioxidant activity applying an improved ABTS radical cation decolorization assay. *Free Radic. Biol. Med.* **1999**, *26*, 1231–1237. [CrossRef]

18. Decker, E.A.; Welch, B. Role of ferritin as a lipid oxidation catalyst in muscle food. *J. Agric. Food Chem.* **1990**, *38*, 674–677. [CrossRef]

19. Ellman, G.L.; Courtney, K.D.; Andres Jr, V.; Featherstone, R.M. A new and rapid colorimetric determination of acetylcholinesterase activity. *Biochem. Pharmacol.* **1961**, *7*, 88–95. [CrossRef]

20. Arruda, M.; Viana, H.; Rainha, N.; Neng, N.R.; Rosa, J.S.; Nogueira, J.M.; Barreto, M.C. Anti-acetylcholinesterase and antioxidant activity of essential oils from *Hedychium gardnerianum* Sheppard ex Ker-Gawl. *Molecules* **2012**, *17*, 3082–3092. [CrossRef] [PubMed]

21. Shimizu, K.; Kondo, R.; Sakai, K.; Lee, S.H.; Sato, H. The inhibitory components from *Artocarpus incisus* on melanin biosynthesis. *Planta Med.* **1998**, *64*, 408–412. [CrossRef]

22. Manosroi, A.; Jantrawut, P.; Akihisa, T.; Manosroi, W.; Manosroi, J. In vitro anti-aging activities of *Terminalia chebula* gall extract. *Pharm. Biol.* **2010**, *48*, 469–481. [CrossRef] [PubMed]

23. Mandl, I.; MacLennan, J.D.; Howes, E.L.; De Bellis, R.H.; Sohler, A. Isolation and characterization of proteinase and collagenase from *Cl. histolyticum*. *J. Clin. Investig.* **1953**, *32*, 1323–1329. [CrossRef] [PubMed]

24. Ndlovu, G.; Fouche, G.; Tselanyane, M.; Cordier, W.; Steenkamp, V. In vitro determination of the anti-aging potential of four southern African medicinal plants. *BMC Complement. Altern. Med.* **2013**, *13*, 304. [CrossRef] [PubMed]

25. Halket, J.M.; Zaikin, V.G. Derivatization in mass spectrometry-1. Silylation. *Eur. J. Mass Spectro.* **2003**, *9*, 1–21. [CrossRef] [PubMed]

26. Seca, A.M.L.; Gouveia, V.L.M.; Barreto, M.C.; Silva, A.M.S.; Pinto, D.C.G.A. Comparative study by GC-MS and chemometrics on the chemical and nutritional profile of *Fucus spiralis* L. juvenile and mature life-cycle phases. *J. Appl. Phycol.* **2018**, *30*, 2539–2548. [CrossRef]

27. Genovese, G.; Tedone, L.; Hamann, M.T.; Morabito, M. The Mediterranean red alga *Asparagopsis*: A source of compounds against *Leishmania*. *Mar. Drugs* **2009**, *7*, 361–366. [CrossRef]

28. Gribble, G.W. Biological activity of recently discovered halogenated marine natural products. *Mar. Drugs* **2015**, *13*, 4044–4136. [CrossRef]

29. Rocha, D.H.A.; Seca, A.M.L.; Pinto, D.C.G.A. Seaweed secondary metabolites in vitro and in vivo anticancer activity. *Mar. Drugs* **2018**, *16*, 410. [CrossRef] [PubMed]

30. Mandrekar, V.K.; Gawas, U.B.; Majik, M.S. Brominated molecules from marine algae and their pharmacological importance. In *Studies in Natural Products Chemistry, Atta-ur-Rahman*; Elsevier: Amsterdam, The Netherlands, 2019; Volume 61, pp. 461–490.

31. Pereira, R.B.; Pinto, D.C.G.A.; Pereira, D.M.; Gomes, N.G.M.; Silva, A.M.S.; Andrade, P.B.; Valentão, P. UHPLC-MS/MS profiling of *Aplysia depilans* and assessment of its potential therapeutic use: Interference on iNOS expression in LPS-stimulated RAW 264.7 macropheges and caspase-mediated pro-apoptotic effect on SH-SY5Y cells. *J. Func. Foods* **2017**, *37*, 164–175. [CrossRef]

32. Zhong, B.; Robinson, N.A.; Warner, R.D.; Barrow, C.J.; Dunshea, F.R.; Suleria, H.A.R. LC-ESI-QTOF-MS/MS Characterization of seaweed phenolics and their antioxidant potential. *Mar. Drugs* **2020**, *18*, 331. [CrossRef] [PubMed]

33. Harvey, D.J.; Vouros, P. Mass spectrometric fragmentation of trimethylsilyl and related alkylsilyl derivatives. *Mass Spectro. Rev.* **2020**, *39*, 105–211. [CrossRef] [PubMed]

34. Greff, S.; Zubia, M.; Genta-Jouve, G.; Massi, L.; Perez, T.; Thomas, O.P. Mahorones, highly brominated cyclopentenones from the red alga *Asparagopsis taxiformis*. *J. Nat. Prod.* **2014**, *77*, 1150–1155. [CrossRef]

35. Paul, N.A.; De Nys, R.; Steinberg, P.D. Chemical defence against bacteria in the red alga. *Mar. Ecol. Prog. Ser.* **2006**, *306*, 87–101. [CrossRef]
36. Mata, L.; Wright, E.; Owens, L.; Paul, N.; de Nys, R. Water-soluble natural products from seaweed have limited potential in controlling bacterial pathogens in fish aquaculture. *J. Appl. Phycol.* **2013**, *25*, 1963–1973. [CrossRef]
37. Harrison, L.M.; Balan, K.V.; Babu, U.S. Dietary fatty acids and immune response to food-borne bacterial infections. *Nutrients* **2013**, *5*, 1801–1822. [CrossRef] [PubMed]
38. Pereira, H.; Barreira, L.; Figueiredo, F.; Custódio, L.; Vizetto-Duarte, C.; Polo, C.; Rešek, E.; Engelen, A.; Varela, J. Polyunsaturated fatty acids of marine macroalgae: Potential for nutritional and pharmaceutical applications. *Mar. Drugs* **2012**, *10*, 1920–1935. [CrossRef] [PubMed]
39. Mancini, A.; Imperlini, E.; Nigro, E.; Montagnese, C.; Daniele, A.; Orrù, S.; Buono, P. Biological and nutritional properties of palm oil and palmitic acid: Effects on health. *Molecules* **2015**, *20*, 17339–17361. [CrossRef] [PubMed]
40. Agostoni, C.; Moreno, L.; Shamir, R. Palmitic acid and health: Introduction. *Crit. Rev. Food Sci. Nutr.* **2016**, *56*, 1941–1942. [CrossRef] [PubMed]
41. Bazes, A.; Silkina, A.; Douzenel, P.; Faÿ, F.; Kervarec, N.; Morin, D.; Berge, J.-P.; Bourgougnon, N. Investigation of the antifouling constituents from the brown alga *Sargassum muticum* (Yendo) Fensholt. *J. Appl. Phycol.* **2009**, *21*, 395–403. [CrossRef]
42. Broadgate, W.J.; Malin, G.; Küpper, F.C.; Thompson, A.; Liss, P.S. Isoprene and other non-methane hydrocarbons from seaweeds: A source of reactive hydrocarbons to the atmosphere. *Mar. Chem.* **2004**, *88*, 61–73. [CrossRef]
43. Albuquerque, H.M.T.; Santos, C.M.M.; Silva, A.M.S. Cholesterol-based compounds: Recent advances in synthesis and applications. *Molecules* **2019**, *24*, 116. [CrossRef]
44. Burreson, B.J.; Moore, R.E.; Roller, P.P. Volatile halogen compounds in the alga *Asparagopsis taxiformis* (Rhodophyta). *J. Agric. Food. Chem.* **1976**, *24*, 856–861. [CrossRef]
45. Máximo, P.; Ferreira, L.M.; Branco, P.; Lima, P.; Lourenço, A. Secondary metabolites and biological activity of invasive macroalgae of Southern Europe. *Mar. Drugs* **2018**, *16*, 265. [CrossRef] [PubMed]

**applied
sciences**

Article

Extraction of Antioxidant Compounds and Pigments from Spirulina (*Arthrospira platensis*) Assisted by Pulsed Electric Fields and the Binary Mixture of Organic Solvents and Water

Francisco J. Martí-Quijal [1], Francesc Ramon-Mascarell [1], Noelia Pallarés [1], Emilia Ferrer [1], Houda Berrada [1], Yuthana Phimolsiripol [2,3] and Francisco J. Barba [1,4,*]

[1] Department of Preventive Medicine and Public Health, Food Science, Toxicology and Forensic Medicine, Faculty of Pharmacy, Universitat de València, Avda. Vicent Andrés Estellés s/n, 46100 Burjassot, Spain; francisco.j.marti@uv.es (F.J.M.-Q.); ramas@alumni.uv.es (F.R.-M.); noelia.pallares@uv.es (N.P.); emilia.ferrer@uv.es (E.F.); houda.berrada@uv.es (H.B.)
[2] Faculty of Agro-Industry, Chiang Mai University, Chiang Mai 50100, Thailand; yuthana.p@cmu.ac.th
[3] Cluster of Agro Bio-Circular-Green Industry (Agro-BCG), Chiang Mai University, Chiang Mai 50200, Thailand
[4] Nutrition and Bromatology Group, Department of Analytical and Food Chemistry, Faculty of Food Science and Technology, Ourense Campus, University of Vigo, 32004 Ourense, Spain
* Correspondence: francisco.barba@uv.es; Tel.: +34-963-544-972

Citation: Martí-Quijal, F.J.; Ramon-Mascarell, F.; Pallarés, N.; Ferrer, E.; Berrada, H.; Phimolsiripol, Y.; Barba, F.J. Extraction of Antioxidant Compounds and Pigments from Spirulina (*Arthrospira platensis*) Assisted by Pulsed Electric Fields and the Binary Mixture of Organic Solvents and Water. *Appl. Sci.* **2021**, *11*, 7629. https://doi.org/10.3390/app11167629

Academic Editors: Claudio Medana, Ana M. L. Seca and Eugenia Gallardo

Received: 16 July 2021
Accepted: 17 August 2021
Published: 19 August 2021

Abstract: The application of pulsed electric fields (PEF) is an innovative extraction technology promoting cell membrane electroporation, thus allowing for an efficient recovery, from an energy point of view, of antioxidant compounds (chlorophylls, carotenoids, total phenolic compounds, etc.) from microalgae. Due to its selectivity and high extraction yield, the effects of PEF pre-treatment (3 kV/cm, 100 kJ/kg) combined with supplementary extraction at different times (5–180 min) and with different solvents (ethanol (EtOH)/H_2O, 50:50, v/v; dimethyl sulfoxide (DMSO)/H_2O, 50:50, v/v) were evaluated in order to obtain the optimal conditions for the extraction of different antioxidant compounds and pigments. In addition, the results obtained were compared with those of a conventional treatment (without PEF pre-treatment but with constant shaking). After carrying out the different experiments, the best extraction conditions to recover the different compounds were obtained after applying PEF pre-treatment combined with the binary mixture EtOH/H_2O, 50:50, v/v, for 60–120 min. PEF extraction was more efficient throughout the study, especially at short extraction times (5–15 min). In this sense, recovery of 55–60%, 85–90%, and 60–70% was obtained for chlorophylls, carotenoids, and total phenolic compounds, respectively, compared to the maximum total extracted amount. These results show that PEF improves the extraction yield of antioxidant bioactive compounds from microalgae and is a promising technology due to its profitability and environmental sustainability.

Keywords: pulsed electric fields; green extraction; microalgae; antioxidants; pigments

1. Introduction

Over the last decade, several research studies have evaluated the use of microalgae as a source of nutrients and bioactive compounds. This is due to a growing interest in the development of new foods that provide health benefits and that meet basic energy and nutritional requirements [1,2]. It is preferred that functional foods have a natural origin, such as from plants, algae and/or microalgae. In this sense, proteins obtained from microalgae have been used to replace proteins of animal origin in meat-like preparations such as turkey burgers [3] or in the fortification of vegan foods such as kefir produced from soy and almond based beverages [4].

Microalgae are becoming increasingly relevant, especially for their composition, since they are a source of high-added-value compounds [5], such as carotenoids, chlorophylls, and other pigments (antioxidants) [6], and polyunsaturated fatty acids [7,8].

Conventional extraction of antioxidant bioactive compounds from microalgae is often carried out using solvents and using dry biomass [9]. However, this conventional extraction method is very slow, involves the extraction of unwanted compounds, and can promote the degradation of some thermolabile compounds. Therefore, there is a need for innovative approaches such as pulsed electric fields (PEF) that affect the quality of the extracted compounds to a lesser extent and can be applied in a continuous flow to obtain higher extraction efficiency rates, minimize the use of solvents, and thus be a more efficient and sustainable extraction alternative [10,11].

In recent years, several studies have applied PEF technology on microalgae for extraction purposes [12–15]. This non-thermal technique consists of applying high-voltage electrical pulses between two electrodes in the treatment chamber [16]. Short electric pulses (1–100 µs) at field intensities of 0.1–1 kV cm^{-1} are employed for reversible permeabilization in plant cells for stress induction, at 0.5–3 kV cm^{-1} for irreversible permeabilization of plant and animal tissues, and at 15–40 kV cm^{-1} for irreversible permeabilization of microbial cells.

The external electric field increases the transmembrane potential and promotes membrane pore formation of the biological cell. High-intensity electric pulses can be generated by the switched discharge of a suitable capacitor bank. The properties of the discharge circuit determine the shape of the time-dependent potential at the treatment chamber. Parallel plate electrodes or colinear type treatment chambers constitute the most commonly used circuits [17]. The PEF technique can be effectively applied in many food processing applications such as microorganism/enzyme inactivation, recovery of bioactive compounds, drying and freezing processes, and to promote the enhancement of some selected properties of food macromolecules and some chemical reactions [16]. It has several advantages compared to conventional techniques as it favours the extraction of bioactive compounds by generating reversible or irreversible micropores in the plasma membrane of the cells, promoting the migration of interesting compounds into the cytoplasm through the membrane with high selectivity/purity, no thermal effect, and short extraction times [18]. The efficiency of PEF to permeabilize cell membranes differs according to process parameters such as electric field strength, treatment time, specific energy, pulse shape, pulse width, frequency, temperature, the properties of the treated food sample such as its pH and conductivity, and the characteristics of the target cells [19].

An example of its usefulness is observed in the extraction of lipids from microalgae. In this regard, a first PEF-assisted extraction of water-soluble bioactive compounds that cross the membrane through the formed micropores can be performed and then a second extraction with ethanol (EtOH) can be applied to obtain lipids or lipid-soluble metabolites that have remained inside the cell, which provides advantages in subsequent purification processing [20]. The use of PEF is possible in aqueous solutions with a low dry matter content, facilitating the more energy-efficient isolation of high-added-value compounds extracted from microalgae directly from the culture without dehydration or drying [20,21].

Most of the studies evaluating the application of PEF focus on the use of aqueous suspensions of microalgae, which mainly allows for the extraction of water-soluble compounds [22,23], while the extraction of non-polar pigments (e.g., chlorophylls and carotenoids) is very low under these conditions due to their low solubility in water [24]. Some previous studies have evaluated the use of complementary extractions with EtOH (96%), obtaining higher yields in the extraction of pigments from microalgae pre-treated with PEF [25]. Studies combining the use of PEF in suspensions of microalgae *(Nannochloropsis)* in water and the subsequent addition of an organic solvent to improve pigment extraction efficiency have been also carried out [12,13].

PEF-assisted extraction yields of nutritionally valuable compounds (lipids, pigments, and proteins) can vary depending on the microalgae used [26]. Therefore, specific studies are needed to evaluate the extraction in different microalgae species at different times and with different solvents to obtain the necessary information and thus be able to scale up the process to an industrial level in an efficient, sustainable way and with a high extraction yield.

Several studies are currently evaluating PEF-assisted extraction from microalgae biomass such as Chlorella and *Nannochloropsis* [12,22,25,27]. However, there is a lack of data on the impact of PEF on the extraction of high-added-value compounds from other microalgae species such as spirulina (*Arthrospira platensis*). Spirulina is an undifferentiated filamentous cyanobacterium [28,29] whose cells are 3–12 µm wide and can reach 16 µm [30]; it has been used as food for centuries by various cultures as it has a biochemical profile rich in bioactive molecules and there are studies that support its benefits for human health [31].

However, considering the gap that exists regarding different extraction levels of antioxidant compounds according to the different microalgae species, the main aim of this study was to evaluate how PEF pre-treatment combined with the binary mixtures of ethanol (EtOH)/H$_2$O, 50:50, v/v and dimethyl sulfoxide (DMSO)/H$_2$O, 50:50, v/v at different extraction times can affect to the recovery of antioxidant bioactive compounds from spirulina.

2. Materials and Methods

2.1. Sample

The spirulina, in noodle form, was produced by the company Ecospirulina (Serra, Comunitat Valenciana, Spain). The cultured biomass comes from the species *Artrospira platensis* (more recently *Limnospira platensis*), strain *paracas 15016*. The Paracas reference refers to the lake from which it originated, Lake Paracas, south of Lima, Peru.

In Ecospirulina, cultivation is carried out in ponds in a greenhouse under natural sunlight and without the use of artificial light. Shading is applied to partially cover the culture ponds, which allows pigment production to be controlled. During the production of the sample used in this trial, the average temperature during the day was 32 °C, while the average temperature during the night was 24 °C. The pH of the culture varied between 9.8 and 10.4, being regulated by the addition of CO$_2$ at the time of each harvest.

The biomass was filtered through a drum filter with 30-micron mesh. The culture substrates were returned to the culture pond, while the biomass was vacuum pressed and then converted into noodle form. In this format, it was air-dried at low temperature (40 ± 2 °C) to reduce the degradation of poorly resistant bioactive compounds at higher temperatures.

2.2. Extraction Procedure

Four samples of 2% (w/v) aqueous suspensions were made from the spirulina dry matter (DM). For this, 198 mL of deionised water were added to 4 g of dry biomass. In two of four samples, the same volume (198 mL) of EtOH or DMSO was subsequently added to finish with a 1% suspension (Figure 1). These two mixtures only received a conventional shaking treatment under stirring. The other 2% suspensions in water were treated with PEF before mixing with the organic solvents. As can be observed in Figure 1, the extracts were obtained from the 1% suspensions at different times and centrifuged to obtain the supernatant to be analysed. A total of 24 extracts were obtained, 12 of which were obtained by conventional extraction and the other 12 by PEF-assisted extraction.

2.2.1. Pulsed Electric Fields (PEF) Extraction

For the PEF pre-treatment of the spirulina 2% (w/v) solution was used with the PEF-Cellcrack III equipment (German Institute for Food Technology (DIL)) (ELEA, Germany). A treatment chamber of 900 mL capacity was used, the distance between the electrodes was fixed at 10 cm, and the total mass added to the treatment chamber was 202 g (198 g of water + 4 g of microalgae). A 100 kJ/kg treatment was applied with an electric field of 3 kV/cm according to a previous work [24]. Before and after the treatment, the temperature and conductivity of each sample was measured with a portable conductivity meter, ProfiLine Cond 3310 (WTW, Xylem Analytics, Weilheim in Oberbayern, Germany). The minimum electric field strength needed to produce changes in the cells is 1 kV/cm, and it has been shown that with a pulse duration of milliseconds, an electric field of 3–4 kV/cm is sufficient to create electroporation [14,32].

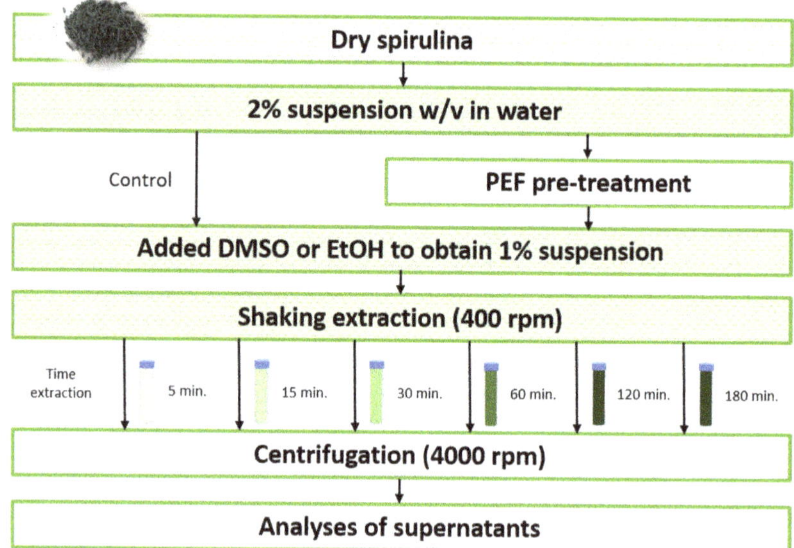

Figure 1. A schematic representation of the extraction process.

2.2.2. Solvent Extractions

First, conventional extraction (Control) was performed on one set of samples. The solvents dimethyl sulfoxide (DMSO) or ethanol (EtOH) (1:1, *v/v*) in deionised water were added up to a final volume of 400 mL to the dry spirulina samples for nutrient extraction. Once the solvents were added, the samples were shaken at 400 rpm for 5, 15, 30, 60, 60, 120, or 180 min at room temperature to test the effect of shaking time on the extraction of compounds from the processed biomasses. Subsequently, the samples were centrifuged for 10 min at 4000 rpm using a 5810R centrifuge (Eppendorf Ibérica, Madrid, Spain). The extract obtained was kept at −20 °C for further analysis.

Next, the remaining samples were pre-treated with PEF under the conditions described above, and then an extraction process was carried out following the same methodology used for the conventional method.

Figure 2 shows the extracts obtained after an extraction using (a) DMSO/H_2O or (b) EtOH/H_2O at different extraction times (5–180 min) and the extracts obtained after pre-treatment with PEF using (c) DMSO/H_2O and (d) EtOH/H_2O.

2.3. Chemical Analysis

2.3.1. Total Phenolic Compounds (TPC)

For the determination of total phenolic compounds (TPC) (mg gallic acid equivalents (GAE)/g DM), the Folin–Ciocalteu method was used, using the procedure described by Parniakov et al. [12]. This technique is based on the property of phenols to react against oxidising agents. The Folin–Ciocalteu reagent contains molybdate and sodium tungstate, which react with the phenolic compounds present in the medium to form phosphomolybdic and phosphotungstic complexes. In a basic medium, electron transfer reduces these complexes to tungsten oxide (W_8O_{23}) and molybdenum oxide (Mo_8O_{23}), which are chromogens with an intense blue colour, proportional to the concentration of phenolic groups present in the sample. Gallic acid (Sigma-Aldrich, Steinheim, Germany) was used as a standard. First, Folin–Ciocalteu reagent at 50% *v/v*, 2% Na_2CO_3, and diluted gallic acid standards were prepared. To carry out the analysis, 3 mL of Na_2CO_3 was added to a test tube, then 100 µL of standard or sample extract was added, and finally 100 µL of Folin–Ciocalteu reagent was added to this mixture. The samples were incubated for 60 min at room temperature under darkness. Finally, the samples were measured at 750 nm wavelength using a Perkin-Elmer UV/Vis Lambda 2 spectrophotometer (Perkin-Elmer, Jügesheim, Germany). All analyses were performed in triplicate.

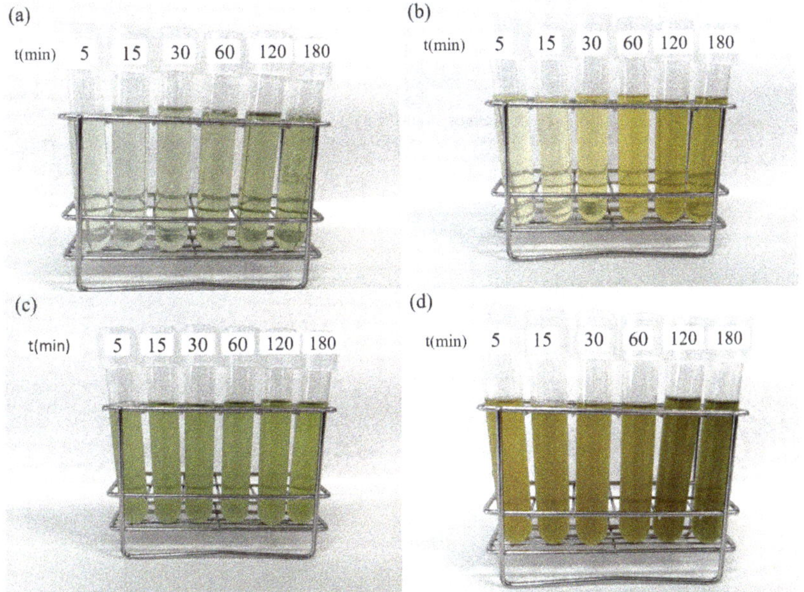

Figure 2. Conventional extraction with (**a**) DMSO/H_2O or (**b**) EtOH/H_2O and their respective extracts using PEF pre-treatment with (**c**) DMSO/H_2O and (**d**) EtOH/H_2O.

2.3.2. Trolox Equivalent Antioxidant Capacity (TEAC)

To determine the total antioxidant capacity (TAC), the Trolox equivalent antioxidant capacity (TEAC) assay was used. The TEAC value (micromolar Trolox equivalents, µM TE) measures the antioxidant capacity of a substance, compared to the standard, Trolox (Sigma-Aldrich, Steinheim, Germany). The TEAC value was measured using the method described by Safafar et al. [33] based on the discolouration of the ABTS radical.

The radical cation $ABTS^{\bullet+}$ (chromophore) was produced by reacting a 7 mM ABTS stock solution with 440 µL of an oxidant such as 140 mM potassium persulphate ($K_2S_2O_8$). The mixture was kept in the dark at room temperature for 16 h before use. The solution was then diluted with 96% EtOH until its absorbance at 734 nm was 0.70 ± 0.02 at 30 °C.

Once the initial absorbance was reached, 2 mL of $ABTS^{\bullet+}$ was mixed with 100 µL of extract and the sample was measured after 3 min. The reaction produced a discolouration due to neutralisation of the radical cation of $ABTS^{\bullet+}$, which depended on the concentration of antioxidants in the sample. The absorbance was measured at a wavelength 734 nm in a Perkin-Elmer UV/Vis Lambda 2 spectrophotometer (Perkin-Elmer, Jügesheim, Germany). All analyses were performed in triplicate.

2.3.3. Oxygen Radical Absorbance Capacity (ORAC)

The method used was previously described by Khawli et al. [34]. This assay measures the oxidative degradation of a fluorescent molecule such as fluorescein (Sigma-Aldrich, St. Louis, MO, USA) after the addition of a free radical generator such as 2,2′-azobis(2-

aminodinopropane) dihydrochloride (AAPH). The determination measures the degree of antioxidant protection of the sample with respect to the Trolox standard and is expressed in micromolar Trolox equivalents (µM TE).

The automated ORAC assay was performed on a Wallac 1420 VICTOR 2 plate reader (Perkin-Elmer, Jügesheim, Germany). The measurements were carried out in 96-well plates in which only the inner 60 wells were used. For the determination, a phosphate buffer solution (7.5 mM and pH 7–7.4) was prepared. For this purpose, 22.72 g of Na_2HPO_4 and 22.16 g of KH_2PO_4 were weighed and then each dissolved in 200 mL of deionised water. A volume of 61.6 mL of the first solution and 38.9 mL of the second solution were mixed and made up to 1000 mL with deionised water.

The standard (Trolox 100 µM) was prepared each day by adding 12.5 mg of Trolox to 50 mL of the previously prepared phosphate buffer. From this solution, 1 mL was taken and made up to 10 mL with the phosphate buffer to obtain the desired concentration. For fluorescein, 44 mg were weighed and made up to 100 mL with phosphate buffer. The working solution of 78 nM fluorescein was prepared daily. For this, 0.167 mL of the first solution was taken and made up to 25 mL with phosphate buffer.

Finally, the working dilution of the 221 mM AAPH radical was prepared daily by taking 600 mg of AAPH and bringing it up to 10 mL with phosphate buffer. As the ORAC assay is extremely sensitive, the microalgae extracts were adequately diluted prior to analysis to avoid interferences. In this case, the microalgae samples were diluted between 1:100 and 1:200, v/v.

For plate preparation, 50 µL of fluorescein (78 nM) and 50 µL of sample, blank (phosphate buffer), or standard (Trolox, 100 µM) were placed in each well and finally 25 µL of AAPH (221 mM) was added. As measurement variations may occur from well to well due to the low conductivity of polypropylene plates, the plates were pre-warmed at 37 °C for 10 min after the addition of fluorescein and before the addition of AAPH to avoid this problem. The plates were analysed immediately after the addition of AAPH and measurements were taken every 5 min until the relative fluorescence intensity of the standard (Trolox) was less than 5% of the initial reading value. All analyses were performed in triplicate.

2.3.4. Chlorophyll a, Chlorophyll b, and Carotenoids

Chlorophyll a, chlorophyll b, and carotenoids contents were estimated spectrophotometrically according to the study by Parniakov et al. [35]. They were calculated using the equations of Lichtenthaler and Wellburn [36] for EtOH and Wellburn [37] for DMSO. This method is based on the determination of carotenoid and chlorophyll content based on the maximum absorbances of chlorophyll a (C_a), chlorophyll b (C_b), and total carotenoids (C_{x+c}). Using EtOH as a solvent, the maximum absorbances were found at the wavelengths of 664 nm, 648 nm, and 470 nm for chlorophyll a (C_a), chlorophyll b (C_b) and total carotenoids (C_{x+c}), respectively. For DMSO, the wavelengths were 665 nm, 649 nm, and 480 nm, respectively. Aliquots of the extracts obtained were diluted and the absorbances (A) were measured at the wavelengths listed above according to the solvent used. All analyses were performed in triplicate.

EtOH equations:

$$C_a \, (\mu g/mL) = 13.36 \, A_{664} - 5.19 \, A_{648} \tag{1}$$

$$C_b \, (\mu g/mL) = 27.43 \, A_{648} - 8.12 \, A_{664} \tag{2}$$

$$C_{x+c} \, (\mu g/mL) = (1000 \, A_{470} - 2.13 \, C_a - 97.64 \, C_b)/209 \tag{3}$$

DMSO equations:

$$C_a \, (\mu g/mL) = 12.47 \, A_{665} - 3.62 \, A_{649} \tag{4}$$

$$C_b \, (\mu g/mL) = 25.06 \, A_{649} - 6.5 \, A_{665} \tag{5}$$

$$C_{x+c} \, (\mu g/mL) = (1000 \, A_{480} - 1.29 \, C_a - 53.78 \, C_b)/220 \tag{6}$$

2.4. Statistical Analysis

Data were analysed using analysis of variance (ANOVA), where PEF pre-treatment, solvents, and extraction time were the factors and chlorophyll, carotenoids, TPC, TEAC, and ORAC concentrations were the variables. Data were expressed as mean $\pm$ standard deviation in all cases. A value of $p < 0.05$ was considered significant. In addition, the LSD (Least Significant Differences) test was performed to determine the differences between the means of the values obtained. All analyses were performed with STATGRAPHICS Centurion XVI 16.1.03 (Statgraphics Technologies Inc., Princeton, NJ, USA).

3. Results and Discussion

In the present study, the effect of extraction time, the use of polar solvents such as DMSO and EtOH (aprotic and protic, respectively) combined with H_2O (50:50, v/v), and the application of a PEF pre-treatment to improve the extraction yield and efficiency of pigments and antioxidant compounds were evaluated.

3.1. Conventional Extraction

3.1.1. Chlorophyll a, Chlorophyll b, and Carotenoids

Figure 3 shows the chlorophyll a, chlorophyll b, and carotenoid content of the extracts obtained after conventional extraction with respect to the extraction time and solvent used. After performing a 2-factor analysis of variance (ANOVA) (time and solvent), it was observed that both extraction time and solvent had a significant effect ($p < 0.05$) on the extraction of chlorophyll a, chlorophyll b, and carotenoids, observing a higher extraction of all these compounds with the longer extraction times, and in general, observing higher values of chlorophyll a, chlorophyll b, and carotenoids when an $EtOH/H_2O$ mixture was used compared to that seen with the $DMSO/H_2O$ mixture. Thus, at 180 min, the highest values of chlorophyll a (0.57 $\pm$ 0.01 mg/g DM), chlorophyll b (0.55 $\pm$ 0.01 mg/g DM), and carotenoids (0.50 $\pm$ 0.01 mg/g DM) were obtained after using the $EtOH/H_2O$ mixture, representing an increase of 34%, 54%, and 84.2%, respectively, compared to those obtained with the $DMSO/H_2O$ solvent at equivalent extraction times.

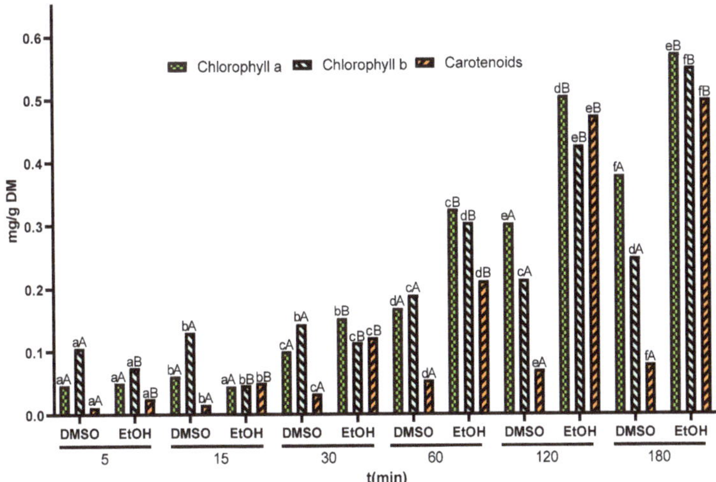

Figure 3. Chlorophyll a, chlorophyll b, and total carotenoids. Conventional extraction (5–180 min) using two binary mixtures (EtOH or DMSO in 50% deionised water). Different lower-case letters in the same parameter and solvent indicate statistical differences as a function of extraction time. Different capital letters in the same parameter and time indicate statistical differences as a function of solvent.

However, it should be noted that this was not verified for all the compounds evaluated. For example, when using the solvent EtOH/H$_2$O, no significant differences were found in the chlorophyll a content after 5 and 15 min of extraction. Moreover, no statistically significant differences in chlorophyll b content were observed after 15 and 30 min of extraction with the solvent DMSO/H$_2$O. When analysing the statistical results between solvents, no differences were found in chlorophyll a extraction after 5 and 15 min of extraction.

These results are in agreement with those obtained by other authors, who observed a higher extraction of chlorophyll a, chlorophyll b, and carotenoids when the EtOH/H$_2$O mixture was used [13]. They attributed this effect to a change in the polarity of the medium, which made it easier for the compounds to cross the lipid membrane of the cell, favouring the extraction of the compounds studied.

3.1.2. Total Phenolic Compounds (TPC) and Total Antioxidant Capacity (TAC)

Figure 4 shows the results obtained for total phenolic compounds (TPC) and total antioxidant capacity (TAC) determined by TEAC and ORAC. After performing a 2-factor analysis of variance (ANOVA) (time and solvent), it was observed that both time and solvent had a significant effect ($p < 0.05$) on the extraction of phenolic compounds.

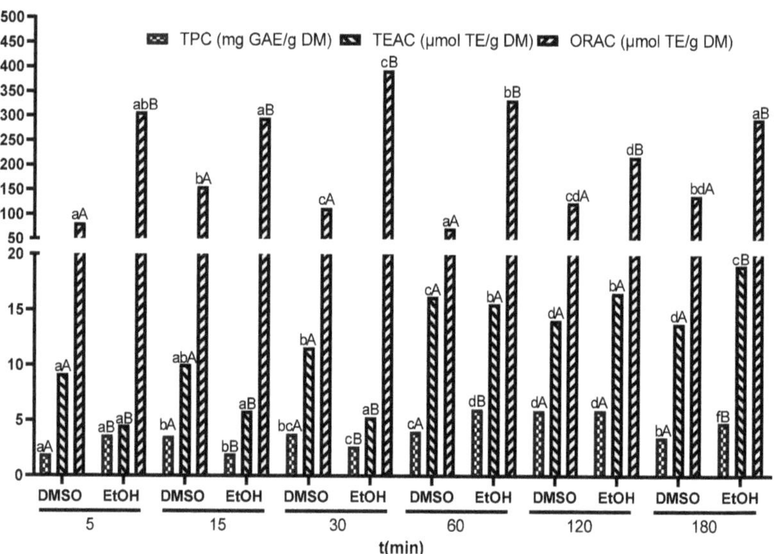

Figure 4. Total phenolic compounds, TEAC and ORAC. Conventional extraction (5–180 min) using two binary mixtures (EtOH or DMSO in 50% deionised water). Different lower-case letters in the same parameter and solvent indicate statistical differences as a function of extraction time. Different capital letters in the same parameter and time indicate statistical differences as a function of solvent.

ORAC values ranged from 295.87 ± 19.01 to 393.24 ± 15.28 µmol TE/g DM after using EtOH/H$_2$O and from 72.45 ± 5.95 to 155.95 ± 10.78 µmol TE/g DM when the mixture DMSO/H$_2$O was used. In addition, a greater effect on the extraction of antioxidant compounds was observed at 30 and 180 min, respectively. On the other hand, the TEAC values ranged from 4.52 ± 0.02 to 19.05 ± 1.25 µmol TE/g DM after using the EtOH/H$_2$O mixture and from 9.19 ± 0.62 to 16.19 ± 1.37 µmol TE/g DM with the DMSO/H$_2$O mixture. The maximum value for the TEAC assay (19.05 µmol TE/g DW) was obtained after 180 min extraction and using EtOH 50% as the solvent. The highest ORAC value was also obtained after using EtOH, but after 30 min of extraction (393.24 µmol TE/g DW). Regarding DMSO, the best values for TEAC and ORAC were obtained at 60 min (16.19 µmol TE/g DW) and 15 min (155.95 µmol TE/g DW), respectively. The higher ORAC values are due to other antioxidant compounds not measured in this study, especially fat-soluble compounds such as vitamin E that may have an impact on the antioxidant capacity [38–40]. Moreover, ORAC sensitivity for other antioxidant compounds should also be taken into account [34].

In the determination of total phenolic compounds (TPC), concentrations between 3.64 ± 0.07 and 6.04 ± 0.28 mg GAE/g DM were obtained for the EtOH/H_2O extraction and between 1.91 ± 0.07 and 5.921 ± 0.175 mg GAE/g DM for the DMSO/H_2O extraction and 5.921 ± 0.175 mg GAE/g DM in the extraction with DMSO/H_2O: For both solvents, the highest values were obtained at 120 min, the phenolic content at that time was similar to that obtained by Shanti et al. [41] for spirulina.

Regarding the mean values of TAC (TEAC + ORAC), it is worth noting that the values obtained with DMSO/H_2O are relatively lower compared to those of the EtOH/H_2O mixture, independent of the extraction time (Figure 5). This may be due to a lower extraction of chlorophylls, carotenoids, and phenolic compounds, as seen above, suggesting a clear contribution of these compounds to TAC. This same correlation has been previously observed by other authors after evaluating the use of these solvents for the extraction of these compounds from the microalgae *Nannochloropsis* [13].

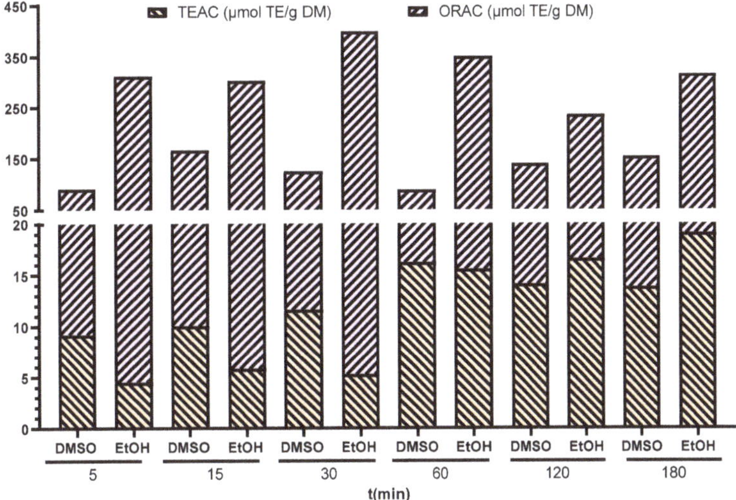

Figure 5. Total antioxidant capacity (TAC) values obtained by adding the mean values of TEAC and ORAC for each time and solvent.

To evaluate the possible correlations between the different antioxidant compounds (chlorophyll a, chlorophyll b, carotenoids, and total phenolic compounds) and the determination methods used for total antioxidant capacity (TEAC and ORAC), a Pearson's test was performed for 36 samples (Table 1). The main correlations were observed between TEAC and carotenoids (R = 0.9094, $p < 0.05$), chlorophyll a (R = 0.7986, $p < 0.05$), chlorophyll b (R = 0.8419, $p < 0.05$), and total phenolic compounds (R = 0.7114, $p < 0.05$), obtaining moderate to high positive correlation coefficients. These results are in agreement with those of other authors [42,43] who evaluated the existing correlations between the different antioxidant compounds and the methods used for the joint evaluation of the total antioxidant capacity.

Table 1. Correlation between antioxidant compounds and total antioxidant capacity (TEAC and ORAC).

	TPC	TEAC	ORAC	Chlorophyll a	Chlorophyll b	Carotenoids
TPC		0.7114 (36) 0.0000	0.0301 (36) 0.8618	0.7016 (36) 0.0000	0.6827 (36) 0.0000	0.5727 (36) 0.0003
TEAC	0.7114 (36) 0.0000		−0.3054 (36) 0.0701	0.7986 (36) 0.0000	0.8419 (36) 0.0000	0.6306 (36) 0.0000
ORAC	0.0301 (36) 0.8618	−0.3054 (36) 0.0701		0.1415 (36) 0.4103	0.1164 (36) 0.4991	0.3586 (36) 0.0317
Chlorophyll a	0.7016 (36) 0.0000	0.7986 (36) 0.0000	0.1415 (36) 0.4103		0.9511 (36) 0.0000	0.8735 (36) 0.0000
Chlorophyll b	0.6827 (36) 0.0000	0.8419 (36) 0.0000	0.1164 (36) 0.4991	0.9511 (36) 0.0000		0.9169 (36) 0.0000
Carotenoids	0.5727 (36) 0.0003	0.6306 (36) 0.0000	0.3586 (36) 0.0317	0.8735 (36) 0.0000	0.9169 (36) 0.0000	

TPC: Total phenolic compounds; TEAC: Trolox equivalent antioxidant capacity; ORAC: Oxygen radical antioxidant capacity.

It should be noted that there was no correlation between antioxidant compounds (chlorophyll a and chlorophyll b) and ORAC. This is because the ORAC method is a more sensitive technique and some authors consider it better than TEAC [44] when measuring total antioxidant compounds, as ORAC measures the antioxidant capacity of compounds other than phenolics. A significant correlation was observed, but with a low correlation coefficient between ORAC and carotenoids (R = 0.3586, $p < 0.05$).

3.2. Pulsed Electric Fields (PEF)-Assisted Extraction

3.2.1. Ethanol/Water

Figure 6 shows the results obtained after applying both treatments (conventional and PEF) using the mixture EtOH/H_2O. After performing a two-way ANOVA analysis (time and treatment), it was observed that both the extraction time and the use of the PEF pre-treatment had a significant effect ($p < 0.05$) on the extraction of chlorophylls, carotenoids, (Figure 6a), and total phenolic compounds (Figure 6b). This effect is due to the electroporation produced by the PEF, which facilitates the extraction of these compounds more efficiently and with less agitation time.

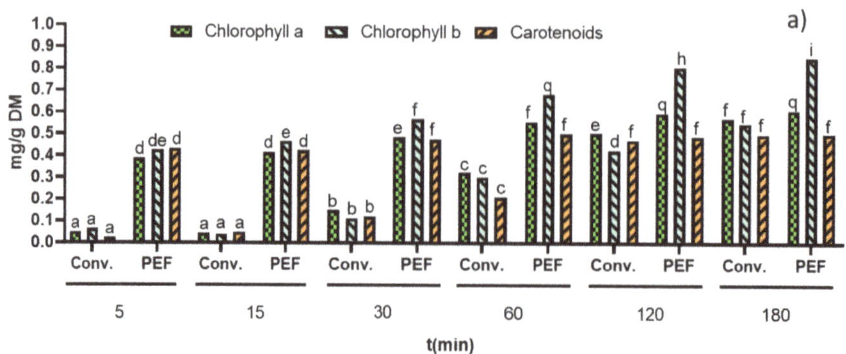

Figure 6. *Cont.*

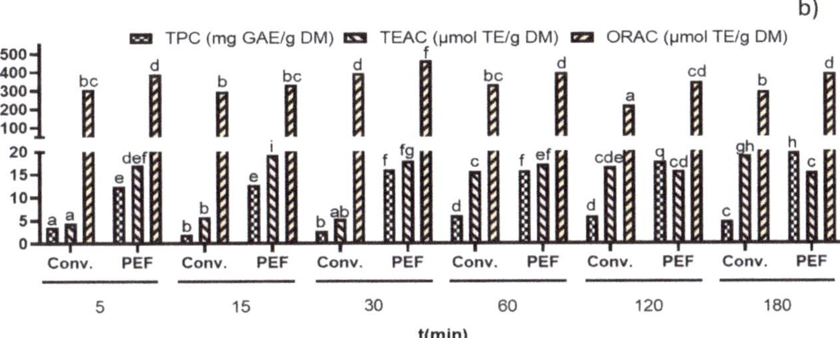

Figure 6. Chlorophyll a, chlorophyll b, and carotenoid content (**a**); TPC, TEAC, and ORAC (**b**) after conventional extraction and PEF-assisted extraction with EtOH/H$_2$O. Different lower-case letters in the same parameter indicate statistical differences depending on the extraction time or treatment used.

It should be noted that no significant differences were observed regarding the maximum carotenoid content obtained after applying PEF pre-treatment for 60 min (0.50 ± 0.01 mg/g DM) or after conventional treatment without PEF for 180 min (0.50 ± 0.01 mg/g DM), which showed that PEF is an effective tool to reduce carotenoid extraction time, being 3 times faster than conventional extraction. A similar effect was observed for chlorophyll a: We obtained similar values after applying PEF and supplementary extraction for 120 min (0.60 ± 0.01 mg/g DM) as those obtained with conventional treatment for 180 min (0.57 ± 0.04 mg/g DM), observing a reduction of 60 min in the time to obtain the maximum chlorophyll a content. Moreover, PEF pre-treatment increased the extraction of chlorophyll b throughout all the extraction times compared to that of the control sample.

After analysing the statistical data of the TPC values, it was found that the pre-treatment with PEF also had a very positive effect on the extraction, obtaining significant differences ($p < 0.05$) compared to those of the conventional treatment, independently of the extraction time. Moreover, it was also observed that after PEF extraction, the TPC values were 2–3-fold higher. For example, after 180 min of conventional extraction, TPC values of 4.84 ± 0.48 mg GAE/g DM were obtained, while the value obtained at the same time after PEF pre-treatment was 19.75 ± 0.50 mg GAE/g DM, representing a 75% increase. The increase in chlorophylls, carotenoids, total phenolic compounds, and TAC content obtained after PEF application compared to those of conventional extraction is in agreement with the results obtained by other authors after similar experiments with the microalgae *Nannochloropsis* spp. [13].

3.2.2. Dimethyl Sulfoxide/Water

Figure 7 shows the results obtained for chlorophyll, carotenoids, TPC, TEAC, and ORAC content after conventional and PEF-assisted extraction using DMSO/H$_2$O as a solvent. After performing a two-way ANOVA (time and treatment), it was observed that both extraction time and treatment had a significant effect ($p < 0.05$) on the extraction of chlorophylls, carotenoids (Figure 7a), and TPC (Figure 7b). This effect was less than that observed after using the EtOH/H$_2$O mixture, mainly due to a lower solvent extraction capacity, as could be observed after using the conventional extraction method (Figures 3 and 4).

Lower TEAC values were also observed due to a decreased extraction of antioxidant compounds (chlorophylls, carotenoids, and phenolic compounds) (Table 1). However, the ORAC values were not significantly affected, although a non-significant increase was found when longer extraction times were used.

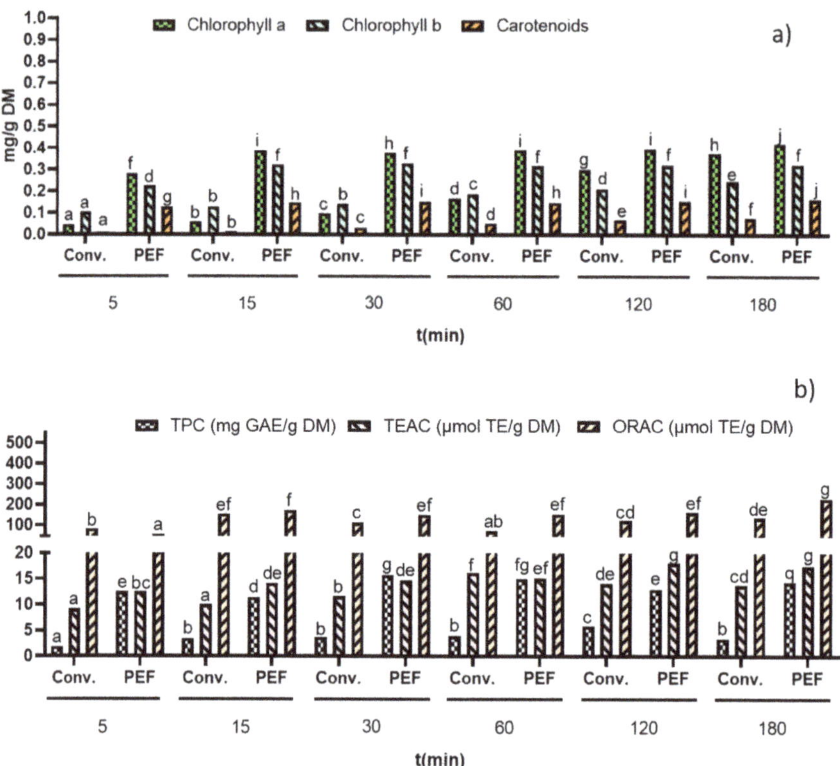

Figure 7. Chlorophyll a, chlorophyll b, and carotenoid content (**a**); TPC, TEAC, and ORAC (**b**) after conventional extraction and PEF-assisted extraction with DMSO/H$_2$O. Different lower-case letters in the same parameter indicate statistical differences depending on the extraction time or treatment used.

It should be noted that the maximum extraction of chlorophyll a and chlorophyll b occurred 15 min after applying the PEF pre-treatment, with values of 0.39 ± 0.01 mg/g DM and 0.32 ± 0.01 mg/g DM, respectively. However, after applying the conventional treatment, these values were not obtained until 180 min. PEF pre-treatment reduced the extraction time by 165 min. Regarding carotenoids, a similar effect was found, with the maximum carotenoid extraction (0.15 ± 0.01 mg/g DM) observed after applying the PEF pre-treatment and subsequent extraction for 15 min, obtaining lower values (0.08 ± 0.00 mg/g DM) than after using conventional extraction for 180 min.

The TPC content was also increased by PEF pre-treatment, for example, the maximum content extracted at 5 min after PEF pre-treatment (12.53 ± 0.31 mg AGE/g DM) was much higher than that obtained at 180 min after conventional treatment (4.84 ± 0.48 mg AGE/g DM).

3.3. Pulsed Electric Fields (PEF) Efficiency

In order to better evaluate the effect of PEF on the extraction of compounds and to compare it with conventional extraction, the Y$_{PEF}$ efficiency coefficient was introduced. This coefficient is defined as the ratio between the values obtained for chlorophylls, carotenoids (Figure 8a), TPC, TEAC, and ORAC (Figure 8b) with PEF-assisted extraction and those same values obtained with conventional extraction.

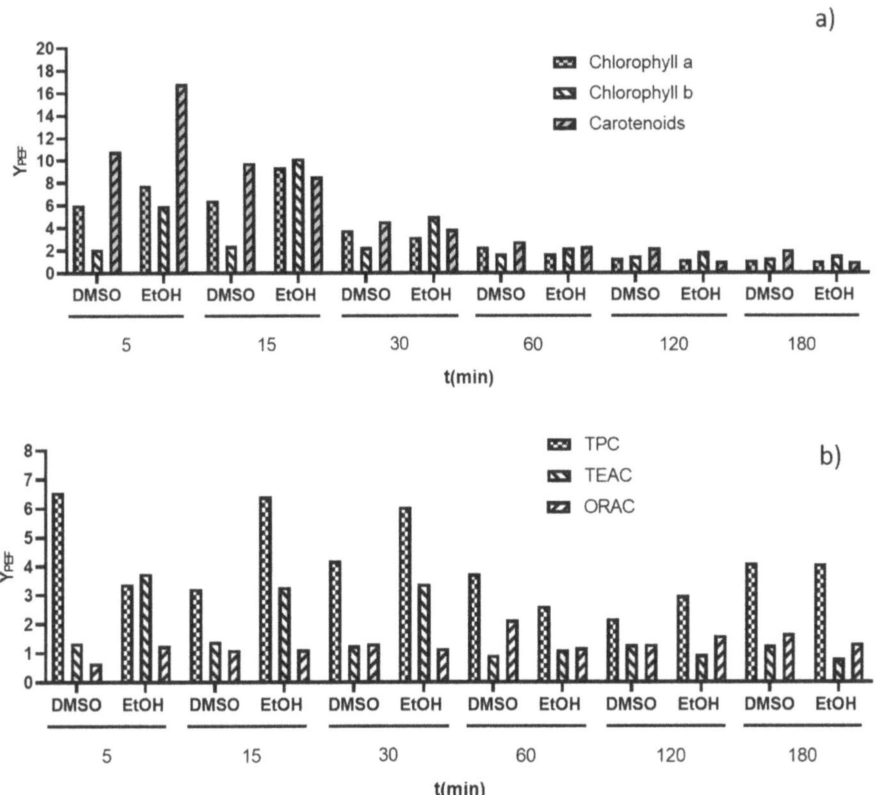

Figure 8. Efficiency ratio (Y_{PEF}) for chlorophyll a, chlorophyll b and carotenoids (**a**), and TPC, TEAC, and ORAC (**b**) vs. time.

The value of Y_{PEF} corresponds to $\dfrac{Ca\left(\frac{mg}{g}\ MS\right)PEF}{Ca\left(\frac{mg}{g}\ MS\right)conv.}$ (for chlorophyll a), $\dfrac{Cb\left(\frac{mg}{g}\ MS\right)PEF}{Cb\left(\frac{mg}{g}\ MS\right)conv.}$ (for chlorophyll b), etc.

It is important to note that the maximum Y_{PEF} values were observed at the optimum extraction times of 5 and 15 min, independently of the solvent mixture, obtaining a greater effect in the extraction of carotenoids and TPC. From 30 min onwards, Y_{PEF} values decreased rapidly, which gives an idea of the effectiveness of the PEF in extracting these compounds, suggesting that in a reduced time the extraction can be increased, and the cost reduced. In this regard, a recovery of 55–60% for chlorophylls, 85–90% for carotenoids, and 60–70% for total phenolic compounds was obtained with respect to the maximum total content extracted.

The Y_{PEF} values for TEAC were also increased, as these are directly related to the extraction of the antioxidant compounds. However, this effect was not observed for the ORAC method, as no direct relation between higher extraction and ORAC values was obtained.

4. Conclusions

From the results obtained in this study it can be concluded that both time and solvent have a significant impact on the recovery of antioxidant compounds when conventional extraction is used, obtaining the highest values of phenolic compounds, chlorophyll a, chlorophyll b, and carotenoids when the mixture EtOH/H_2O was used for 180 min. Moreover, a strong relationship of TEAC values with total phenolic compounds, carotenoids, and chlorophylls was found while ORAC values were positively correlated with carotenoids. Both PEF treatment and extraction time had a statistically significant effect on the recovery of antioxidant compounds when the EtOH/H_2O mixture was used, showing a considerable reduction in the extraction time required to recover polyphenols, carotenoids, and chlorophylls compared to those of conventional treatment. When the impact of PEF and the DMSO/H_2O mixture was evaluated, it was found that both the treatment and the extraction time had a statistically significant effect, reducing the extraction times compared to the conventional treatment; however, in this case the maximum content of antioxidant compounds was lower than that observed for EtOH/H_2O. TAC values were also increased after PEF treatment, mainly due to an increase in the extraction of antioxidant compounds. In addition, the maximum efficiency values were observed at 5 and 15 min for the two solvents, with a greater effect on the extraction of carotenoids and total phenolic compounds. In the future, the implementation of PEF in the extraction of antioxidant bioactive compounds from microalgae would be interesting, as it could be a promising and environmentally sustainable technology.

Author Contributions: Conceptualization, E.F., H.B. and F.J.B.; methodology, F.R.-M., F.J.M.-Q. and N.P.; formal analysis, F.R.-M., F.J.M.-Q. and N.P.; software, F.R.-M. and F.J.M.-Q.; resources, F.J.B., E.F. and H.B.; writing—original draft preparation, F.R.-M., F.J.M.-Q., N.P., Y.P. and F.J.B.; writing—review and editing, F.J.B., E.F. and H.B.; supervision, F.J.B., E.F. and H.B.; funding acquisition, F.J.B., E.F. and H.B. All authors have read and agreed to the published version of the manuscript.

Funding: This research was funded by the University of Valencia through the project OTR2021-21736INVES, supported by the University of Vigo. Moreover, it was partially funded by the EU Commission and BBI-JU Horizon H2020, through the AQUABIOPRO-FIT project (aquaculture and agriculture biomass side stream proteins and bioactives for feed, fitness and health promoting nutritional supplements) grant number 790956.

Acknowledgments: The authors would like to acknowledge Nicolas Mazurier (Ecospirulina company) for providing the microalgae samples to carry out the experiments. They also thank Generalitat Valenciana for financial support (IDIFEDER/2018/046—Procesos innovadores de extracción y conservación: pulsos eléctricos y fluidos supercríticos) through the European Union ERDF funds (European Regional Development Fund). Moreover, F.J.M.-Q. also would like to acknowledge the pre-PhD scholarship program of University of Valencia "Atracció de Talent".

Conflicts of Interest: The authors declare no conflict of interest.

References

1. Granato, D.; Nunes, D.S.; Barba, F.J. An integrated strategy between food chemistry, biology, nutrition, pharmacology, and statistics in the development of functional foods: A proposal. *Trends Food Sci. Technol.* **2017**, *62*, 13–22. [CrossRef]
2. Granato, D.; Barba, F.J.; Kovačević, D.B.; Lorenzo, J.M.; Cruz, A.G.; Putnik, P. Functional foods: Product development, technological trends, efficacy testing, and safety. *Ann. Rev. Food Sci. Technol.* **2020**, *11*, 93–118. [CrossRef] [PubMed]
3. Marti-Quijal, F.J.; Zamuz, S.; Tomašević, I.; Rocchetti, G.; Lucini, L.; Marszałek, K.; Barba, F.J.; Lorenzo, J.M. A chemometric approach to evaluate the impact of pulses, Chlorella and Spirulina on proximate composition, amino acid, and physicochemical properties of turkey burgers. *J. Sci. Food Agric.* **2019**, *99*, 3672–3680. [CrossRef]
4. Atik, D.S.; Gürbüz, B.; Bölük, E.; Palabıyık, İ. Development of vegan kefir fortified with *Spirulina platensis*. *Food Biosci.* **2021**, *42*, 101050. [CrossRef]
5. Barba, F.J. Microalgae and seaweeds for food applications: Challenges and perspectives. *Food Res. Int.* **2017**, *99*, 969–970. [CrossRef] [PubMed]
6. Reddy, M.; Bhat, V.B.; Kiranmai, G.; Reddy, M.; Reddanna, P.; Madyastha, K. Selective inhibition of cyclooxygenase-2 by c-phycocyanin, a biliprotein from *Spirulina platensis*. *Biochem. Biophys. Res. Commun.* **2000**, *277*, 599–603. [CrossRef]
7. Cohen, Z.; Vonshak, A. Fatty acid composition of Spirulina and spirulina-like cyanobacteria in relation to their chemotaxonomy. *Phytochemistry* **1991**, *30*, 205–206. [CrossRef]
8. Golmakani, M.-T.; Rezaei, K.; Mazidi, S.; Razavi, S.H. γ-Linolenic acid production by *Arthrospira platensis* using different carbon sources. *Eur. J. Lipid Sci. Technol.* **2012**, *114*, 306–314. [CrossRef]
9. Ambrozova, J.V.; Misurcova, L.; Vicha, R.; Machu, L.; Samek, D.; Baron, M.; Mlcek, J.; Sochor, J.; Jurikova, T. Influence of extractive solvents on lipid and fatty acids content of edible freshwater algal and seaweed products, the green microalga *Chlorella kessleri* and the cyanobacterium *Spirulina platensis*. *Molecules* **2014**, *19*, 2344–2360. [CrossRef]

10. Günerken, E.; D'Hondt, E.; Eppink, M.; Garcia-Gonzalez, L.; Elst, K.; Wijffels, R. Cell disruption for microalgae biorefineries. *Biotechnol. Adv.* **2015**, *33*, 243–260. [CrossRef]

11. Carullo, D.; Abera, B.D.; Casazza, A.A.; Donsì, F.; Perego, P.; Ferrari, G.; Pataro, G. Effect of pulsed electric fields and high pressure homogenization on the aqueous extraction of intracellular compounds from the microalgae *Chlorella vulgaris*. *Algal Res.* **2018**, *31*, 60–69. [CrossRef]

12. Parniakov, O.; Barba, F.J.; Grimi, N.; Marchal, L.; Jubeau, S.; Lebovka, N.; Vorobiev, E. Pulsed electric field and pH assisted selective extraction of intracellular components from microalgae *Nannochloropsis*. *Algal Res.* **2015**, *8*, 128–134. [CrossRef]

13. Parniakov, O.; Barba, F.J.; Grimi, N.; Marchal, L.; Jubeau, S.; Lebovka, N.; Vorobiev, E. Pulsed electric field assisted extraction of nutritionally valuable compounds from microalgae *Nannochloropsis* spp. using the binary mixture of organic solvents and water. *Innov. Food Sci. Emerg. Technol.* **2015**, *27*, 79–85. [CrossRef]

14. Martínez, J.M.; Gojkovic, Z.; Ferro, L.; Maza, M.; Álvarez, I.; Raso, J.; Funk, C. Use of pulsed electric field permeabilization to extract astaxanthin from the Nordic microalga *Haematococcus pluvialis*. *Bioresour. Technol.* **2019**, *289*, 121694. [CrossRef] [PubMed]

15. Lai, Y.S.; Parameswaran, P.; Li, A.; Baez, M.; Rittmann, B.E. Effects of pulsed electric field treatment on enhancing lipid recovery from the microalga *Scenedesmus*. *Bioresour. Technol.* **2014**, *173*, 457–461. [CrossRef] [PubMed]

16. Puértolas, E.; Koubaa, M.; Barba, F.J. An overview of the impact of electrotechnologies for the recovery of oil and high-value compounds from vegetable oil industry: Energy and economic cost implications. *Food Res. Int.* **2016**, *80*, 19–26. [CrossRef]

17. Knorr, D.; Froehling, A.; Jaeger, H.; Reineke, K.; Schlueter, O.; Schoessler, K. Emerging technologies in food processing. *Annu. Rev. Food Sci. Technol.* **2011**, *2*, 203–235. [CrossRef]

18. Poojary, M.M.; Barba, F.J.; Aliakbarian, B.; Donsì, F.; Pataro, G.; Dias, D.A.; Juliano, P. Innovative alternative technologies to extract carotenoids from microalgae and seaweeds. *Mar. Drugs* **2016**, *14*, 214. [CrossRef]

19. Arshad, R.N.; Abdul-Malek, Z.; Munir, A.; Buntat, Z.; Ahmad, M.H.; Jusoh, Y.M.; Bekhit, A.E.-D.; Roobab, U.; Manzoor, M.F.; Aadil, R.M. Electrical systems for pulsed electric field applications in the food industry: An engineering perspective. *Trends Food Sci. Technol.* **2020**, *104*, 1–13. [CrossRef]

20. Eing, C.; Goettel, M.; Straessner, R.; Gusbeth, C.; Frey, W. Pulsed electric field treatment of microalgae—Benefits for microalgae biomass processing. *IEEE Trans. Plasma Sci.* **2013**, *41*, 2901–2907. [CrossRef]

21. Käferböck, A.; Smetana, S.; de Vos, R.; Schwarz, C.; Toepfl, S.; Parniakov, O. Sustainable extraction of valuable components from Spirulina assisted by pulsed electric fields technology. *Algal Res.* **2020**, *48*, 101914. [CrossRef]

22. Grimi, N.; Dubois, A.; Marchal, L.; Jubeau, S.; Lebovka, N.; Vorobiev, E. Selective extraction from microalgae *Nannochloropsis* sp. using different methods of cell disruption. *Bioresour. Technol.* **2014**, *153*, 254–259. [CrossRef] [PubMed]

23. Martínez, J.M.; Delso, C.; Álvarez, I.; Raso, J. Pulsed electric field-assisted extraction of valuable compounds from microorganisms. *Compr. Rev. Food Sci. Food Saf.* **2020**, *19*, 530–552. [CrossRef] [PubMed]

24. Kokkali, M.; Martí-Quijal, F.J.; Taroncher, M.; Ruiz, M.-J.; Kousoulaki, K.; Barba, F.J. Improved extraction efficiency of antioxidant bioactive compounds from *Tetraselmis chuii* and *Phaedoactylum tricornutum* using pulsed electric fields. *Molecules* **2020**, *25*, 3921. [CrossRef]

25. Luengo, E.; Condon_Abanto, S.; Alvarez, I.; Raso, J. Effect of pulsed electric field treatments on permeabilization and extraction of pigments from *Chlorella vulgaris*. *J. Membr. Biol.* **2014**, *247*, 1269–1277. [CrossRef]

26. Barba, F.J.; Grimi, N.; Vorobiev, E. New approaches for the use of non-conventional cell disruption technologies to extract potential food additives and nutraceuticals from microalgae. *Food Eng. Rev.* **2015**, *7*, 45–62. [CrossRef]

27. Lam, G.T.; Postma, P.; Fernandes, D.; Timmermans, R.; Vermuë, M.; Barbosa, M.; Eppink, M.; Wijffels, R.; Olivieri, G. Pulsed electric field for protein release of the microalgae *Chlorella vulgaris* and *Neochloris oleoabundans*. *Algal Res.* **2017**, *24*, 181–187. [CrossRef]

28. Rippka, R.; Stanier, R.Y.; Deruelles, J.; Herdman, M.; Waterbury, J.B. Generic assignments, strain histories and properties of pure cultures of cyanobacteria. *Microbiology* **1979**, *111*, 1–61. [CrossRef]

29. Markou, G.; Angelidaki, I.; Nerantzis, E.; Georgakakis, D. Bioethanol production by carbohydrate-enriched biomass of *Arthrospira* (Spirulina) *platensis*. *Energies* **2013**, *6*, 3937–3950. [CrossRef]

30. Ramírez-Moreno, L.; Olvera-Ramírez, R. Traditional and present use of Spirulina sp. (*Arthrospira* sp.). *Interciencia* **2006**, *31*, 657–663.

31. De la Jara, A.; Ruano, C.; Polifrone, M.; Assunção, P.; Casillas, Y.B.; Wägner, A.; Serra-Majem, L. Impact of dietary *Arthrospira* (Spirulina) biomass consumption on human health: Main health targets and systematic review. *J. Appl. Phycol.* **2018**, *30*, 2403–2423. [CrossRef]

32. Luengo, E.; Raso, J. Pulsed Electric Field-Assisted Extraction of Pigments from Chlorella vulgaris. In *Handbook of Electroporation*; Springer Science and Business Media LLC: Midtown Manhattan, NY, USA, 2017; pp. 2939–2954.

33. Safafar, H.; Van Wagenen, J.; Møller, P.; Jacobsen, C. Carotenoids, phenolic compounds and tocopherols contribute to the antioxidative properties of some microalgae species grown on industrial wastewater. *Mar. Drugs* **2015**, *13*, 7339–7356. [CrossRef] [PubMed]

34. Khawli, F.A.; Martí-Quijal, F.J.; Pallarés, N.; Barba, F.J.; Ferrer, E. Ultrasound extraction mediated recovery of nutrients and antioxidant bioactive compounds from *Phaeodactylum tricornutum* microalgae. *Appl. Sci.* **2021**, *11*, 1701. [CrossRef]

35. Parniakov, O.; Apicella, E.; Koubaa, M.; Barba, F.J.; Grimi, N.; Lebovka, N.; Pataro, G.; Ferrari, G.; Vorobiev, E. Ultrasound-assisted green solvent extraction of high-added value compounds from microalgae *Nannochloropsis* spp. *Bioresour. Technol.* **2015**, *198*, 262–267. [CrossRef]

36. Lichtenthaler, H.K.; Wellburn, A.R. Determinations of total carotenoids and chlorophylls a and b of leaf extracts in different solvents. *Biochem. Soc. Trans.* **1983**, *11*, 591–592. [CrossRef]
37. Wellburn, A.R. The spectral determination of chlorophylls a and b, as well as total carotenoids, using various solvents with spectrophotometers of different resolution. *J. Plant Physiol.* **1994**, *144*, 307–313. [CrossRef]
38. Banskota, A.H.; Sperker, S.; Stefanova, R.; McGinn, P.J.; O'Leary, S.J.B. Antioxidant properties and lipid composition of selected microalgae. *Environ. Boil. Fishes* **2018**, *31*, 309–318. [CrossRef]
39. Balboa, E.M.; Conde, E.; Moure, A.; Falqué, E.; Domínguez, H. In vitro antioxidant properties of crude extracts and compounds from brown algae. *Food Chem.* **2013**, *138*, 1764–1785. [CrossRef] [PubMed]
40. Lv, J.; Yang, X.; Ma, H.; Hu, X.; Wei, Y.; Zhou, W.; Li, L. The oxidative stability of microalgae oil (*Schizochytrium aggregatum*) and its antioxidant activity after simulated gastrointestinal digestion: Relationship with constituents. *Eur. J. Lipid Sci. Technol.* **2015**, *117*, 1928–1939. [CrossRef]
41. Shanthi, G.; Premalatha, M.; Anantharaman, N. Effects of l-amino acids as organic nitrogen source on the growth rate, biochemical composition and polyphenol content of *Spirulina platensis*. *Algal Res.* **2018**, *35*, 471–478. [CrossRef]
42. Amrani-Allalou, H.; Boulekbache-Makhlouf, L.; Mapelli-Brahm, P.; Sait, S.; Tenore, G.C.; Benmeziane, A.; Kadri, N.; Madani, K.; Martínez, A.J.M. Antioxidant activity, carotenoids, chlorophylls and mineral composition from leaves of *Pallenis spinosa*: An Algerian medicinal plant. *J. Complement. Integr. Med.* **2019**, *17*, 17. [CrossRef] [PubMed]
43. Carbonell-Capella, J.M.; Barba, F.J.; Esteve, M.J.; Frígola, A. Quality parameters, bioactive compounds and their correlation with antioxidant capacity of commercial fruit-based baby foods. *Food Sci. Technol. Int.* **2013**, *20*, 479–487. [CrossRef] [PubMed]
44. Silva, E.M.; Souza, J.N.S.; Rogez, H.; Rees, J.F.; Larondelle, Y. Antioxidant activities and polyphenolic contents of fifteen selected plant species from the Amazonian region. *Food Chem.* **2007**, *101*, 1012–1018. [CrossRef]

MDPI

Article

Identification and Application of Bioactive Compounds from *Garcinia xanthochymus* Hook. for Weed Management

Md. Mahfuzur Rob [1,2,3], Kawsar Hossen [1,3], Mst. Rokeya Khatun [1,3], Keitaro Iwasaki [4], Arihiro Iwasaki [4], Kiyotake Suenaga [4] and Hisashi Kato-Noguchi [1,3,*]

[1] Department of Applied Biological Science, Faculty of Agriculture, Miki, Kagawa University, Kagawa 761-0795, Japan; mahfuzrob.hort@sau.ac.bd (M.M.R.); kawsar.ag@nstu.edu.bd (K.H.); rokeya.entom@bau.edu.bd (M.R.K.)

[2] Department of Horticulture, Faculty of Agriculture, Sylhet Agricultural University, Sylhet 3100, Bangladesh

[3] The United Graduate School of Agricultural Sciences, Ehime University, 3-5-7 Tarumi, Matsuyama, Ehime 790-8566, Japan

[4] Department of Chemistry, Faculty of Science and Technology, Keio University, 3-14-1 Hiyoshi, Kohoku, Yokohama 223-8522, Japan; keitalo@keio.jp (K.I.); a.iwasaki@chem.keio.ac.jp (A.I.); suenaga@chem.keio.ac.jp (K.S.)

* Correspondence: kato.hisashi@kagawa-u.ac.jp

Citation: Rob, M..M.; Hossen, K.; Khatun, M..R.; Iwasaki, K.; Iwasaki, A.; Suenaga, K.; Kato-Noguchi, H. Identification and Application of Bioactive Compounds from *Garcinia xanthochymus* Hook. for Weed Management. *Appl. Sci.* **2021**, *11*, 2264. https://doi.org/10.3390/app11052264

Academic Editors: Ana M. L. Seca and Eugenia Gallardo

Received: 1 February 2021
Accepted: 28 February 2021
Published: 4 March 2021

Publisher's Note: MDPI stays neutral with regard to jurisdictional claims in published maps and institutional affiliations.

Abstract: The allelopathic potential of plant species and their related compounds has been increasingly reported to be biological tools for weed control. The allelopathic potential of *Garcinia xanthochymus* was assessed against several test plant species: lettuce, rapeseed, Italian ryegrass, and timothy. The extracts of *G. xanthochymus* leaves significantly inhibited all the test plants in a concentration- and species-specific manner. Therefore, to identify the specific compounds involved in the allelopathic activity of the *G. xanthochymus* extracts, assay-guided purification was carried out and two allelopathic compounds were isolated and identified as methyl phloretate {3-(4-hydroxyphenyl) propionic acid methyl ester} and vanillic acid (4-hydroxy-3-methoxybenzoic acid). Both of the substances significantly arrested the cress and timothy seedlings growth. I_{50} values (concentrations required for 50% inhibition) for shoots and roots growth of the cress and timothy were 113.6–104.6 and 53.3–40.5 µM, respectively, for methyl phloretate, and 331.6–314.7 and 118.8–107.4 µM, respectively, for vanillic acid, which implied that methyl phloretate was close to 3- and 2-fold more effective than vanillic acid against cress and timothy, respectively. This report is the first on the presence of methyl phloretate in a plant and its phytotoxic property. These observations suggest that methyl phloretate and vanillic acid might participate in the phytotoxicity of *G. xanthochymus* extract.

Keywords: *Garcinia xanthochymus*; growth inhibitory compounds; allelopathy; vanillic acid; methyl phloretate

1. Introduction

With the world population increasing, maximizing agricultural production is essential. Weeds constrain agricultural production by directly competing with crops for growth resources [1]. Using synthetic herbicides may be the most efficient and cost-effective weed management strategy that has contributed to improving crop production over the last few decades [2]. However, non-judicious use of synthetic agrochemicals has had a significant negative effect on the planet and human wellbeing [3]. The evolution of resistant weeds is another consequence of extensive herbicide use [4]. Thus, to achieve safer agriculture, it is essential to replace hazardous synthetic herbicides with eco-friendly weed-management approaches. Natural compounds with diversified structures and modes of action could offer a new way to develop natural bio-herbicides [5]. Plants produce secondary metabolites called allelochemicals, which can negatively influence the process of growth, and development of surrounding plants [6,7]. In recent years, plant-derived natural phytotoxic compounds have been broadly investigated for herbicidal properties and have been shown to suppress weeds. These compounds can then be used as promising templates for standard bio-herbicides [8].

Garcinia xanthochymus Hook. f. ex T. Anderson (Clusiaceae) is a medium-size tree that has straight trunk with spreading type branches rising in a whorl [9]. This plant is generally found in Bangladesh, India, Myanmar, Thailand, and China [10,11]. *G. xanthochymus* is popularly known as false mangosteen because the shape of its fruit is similar to that of mangosteen. The leaves of the plant are light green color during young stage and become dark green when mature [12]. Distinct parts of this tree have traditionally been used for different therapeutic practices for many years in different south Asian countries [13]. In Bangladesh, *G. xanthochymus* is usually treated as folk drug for the treatment of diarrhea, dysentery, and vomiting. This plant is also used to treat worms and food toxins [14].

Earlier reports revealed that *G. xanthochymus* contains several phytochemicals that have antibacterial, anti-inflammatory, and antioxidant activities [15]. Although extensive research on different biological activities of *G. xanthochymus* have been conducted, very little is known about its phytotoxic potential or the constituents responsible for its phytotoxicity. Hence, this study aimed to (i) examine the phytotoxic potential of *G. xanthochymus* and (ii) detect phytotoxic substances which can be used as a potential candidate for bioherbicide.

2. Materials and Methods

2.1. Plant Sample

The leaf samples of *Garcinia xanthochymus* Hook.f. ex T.Anderson were obtained from a North-Eastern district (Netrokona) of Bangladesh (24.8750° N 90.7333° E) in the time of June and July 2017. A voucher specimen (SAUCB 19127) was submitted to the Crop Botany Herbarium at Sylhet Agricultural University, Sylhet-3100, Bangladesh. Collected leaves were thoroughly washed and dried under room temperature. Then, the leaves were powdered in a grinder and kept for further use at 2 °C.

2.2. Model Test Species

Four model test plant species Italian ryegrass (*Lolium multiflorum* Lam.), lettuce (*Lactuca sativa* L.), rapeseed (*Brassica napus* L.) and timothy (*Phleum pratense* L.) were chosen for conducting the phytotoxicity assay.

2.3. Extraction and Bioassay

In total, 100g leaf powder of *G. xanthochymus* was extracted with 500 mL aqueous methanol (70% (*v/v*)) at room temperature for two days. After filtration with single layer filter paper (No. 2, 125 mm; Toyo Ltd., Tokyo, Japan), the residue was re-extracted for one day with equal volume of methanol and filtered again. The two filtrates were then mixed and dried in a rotavapor at 40 °C. The bioassay experiment was conducted with *G. xanthochymus* crude extracts against selected tested plants at different concentrations as 0.001, 0.003, 0.01, 0.03, 0.1, and 0.3 g of dry weight (DW) equivalent extract/mL as described by Rob et al. [16]. To prepare desired concentrates, an aliquot of the extracts was applied to a sheet of filter paper (No. 2) in 28 mm Petri dishes. Then, the solvent of the mixture (methanol) was dried in a draft chamber and 0.6 mL of 0.05% (*v/v*) aqueous Tween 20 (polyoxyethylene sorbitan monolaurate; Nacalai, Kyoto, Japan) solution was added to each Petri dish. Tween 20 was used as a non-toxic surfactant. Then, seeds of test plants were set. The seeds in the Petri dishes treated with Tween 20 without extract were used as a control.

2.4. Purification of The Active Substances

Powder of *Garcinia xanthochymus* leaves (2 kg) was extracted as aforementioned procedure. The obtained extracts were then evaporated with a vacuum evaporator at 40 °C to produce an aqueous residue and adjusted to pH 7.0 with phosphate buffer (1M). This residue was partitioned with an equal volume of ethyl acetate for three times. The unwanted aqueous part in ethyl acetate fraction was removed by adding anhydrous Na_2SO_4. As described above, a bioassay with the generated aqueous and ethyl acetate fractions was conducted against cress. The ethyl acetate fraction imposed stronger phytotoxic activity compared to the aqueous fraction (data not shown). Therefore, ethyl acetate fraction was selected for further bio-guided fractionation by different column chromatographic steps including silica gel, Sephadex LH-20, C_{18} cartridge and reverse-phase HPLC (500×10 mm I.D. ODS AQ-325; YMC Ltd., Kyoto, Japan) monitoring cress phytotoxicity assay in each purification step following the procedure described by Rob et al. (2019) which led to the isolation of two substances, 1 and 2. These substances 1 and 2 were then re-purified with reverse-phase HPLC (4.6×250 mm I.D., S-5 μm, Inertsil$^{®}$ ODS-3; GL Science Inc., Tokyo, Japan) at a flow rate of 0.8 mL min^{-1} with 5% and 30% aqueous methanol and obtained at 10–15 min and 50–60 min retention time, respectively. Finally, the two substances were characterized by spectral data analysis.

2.5. Bioassay of the Isolated Compounds

The identified compounds were dissolved in 2 mL methanol to make different concentrations (1, 3, 10, 30, 100, 300, and 1000 μM). Then, prepared concentrations were applied on cress and timothy to perform bioassays described above.

2.6. Statistical Analysis

Bioassays were replicated thrice with 10 seedlings and repeated twice. The obtained data were subjected to ANOVA followed by Tukey's HSD test with the help of statistical software "SPSS" Version 24 [17]. Concentrations required for the suppression of 50% growth of the tested plants (I_{50} value) in assay experiments were estimated by the logistic regression equation of the concentration-response curves.

3. Results

3.1. Phytotoxic Activity of the G. Xanthochymus Extract

The effect of aqueous methanolic extract obtained from *G. xanthochymus* is presented in Figure 1. All the applied concentrations suppressed the growth of all test plant species, except 0.001 g DW equivalent extract mL^{-1} concentration. At concentration 0.1 g DW equivalent extract mL^{-1}, lettuce growth was completely restricted while at the same treatment, shoot and root length of rapeseed, Italian ryegrass and timothy were restricted to 3.88, 3.41, 3.05% and 2.9, 5.5, 0.0% of control seedlings, respectively. The seedling length of all test plants limited to less than 3% of control, when treated with the concentration obtained from 0.3 g DW equivalent extract mL^{-1}. The I_{50} values of *G. xanthochymus* extract for all tested plants varied from 4.7 to 17.2 mg DW equivalent extract mL^{-1} as shown in Table 1.

3.2. Bioativity of Different Fractions in The Separtion Steps

Different fractions in silica gel column chromatography showed different level of activity at the concentration 0.3 g dry weight equivalent extract/mL of G. xanthochymus. (Figure 2). The highest bioactivity was achieved with the fractions containing 60 and 70% ethyl acetate in n-hexane and inhibited the seedling growth of cress by less than 13 and 2%, respectively. Combined crude of both of these fractions was subjected to separate expected bioactive compounds. In each step, the highest bioactive fraction was selected for the next separation steps (data not shown).

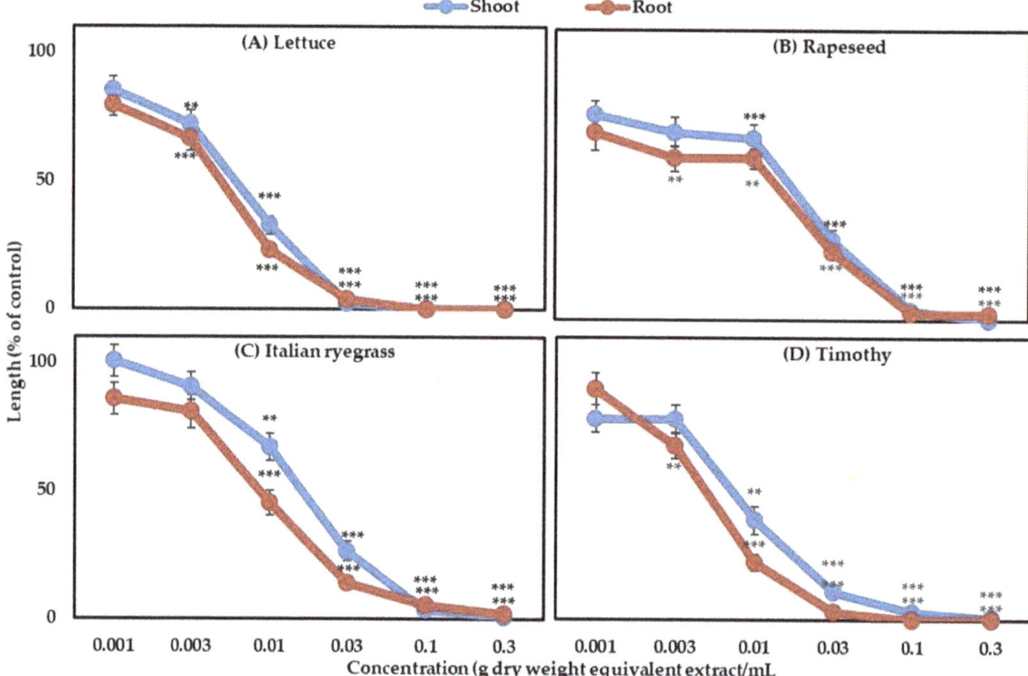

Figure 1. Effect of the *G. xanthochymus* extracts on the shoot and root growth of (**A**) lettuce, (**B**) rapeseed, (**C**) Italian ryegrass, and (**D**) timothy at different concentrations. Mean ± SE was calculated from two independent experiments (Replication = 3 times, number of seedlings/treatments = 10, n = 60). Asterisks denote significant variations between indicated plants (** $p < 0.01$ and *** $p < 0.001$).

Table 1. I_{50} values (concentrations causing 50% growth inhibition) of *Garcinia xanthochymus* leaf extracts (mg dry weight equivalent extract/mL) on lettuce, rapeseed, Italian ryegrass, and timothy ± standard deviation.

Aqueous Methanol Extracts of *G. xanthochymus* (mg Dry Weight Equivalent Extract/mL)		
Test plant Species	Shoot	Root
Lettuce	7.13 ± 0.94	4.73 ± 0.61
Rapeseed	17.20 ± 0.68	14.23 ± 0.85
Italian ryegrass	15.84 ± 0.78	8.53 ± 0.99
Timothy	7.13 ± 0.72	4.81 ± 0.97

3.3. Characterization of The Compounds

Two phytotoxic substances were identified by bio-guided fractionation from the leaf extract of *G. xanthochymys* by spectral analysis. The formula of the substance 1 was assigned as $C_8H_8O_4$ based on HRESIMS at m/z 167.0349 [M–H]- (calcd for $C_8H_7O_4$, 167.0344, Δ = +0.5 mmu); ^{1}H NMR (400 MHz, D_2O) δ_H 7.52 (d, *J* = 7.50 Hz, 1 H, H-2), 7.45 (dd, *J* = 8.5, 2.0 Hz, 1 H, H-6), 6.93 (d, *J* = 8.5 Hz, 1 H, H-5), 3.90 (s, 3 H, H-8); ^{13}C NMR (100 MHz, D_2O) δ_C 173.5 (C-7), 147.9 (C-4), 146.7 (C-3), 128.1 (C-6), 123.2 (C-1), 114.8 (C-5), 113.1 (C-2), 55.8 (C-8). Comparing this spectral data with previously published data substance was recognized as vanillic acid with the systematic name 4-hydroxy-3-methoxy benzoic acid shown in Figure 3 [18].

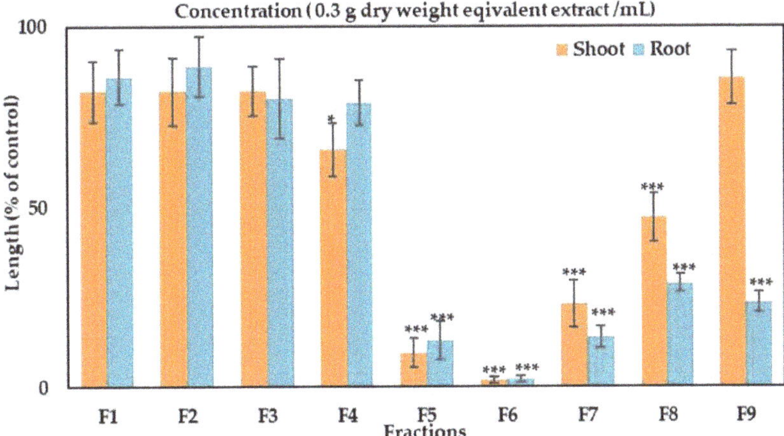

Figure 2. Effect of silica gel column fractions on the seedling growth of cress at the concentration 0.3 g dry weight equivalent extract/mL of *G. xanthochymus*. The column was eluted with raising quantities of the ethyl acetate (10% per step, *v/v*) in n-hexane: F1, F2, F3, F4, F5, F6 and F7 contained 20%, 30%, 40%, 50%, 60%, 70% and 80% ethyl acetate in n-hexane, respectively, F8 (ethyl acetate), F9 (methanol). The values are mean ± SE obtained from two independent experiments. Error bars are standard error of the mean. Asterisks show major variations between treatments and control by least significant difference (LSD) test (* $p < 0.05$ and *** $p < 0.001$).

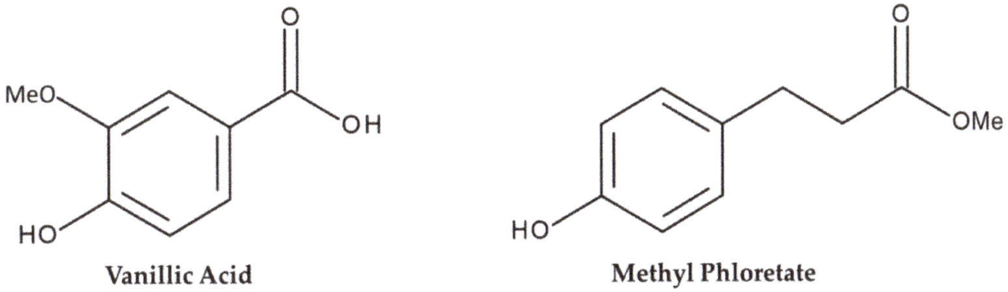

Vanillic Acid　　　　　　**Methyl Phloretate**

Figure 3. Molecular structure of the vanillic acid and the methyl phloretate from the *G. xanthochymus* leaf extract.

The molecular formula of the substance 2 was assigned as $C_{10}H_{12}O_3$ based on HRES-IMS at m/z 181.0830 $[M + H]^+$ (calcd for $C_{10}H_{13}O_3$, 181.0865, $\Delta = -3.5$ mmu); ^{1}H NMR (400 MHz, CDCl$_3$) δ_H 7.07 (d, $J = 8.7$ Hz, 2 H, H5, 9), 6.76 (d, $J = 8.7$ Hz, 2 H, H6, 8), 3.67 (s, 3 H, H10), 2.88 (t, $J = 7.8$ Hz, 2 H, H3), 2.60 (t, $J = 7.8$ Hz, 2 H, H2). The substance was identified as methyl phloretate {3-(4-hydroxyphenyl) propionic acid methyl ester} (Figure 3) by investigating original data with previously documented literature [19].

3.4. Biological Activity of the Isolated Substances

The phytotoxic activity of two identified substances was checked on cress and timothy. Vanillic acid caused significant inhibition on growth of cress and timothy at concentrations 30 and 10 µM, respectively (Figure 4). The I_{50} values of vanillic acid for the shoot and root growth of cress were 331.7 and 314.7 µM, respectively, while those values for timothy were 118.8 and 107.3 µM, respectively (Table 2). Similarly, methyl phloretate also possessed strong phytotoxicity against the test plants with the I_{50} values ranged from 104.7 to 113.7 µM for cress seedlings and 53.4 to 40.6 µM for timothy seedlings (Table 2). It is notable that, the inhibition was concentration dependent and timothy was more susceptible to both compounds compared to cress. Moreover, methyl phloretate was more phytotoxic than vanillic acid concerning the I_{50} values of the compounds (Figure 4).

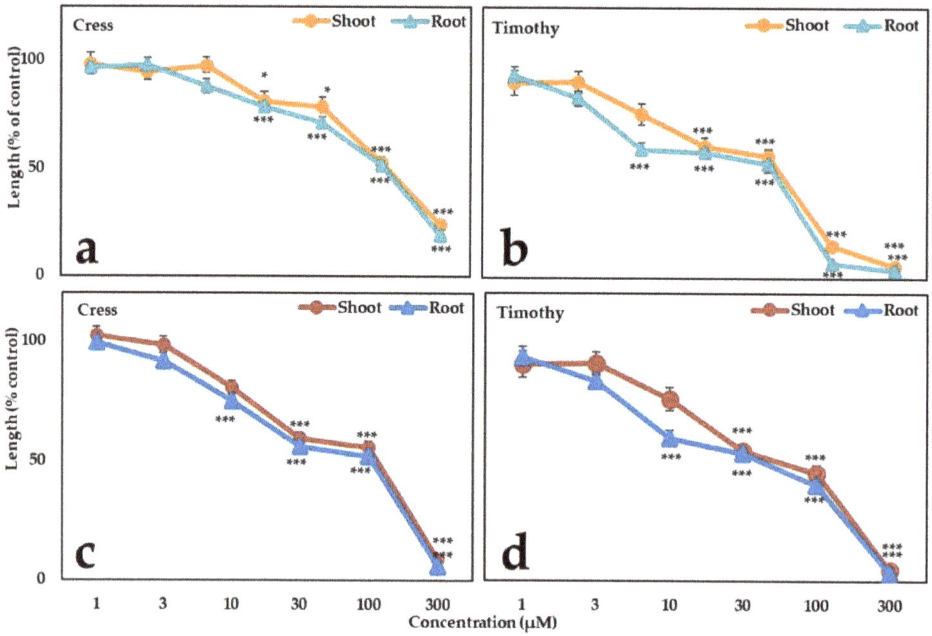

Figure 4. Effects of the vanillic acid (**a**,**b**) and methyl phloretate (**c**,**d**) on cress and timothy. Values represent means $\pm$ SE from three replicates (n = 30). The significant variations between control and treatment are denoted by * $p < 0.05$ and *** $p < 0.001$.

Table 2. The I_{50} value (concentration causing 50% of growth inhibition) of vanillic acid and methyl phloretate (μM) on cress and timothy standard deviation.

Tested Species		Vanillic Acid		Methyl Phloretate	
		Shoot ***	Root ***	Shoot ***	Root ***
dicot	Cress	331.7 ± 7.1	314.7 ± 8.3	113.7 ± 3.8	104.7 ± 2.9
Monocot	Timothy	118.8 ± 4.4	107.3 ± 4.1	53.4 ± 2.8	40.6 ± 1.9

Significant differences between vanillic acid and methyl phloretate represented by *** $p < 0.001$ (paired *t*-test).

4. Discussion

The results revealed that *G. xanthochymus* extract markedly inhibited both the dicot species (lettuce and rapeseed) and the monocot species (Italian ryegrass and timothy). The phytotoxicity of the extract against all the test plants increased with the increase in concentration. Other researchers have also documented such concentration-dependent inhibitory activity of different plant extracts [20–26]. Our previous experiment with *Garcinia pedunculata* also showed strong inhibition against several test species [27]. In addition, the phytotoxic activity of the extracts varied against different test plants. Tuyen et al. [28] also reported species specificity of *Castanea crenata* extracts against radish, lettuce, and barnyard grass. These concentration-dependent and species-specific phytotoxicity of the *G. xanthochymus* leaf extract led us to assume that the extract contains potential phytotoxic substances.

Bio-guided isolation resulted in obtaining two phytotoxic substances from the *G. xanthochymus* leaf extract, which were identified as vanillic acid, and methyl phloretate through spectral analysis. Vanillic acid is one of the most common phenolic compounds found in different plant parts. It was previously identified in extracts from different plants such as *Alnus japonica, Gossypium mexicanum, Rosa canina, Panax ginseng* [29], and *Chenopodium murale* [30]. Vanillic acid can act as an antimicrobial, antioxidant [31], anti-inflammatory, and antidiabetic agent [32]. However, phytotoxic effects of the vanillic acid have also been narrated by many researchers [33]. Methyl phloretate is a methyl ester of phenylpropanoid phenol, phloretic acid [34], which is a naturally occurring phenolic compound that can be obtained from *p*-coumaric acid hydrogenation or synthesized from phloretin, a secondary metabolite of apple leaves [35]. In phloretic acid, there is a propionic acid side chain that is suitable for esterification, leading to production of methyl phloretate [36]. Methyl phloretate is a potential synthetic intermediate and can be used to prepare antidiabetic agents [37].

In our study, the inhibitory activity of vanillic acid and methyl phloretate depended on the concentration and species. In previous reports such inhibitory activities were noted from different phytotoxic substances [20,38]. Piyatida et al. reported that vanillic acid has strong growth-inhibitory activity against cress and timothy, and the inhibition is more pronounced against timothy than cress [39]. This species specificity of allelochemicals might be because of different physical and physiological features of receptor seeds including thickness of seed coat, cell membrane permeability, and sensitive enzymes in seeds [40]. In our experiment, timothy was much more affected by both the allelochemicals compared with cress. Our finding is in line with that by Pèrez et al. [41], who reported that small-seeded tested plants were usually more susceptible to phytochemicals, because the concentration of the phytochemicals necessary to produce suppression is influenced by seed size. Taking into consideration the 1000 seed weight, the weight of timothy is around 5-times lower than cress, which is why timothy is more sensitive to the allelochemicals [42,43].

The data from our experiment also revealed that methyl phloretate has much more inhibitory activity than vanillic acid against tested plant species. The allelochemicals affect plant growth through different chemical reactions. The toxicity of phytochemicals is regulated by the various functional groups in the structure of compound, which act on different positions of the enzymes, and affect their activity [40,44]. Vanillic acid is a mono hydroxy benzoic acid with one methoxy group in its aromatic ring. Maffei et al. [45] reported that the methoxy groups on benzoic acid ring increase the phytotoxicity, whereas the hydroxy groups decrease the phytotoxicity of benzoic acid derivatives in cucumber seed germination and early growth. Maffei et al. [45] also showed that the hydroxy group in the C4 position scavenges the inhibitory action of the methoxy group in vanillic acid. Levi-Minzi et al. [46] showed that 4-hydroxy-3-methoxy benzoic acid is much more phytotoxic against wheat growth compared with 4-hydroxy benzoic acid. We therefore conclude that the methoxy group in the C-3 position contributes to the phytotoxicity of vanillic acid.

On the other hand, methyl phloretate is a monohydroxy derivative of cinnamic acids. With cinnamic acid derivatives, the more hydroxyl groups in its benzene ring the less the phytotoxicity [47]. Therefore, a single hydroxyl group in benzene ring of methyl phloretate may influence its phytotoxic potential. In addition, some researchers proposed that the hydrophobicity of the compound corresponds with higher toxicity in case of cinnamic acid derivatives [48,49]. Hydrophobic compounds bearing lipophilic properties make them capable of passing through the cell membrane more readily [50]. Some researchers reported that the methyl ester of cinnamic acid derivatives has higher growth-inhibitory activity corresponding to their free acids because esterification of the carboxyl group results in more hydrophobic compounds having more phytotoxic potential [51,52]. Accordingly, esterification of phloretic acid (corresponding free phenolic acid of methyl phloretate) which is already documented to have growth-inhibitory activity, might lead to producing more phytotoxic methyl phloretate. Waśko et al. [52] showed that the allelopathic activity of methyl p-coumarate, which has a similar structure to methyl phloretate, was more inhibitory than its corresponding free phenolic acid, p-coumaric acid.

In general, cinnamic acid derivatives are more phytotoxic than benzoic acid derivatives due to their higher hydrophobicity [47,53]. Our results corroborate this finding because the more phytotoxic methyl phloretate is a derivative of cinnamic acids, while the less phytotoxic vanillic acid is a derivative of benzoic acid.

5. Conclusions

The aqueous methanol extract of *G. xanthochymus* significantly inhibited four test plant species, lettuce, rapeseed, Italian ryegrass, and timothy. Generally, leaves of *G. xanthochymus* are not consumed and are classified as waste material. However, from our experiment, it is evident that leaves of *G. xanthochymus* have the potential to be used to manage weeds. Two identified substances, vanillic acid and methyl phloretate, isolated from the *G. xanthochymus* leaf extract imposed significant growth inhibitory activity on cress and timothy. Methyl phloretate had around two- and three-fold higher inhibitory activity against timothy and cress, respectively, compared with vanillic acid. This report is the first on the existence of methyl phloretate in a plant. Therefore, field experiments could determine the potential of the crude extract of *G. xanthochymus* and methyl phloretate it contains as a biological agent to control weeds in an eco-friendly way.

Author Contributions: Conceptualization, M.M.R., and H.K.-N.; methodology, M.M.R., K.H., M.R.K., K.S., A.I., K.I., and H.K-N; software, M.M.R., and K.H.; validation, K.S., K.I., A.I., and H.K.-N.; formal analysis, M.M.R., and K.H.; investigation, M.M.R., and K.H.; resources, H.K.-N.; data curation, H.K.-N.; writing—original draft preparation, M.M.R., and K.H.; writing—review and editing, H.K.-N.; visualization, M.M.R., and K.H.; supervision, H.K.-N. All the authors have read and agreed with the manuscript. All authors have read and agreed to the published version of the manuscript.

Funding: This research work was supported through a MEXT scholarship (Grant Number MEXT-193490) from the Japan government to conduct the study in Japan.

Institutional Review Board Statement: Not applicable

Informed Consent Statement: Not applicable

Acknowledgments: We thankful to the Government of Japan (MEXT scholarship, Grant number: MEXT-173591) to carry out this research work. We also express gratitude to Professor Dennis Murphy, for editing the English of this manuscript.

Conflicts of Interest: There was no conflict of interest declared by the authors.

References

1. Colbach, N.; Darmency, H.; Fernier, A.; Granger, S.; Corre, V.L.; Messéan, A. Simulating changes in cropping practices in conventional and glyphosate-resistant maize. II. Weed impacts on crop production and biodiversity. *Environ. Sci. Pollut. Res.* **2017**, *24*, 13121–13135. [CrossRef] [PubMed]
2. Feng, G.; Chen, M.; Ye, H.C.; Zhang, Z.K.; Li, H.; Chen, L.L.; Zhang, J. Herbicidal activities of compounds isolated from the medicinal plant *Piper sarmentosum*. *Ind. Crops Prod.* **2019**, *132*, 41–47. [CrossRef]
3. Magnoli, K.; Carranza, C.S.; Aluffi, M.E.; Magnoli, C.E.; Barberis, C.L. Herbicides based on 2, 4-D: Its behavior in agricultural environments and microbial biodegradation aspects. A review. *Environ. Sci. Pollut. Res. Int.* **2020**, *27*, 38501–38512. [CrossRef]
4. Moss, S.; Ulber, L.; Hoed, I.D. A herbicide resistance risk matrix. *Crop Prot.* **2019**, *115*, 13–19. [CrossRef]
5. Hossain, M.M. Recent perspective of herbicide: Review of demand and adoption in world agriculture. *J. Bangladesh Agric. Univ.* **2015**, *13*, 13–24. [CrossRef]
6. Arroyo, A.I.; Pueyo, Y.; Pellissier, F.; Ramos, J.; Espinosa-Ruiz, A.; Millery, A.; Alados, C.L. Phytotoxic effects of volatile and water-soluble chemicals of *Artemisia herba-alba*. *J. Arid Environ.* **2018**, *151*, 1–8. [CrossRef]
7. Rice, E.L. *Allelopathy*; Academic Press: Orlando, FL, USA, 1984.
8. Duke, S.O.; Dayan, F.E.; Romagni, J.G.; Rimando, A.M. Natural products as sources of herbicides: Current status and future trends. *Weed Res.* **2000**, *40*, 99–111. [CrossRef]
9. Hooker, J.D. *The Flora of British India*; L. Reeve and Co.: London, UK, 1874.
10. Verheij, E.W.M.; Coronel, R.E. Edible fruits and nuts. In *Plant Resources of South-East Asia (PROSEA)*; Backhuys Publishers: Kerkwerve, The Netherlands, 1991; p. 175.
11. Zhong, F.F.; Chen, Y.; Song, F.J.; Yang, G.Z. Three new xanthones from *Garcinia xanthochymus*. *Acta Pharm. Sinica*. **2008**, *43*, 938–941.
12. Joseph, K.S.; Dandin, V.S.; Hosakatte, N.M. Chemistry and biological activity of *Garcinia xanthochymus*: A review. *J. Biol. Act. Prod. Nat.* **2016**, *6*, 173–194. [CrossRef]

13. Perry, L.M.; Metzger, J. *Medicinal Plants of East and Southeast Asia: Attributed Properties and Uses*; MIT Press: Cambridge, MA, USA, 1980.
14. Trisuwan, K.; Boonyaketgoson, S.; Rukachaisirikul, V.; Phongpaichit, S. Oxygenated xanthones and bioflavonoids from the twigs of *Garcinia xanthochymus*. *Tetrahedron Lett.* **2014**, *55*, 3600–3602. [CrossRef]
15. Hassan, N.K.N.C.; Taher, M.; Susanti, D. Phytochemical constituents and pharmacological properties of *Garcinia xanthochymus*—A review. *Biomed. Pharmacother.* **2018**, *106*, 1378–1389. [CrossRef]
16. Rob, M.M.; Hossen, K.; Iwasaki, A.; Suenaga, K.; Kato-Noguchi, H. Phytotoxic activity and identification of phytotoxic substances from *Schumannianthus dichotomus*. *Plants* **2020**, *9*, 102. [CrossRef] [PubMed]
17. IBM, Corp. *IBM SPSS Statistics for Windows*; Version 16.0; IBM Corp: Armonk, NY, USA, 2016.
18. Stalin, T.; Rajendiran, N. A study on the spectroscopy and photophysics of 4-hydroxy-3-methoxybenzoic acid in different solvents, pH and β-cyclodextrin. *J. Mol. Struct.* **2006**, *794*, 35–45. [CrossRef]
19. Clough, J.M.; Jones, R.V.; McCann, H.; Morris, D.J.; Wills, M. Synthesis and hydrolysis studies of a peptide containing the reactive triad of serine proteases with an associated linker to a dye on a solid phase support. *Org. Biomol. Chem.* **2003**, *1*, 1486–1497. [CrossRef] [PubMed]
20. Bari, I.N.; Kato-Noguchi, H. Phytotoxic effect of *Filicium decipiens* leaf extract. *Am. Eurasian J. Agric. Environ. Sci.* **2017**, *17*, 288–292. [CrossRef]
21. Islam, M.S.; Iwasaki, A.; Suenaga, K.; Kato-Noguchi, H. 2-Methoxystypandrone, a potent phytotoxic substance in *Rumex maritimus* L. *Theor. Exp. Plant Physiol.* **2017**, *29*, 195–202. [CrossRef]
22. Islam, M.S.; Zaman, F.; Iwasaki, A.; Suenaga, K.; Kato-Noguchi, H. Phytotoxic potential of *Chrysopogon aciculatus* (Retz.) Trin. (Poaceae). *Weed Biol. Manag.* **2019**, *19*, 51–58. [CrossRef]
23. Hossen, K.; Das, K.R.; Okada, S.; Iwasaki, A.; Suenaga, K.; Kato-Noguchi, H. Allelopathic potential and active substances from *Wedelia chinensis* (Osbeck). *Foods* **2020**, *9*, 1591. [CrossRef]
24. Hossen, K.; Kato-Noguchi, H. Determination of allelopathic properties of *Acacia catechu* (L.f.) Willd. *Not. Bot. Horti Agrobot. Cluj Napoca* **2020**, *48*, 2050–2059. [CrossRef]
25. Zaman, F.; Iwasaki, A.; Suenaga, K.; Kato-Noguchi, H. Allelopathic potential and identification of two allelopathic substances in *Eleocharis atropurpurea*. *Plant Biosys.* **2020**. [CrossRef]
26. Hossen, K.; Iwasaki, A.; Suenaga, K.; Kato-Noguchi, H. Phytotoxic Activity and Growth Inhibitory Substances from *Albizia richardiana* (Voigt.) King & Prain. *Appl. Sci.* **2021**, *11*, 1455. [CrossRef]
27. Rob, M.M.; Kato-Noguchi, H. Study of the allelopathic activity of *Garcinia pedunculata* Roxb. *Plant Omics.* **2019**, *12*, 31–36. [CrossRef]
28. Tuyen, P.T.; Xuan, T.D.; Anh, T.T.T.; Van, T.M.; Ahmad, A.; Elzaawely, A.A.; Khanh, T.D. Weed suppressing potential and isolation of potent plant growth inhibitors from *Castanea crenata* Sieb. et Zucc. *Molecules* **2018**, *23*, 345. [CrossRef]
29. Khadem, S.; Marles, R.J. Monocyclic phenolic acids; hydroxy-and polyhydroxybenzoic acids: Occurrence and recent bioactivity studies. *Molecules* **2010**, *15*, 7985–8005. [CrossRef]
30. Batish, D.R.; Lavanya, K.; Pal, H.S.; Kohli, R.K. Root-mediated allelopathic interference of nettle-leaved goosefoot (*Chenopodium murale*) on wheat (*Triticum aestivum*). *J. Agron. Crop. Sci.* **2007**, *193*, 37–44. [CrossRef]
31. Kakkar, S.; Bais, S. A review on protocatechuic acid and its pharmacological potential. *ISRN Pharmacol.* **2014**, 1–9. [CrossRef]
32. Bai, F.; Fang, L.; Hu, H.; Yang, Y.; Feng, X.; Sun, D. Vanillic acid mitigates the ovalbumin (OVA)-induced asthma in rat model through prevention of airway inflammation. *Biosci. Biotechnol. Biochem.* **2019**, *83*, 531–537. [CrossRef] [PubMed]
33. Zhang, T.-T.; Zheng, C.-Y.; Hu, W.; Xu, W.-W.; Wang, H.-F. The allelopathy and allelopathic mechanism of phenolic acids on toxic *Microcystis aeruginosa*. *J. Appl. Phycol.* **2010**, *22*, 71–77. [CrossRef]
34. Comi, M.; Lligadas, G.; Ronda, J.C.; Galia, M.; Cadiz, V. Renewable benzoxazine monomers from "lignin-like" naturally occurring phenolic derivatives. *J. Polym. Sci. Part A Polym. Chem.* **2013**, *51*, 4894–4903. [CrossRef]
35. Picinelli, A.; Dapena, E.; Mangas, J.J. Polyphenolic pattern in apple tree leaves in relation to scab resistance. A preliminary study. *J. Agric. Food Chem.* **1995**, *43*, 2273–2278. [CrossRef]
36. Trejo-Machin, A.; Verge, P.; Puchot, L.; Quintana, R. Phloretic acid as an alternative to the phenolation of aliphatic hydroxyls for the elaboration of polybenzoxazine. *Green Chem.* **2017**, *19*, 5065–5073. [CrossRef]
37. Sasaki, S.; Kitamura, S.; Negoro, N.; Suzuki, M.; Tsujihata, Y.; Suzuki, N.; Kobayashi, M. Design, synthesis, and biological activity of potent and orally available G protein-coupled receptor 40 agonists. *J. Med. Chem.* **2011**, *54*, 1365. [CrossRef]
38. Suzuki, M.; Chozin, M.A.; Iwasaki, A.; Suenaga, K.; Kato-Noguchi, H. Phytotoxic activity of Chinese violet (*Asystasia gangetica* (L.) T. Anderson) and two phytotoxic substances. *Weed Biol. Manag.* **2019**, *19*, 3–8. [CrossRef]
39. Piyatida, P.; Suenaga, K.; Kato-Noguchi, H. Allelopathic potential and chemical composition of *Rhinacanthus nasutus* extracts. *Allelopath. J.* **2010**, *26*, 207–216.
40. Macias, F.A.; Galindo, J.C.G.; Massanet, G.M. Potentials allelopathic activity of several sesquiterpene lactone models. *Phytochemistry* **1992**, *31*, 1969–1977. [CrossRef]
41. Pèrez, F.J. Allelopathic effect of hydroxamic acids from cereals on Avena sativa and *A. fatua*. *Phytochemistry* **1990**, *29*, 773–776. [CrossRef]
42. Stanisavljevic, R.; Đjokic, D.; Milenkovic, J.; Đukanovic, L.; Stevovic, V.; Simic, A.; Dodig, D. Seed germination and seedling vigour of Italian ryegrass, cocksfoot and timothy following harvest and storage. *Cienc. Agrotec.* **2011**, *35*, 1141–1148. [CrossRef]

43. Shehzad, M.; Tanveer, A.; Ayub, M.; Mubeen, K.; Sarwar, N.; Ibrahim, M.; Qadir, I. Effect of weed-crop competition on growth and yield of garden cress (*Lepidium sativum* L.). *J. Med. Plants Res.* **2011**, *5*, 6169–6172. [CrossRef]
44. Sanchez-Maldonado, A.F.; Schieber, A.; Ganzle, M.G. Structure-function relationships of the antibacterial activity of phenolic acids and their metabolism by lactic acid bacteria. *J. Appl. Microbiol.* **2011**, *111*, 1176–1184. [CrossRef]
45. Maffei, M.; Bertea, C.M.; Garneri, F.; Scannerini, S. Effect of benzoic acid hydroxy-and methoxy-ring substituents during cucumber (*Cucumis sativus* L.) germination. I. Isocitrate lyase and catalase activity. *Plant Sci.* **1999**, *141*, 139–147. [CrossRef]
46. Levi-Minzi, R.; Saviozzi, A.; Riffaldi, R. Organic acids as seed germination inhibitors. *J. Environ. Sci. Health* **1994**, *29*, 2203–2217. [CrossRef]
47. Pinho, I.A.; Lopes, D.V.; Martins, R.C.; Quina, M.J. Phytotoxicity assessment of olive mill solid wastes and the influence of phenolic compounds. *Chemosphere* **2017**, *185*, 258–267. [CrossRef] [PubMed]
48. Wang, X.; Dong, Y.; Xu, S.; Wang, L.; Han, S. Quantitative structure-activity relationships for the toxicity to the tadpole *Rana japonica* of selected phenols. *Bull. Environ. Contam. Toxicol.* **2000**, *64*, 859–865. [CrossRef]
49. Wang, X.; Yu, J.; Wang, Y.; Wang, L. Mechanism-based quantitative structure–activity relationships for the inhibition of substituted phenols on germination rate of *Cucumis sativus*. *Chemosphere* **2002**, *46*, 241–250. [CrossRef]
50. Jităreanu, A.; Tătărîngă, G.; Zbancioc, A.M.; Stănescu, U. Toxicity of some cinnamic acid derivatives to common bean (*Phaseolus vulgaris*). *Not. Bot. Horti. Agrobo.* **2011**, *39*, 130–134. [CrossRef]
51. Guzman, J.D. Natural cinnamic acids, synthetic derivatives and hybrids with antimicrobial activity. *Molecules* **2014**, *19*, 19292–19349. [CrossRef]
52. Waśko, A.; Szwajgier, D.; Polak-Berecka, M. The role of ferulic acid esterase in the growth of *Lactobacillus helveticus* in the presence of phenolic acids and their derivatives. *Eur. Food Res. Technol.* **2014**, *238*, 299–306. [CrossRef]
53. Reynolds, T. Comparative effects of aromatic compounds on inhibition of lettuce fruit germination. *Ann. Bot.* **1978**, *42*, 419–427. [CrossRef]

Article

Isolation and Identification of Two Potent Phytotoxic Substances from *Afzelia xylocarpa* for Controlling Weeds

Ramida Krumsri [1,2,*], Kaori Ozaki [3], Toshiaki Teruya [4] and Hisashi Kato-Noguchi [1,2]

[1] Department of Applied Biological Science, Faculty of Agriculture, Kagawa University, Miki, Kagawa 761-0795, Japan; kato.hisashi@kagawa-u.ac.jp

[2] The United Graduate School of Agricultural Sciences, Ehime University, 3-5-7 Tarumi, Matsuyama, Ehime 790-8566, Japan

[3] Graduate School of Engineering and Science, University of the Ryukyus, 1 Senbaru, Nishihara, Okinawa 903-0213, Japan; k168239@eve.u-ryukyu.ac.jp

[4] Faculty of Education, University of the Ryukyus, 1 Senbaru, Nishihara, Okinawa 903-0213, Japan; t-teruya@edu.u-ryukyu.ac.jp

* Correspondence: ramidakrumsri@gmail.com

Abstract: Phytotoxic substances released from plants are considered eco-friendly alternatives for controlling weeds in agricultural production. In this study, the leaves of *Afzelia xylocarpa* (Kurz) Craib. were investigated for biological activity, and their active substances were determined. Extracts of *A. xylocarpa* leaf exhibited concentration-dependent phytotoxic activity against the seedling length of *Lepidium sativum* L., *Medicago sativa* L., *Phleum pratense* L., and *Echinochloa crus-galli* (L.) P. Beauv. Bioassay-guided fractionation of the *A. xylocarpa* leaf extracts led to isolating and identifying two compounds: vanillic acid and *trans*-ferulic acid. Both compounds were applied to four model plants using different concentrations. The results showed both compounds significantly inhibited the model plants' seedling length in a species-dependent manner ($p < 0.05$). The phytotoxic effects of *trans*-ferulic acid (IC$_{50}$ = 0.42 to 2.43 mM) on the model plants were much greater than that of vanillic acid (IC$_{50}$ = 0.73 to 3.17 mM) and *P. pratense* was the most sensitive to both compounds. In addition, the application of an equimolar (0.3 mM) mixture of vanillic acid and *trans*-ferulic acid showed the synergistic effects of the phytotoxic activity against the root length of *P. pratense* and *L. sativum*. These results suggest that the leaves of *A. xylocarpa* and its phytotoxic compounds could be used as a natural source of herbicides.

Keywords: allelopathic activity; growth inhibitor; phenolic compounds; bioherbicide; sustainable agriculture

Citation: Krumsri, R.; Ozaki, K.; Teruya, T.; Kato-Noguchi, H. Isolation and Identification of Two Potent Phytotoxic Substances from *Afzelia xylocarpa* for Controlling Weeds. *Appl. Sci.* **2021**, *11*, 3542. https://doi.org/10.3390/app11083542

Academic Editor: Ana M. L. Seca

Received: 8 March 2021
Accepted: 13 April 2021
Published: 15 April 2021

Publisher's Note: MDPI stays neutral with regard to jurisdictional claims in published maps and institutional affiliations.

1. Introduction

Weeds are considered an important problem in agriculture worldwide. They compete with crops throughout the growing period and reduce crop yield and quality [1]. On average, approximately 34% of the total yield loss of crops is due to weeds interference worldwide [2,3]. Farmers mostly rely on synthetic herbicides to control weeds in crop fields because of their easy accessibility and a more rapid return [4]. In recent years, the deleterious effects of synthetic herbicides on the environment have been of concern. The application of synthetic herbicides has led to long-term danger to the environment because of their permanence in nature [5,6]. In particular, the widespread practice of using synthetic herbicides has resulted in the evolution of herbicide-tolerant weed species [7]. Consequently, scientific communities have been paying greater attention to eco-friendly alternatives and sustainable biological solutions for weed management [8–10]. As a result, allelopathic plants with phytotoxic substances are now considered environmentally friendly alternatives for controlling weeds [11,12].

Allelopathic plants release various phytotoxic substances into the environment through different mechanisms [13,14]. These phytotoxic substances (e.g., phenolics, terpenoids, and alkaloids and their derivatives) can affect target plants' physiological functions, such as inhibiting cell membrane permeability and cell division as well as interrupting respiration photosynthesis, nutrient uptake, and enzymatic activities [15–17]. Most of these substances are entirely or partially water-soluble, making them more eco-friendly and easier to apply as natural herbicides [18,19]. Consequently, there has been much research on isolating and identifying phytotoxic substances from different plant species to develop natural herbicides [20–22]. Some of the allelochemicals isolated from plants have been used in developing allelochemical-based herbicides. For instance, cinmethylin herbicide (Cinch®) has been developed based on the chemical structure of 1,4-cineole derived from *Eucalyptus* sp. This herbicide is effectively used as a pre-emergence herbicide to control annual grass weeds that inhibit the enzyme tyrosine aminotransferase activity [23,24].

However, different structural characteristics of the compounds show different modes of action and target sites on the plants [25]. Therefore, it is important to search for and to identify new species with phytotoxic activity to determine specific properties and unique target sites in target plants.

Recently, researchers have been interested in searching for a potential source of phytotoxic substances in the Fabaceae family [26,27]. Fabaceae is a large family of diverse plants known for a high diversity of secondary metabolites with phytotoxic potential [28,29]. Research on their phytotoxic potential suggests future applications as natural herbicides [30,31].

Afzelia xylocarpa (Kurz) Craib. is a dicotyledonous plant belonging to the subfamily Caesalpinioideae in the Fabaceae family. This species is both ecologically and economically important deciduous tree in agroforestry systems in the tropics and is mainly distributed in Thailand, Vietnam, Cambodia, Laos, and Burma [32]. It is used as a traditional medicine against inflammatory ocular diseases, sore throat, and food poisoning [33]. Previous studies have shown different biological activities of this plant, such as antioxidant, antidiabetic, antimicrobial, and anti-inflammatory [34–36]. *A. xylocarpa* is one of the endangered species in the world. This plant suffers from overexploitation as a timber source, and its habitat has also been decreasing [32]. One possible reason for the reduction of the habitat of the species may be its allelopathic potential and/or autotoxicity. Allelopathic effects are more explicated in the forest communities because of the huge biomass of the canopy or released phytotoxic substances through fallen leaves/fruits [37,38]. Various compounds have been reported in *A. xylocarpa* seeds, such as chlorogenic acid, ferulic acid, gallic acid, taxifolin, caffeic acid, rosmarinic acid, daidzein, isovanillic acid, cinnamic acid, vanillin, naringenin, *p*-coumaric acid, cholesterine, campesterol, campestanol, and stigmasterol [39]. Twelve compounds were isolated from *A. xylocarpa* leaves, including kaempferol-7-*O*-β-D-glucopyranoside, friedelin, β-sitosterol, butyl benzoate, stigmas-ta-4, 25-dien-3-one, epifriedelanol, stigmasterol, palmitic acid, linoleic acid, α-linolenic acid, and (3S,5R,6R,7E)-3,5,6-trihydroxy-7-megastigmen-9-one [33]. It was reported that some phenolic acids, such as *p*-coumaric acid, act as phototoxic substances, which are released from different plant parts or decomposed plant residues. These compounds accumulate in the soil and have detrimental effects on the seed germination and growth of the species nearby and autotoxic effects on the species itself [40–42]. Therefore, the understanding of allelopathy is important to preserve this plant species. In addition, the allelopathic substances may potentially be used in biological weed management.

Although some bioactive compounds have been isolated from *A. xylocarpa*, it was not clear which compounds truly act as active substances in this plant. Thus, this study was conducted to determine the phytotoxic activity of *A. xylocarpa* leaves and to isolate active substances by bioassay-guided fractionations. The activities of the identified compounds and a mixture of these compounds were tested on the seedling growth under laboratory conditions. It was envisaged that this study would highlight the effect of active substances isolated from *A. xylocarpa* leaves on the model plants. This information could be useful for assessing the interrelations between active substances and action sites in the target plants. This research may lead to developing a natural herbicide in the future. More bioactive substances produced by this forest species can help better understand their impact on the local environment and natural ecosystem.

2. Materials and Methods

2.1. Plant Material and Model Plants

Leaves of *Afzelia xylocarpa* (Kurz) Craib. were collected from Phitsanulok Province, Thailand, in August 2017. The collected materials were thoroughly washed with tap water, immediately dried under shade to avoid direct sunlight, and finely ground into powder form for further extraction. Four plant species were used as model plants for bioassay experiments. Two dicotyledonous species (*Lepidium sativum* L. and *Medicago sativa* L.) were selected because of their known growth characteristics and sensitivity to allelopathic extracts. Two monocotyledonous species (*Phleum pratense* L. and *Echinochloa crus-galli* (L.) P. Beauv) were selected for their worldwide distribution in crop fields [43,44]. In the isolation process, *L. sativum* was selected as the model plant due to its high sensitivity to phytotoxic substances at low concentrations [45].

2.2. Preparation of the A. xylocarpa Leaf Extracts

The *A. xylocarpa* powder (100 g dry weight) was extracted in 70% aqueous methanol in a 1:5 (*w/v*) ratio at room temperature and then kept under dark conditions to avoid light degradation. After 48 h extraction, the solution was then vacuum filtered through filter paper (No.2, 12.5 cm; Toyo Ltd., Tokyo, Japan). The residue was re-extracted with methanol (500 mL) for another 24 h and filtered. The combined filtrates were concentrated in a vacuum using a rotary evaporator (40 °C) to produce a crude extract (~52 g).

2.3. Bioassay Procedure

To assess the biological activity of the *A. xylocarpa* leaf extracts, concentrations of 1, 3, 10, 30, 100, and 300 mg dry weight (DW) equivalent extract/mL were used. An aliquot of the extracts at assay concentrations (0.6 mL) was transferred into Petri dishes (2.8 cm in diameter) containing a sheet of filter paper (No. 2; 2.8 cm; Toyo Ltd.). After the solvent evaporated, filter paper sheets were moistened with 0.6 mL of 0.05% Tween-20 in distilled water (polyoxyethylene sorbitan monolaurate; Nacalai, Kyoto, Japan). Ten seeds each of *L. sativum* and *M. sativa*, and ten sprouted seeds each of *P. pratense* and *E. crus-galli* were placed into Petri dishes (sprouted in distilled water under darkness at 25 °C for 48 h). A solution of 0.05% Tween-20 served as control. All plates were incubated in a growth chamber under darkness (25 °C), and six replicates were performed for each treatment. After 48 h incubation, the lengths of the shoots and roots of the model plants were measured. The results were expressed as a percentage ratio using the following formula [46]: seedling length (%) = (treated length/length of control) ×100.

2.4. Isolation and Identification of the Active Substances

The *A. xylocarpa* leaves (1400 g) were extracted using the same method described earlier. The aqueous methanol leaf extracts were concentrated in *vacuo* (40 °C) to produce an aqueous solution for further isolation and identification. The isolation methods were modified from those performed by Boonmee et al. [47]. The separated fractions obtained from each isolation step were examined for their activity using an *L. sativum* bioassay as the same assay method described above. After adjusting the pH to 7 using 1 M phosphate buffer, the aqueous solution was partitioned with ethyl acetate at an equal volume and separated into aqueous and ethyl acetate fractions. The ethyl acetate fraction was subjected to silica gel column chromatography (60 g, silica gel 60, 70–230 mesh; Nacalai, Kyoto, Japan), and eluted with stepwise gradient mixtures of *n*-hexane: ethyl acetate (9:1 to 2:8 (*v/v*), 150 mL), and methanol for final elution (300 mL). The most active fraction (F5) was then chromatographed on a Sephadex LH-20 column (100 g; GE Healthcare, Uppsala, Sweden). The mobile phase was used as stepwise water: methanol mixtures (8:2 to 2:8 (*v/v*), 150 mL), and methanol (300 mL). The most active fraction (F2) eluted with 40%

aqueous methanol. Therefore, fraction 2 was loaded onto a reverse-phase C_{18} cartridge (1.2 × 6.5 cm; YMC-Dispo Pack AT ODS-25; YMC Ltd., Kyoto, Japan). The cartridge was eluted step-by-step with water: methanol mixtures (9:1 to 2:8 (*v/v*), 15 mL), and methanol (30 mL). The most active fraction was eluted with 20% aqueous methanol (F1) and finally purified using reverse-phase HPLC (column size 10 mm i.d. × 500 mm S-5 μm; ODS AQ-325, YMC Ltd.), eluted at a flow rate of 1.5 mL/min with 20% aqueous methanol, and detected at a wavelength of 220 nm. Finally, two isolated substances were found, and their chemical structures were assessed using spectral analysis.

2.5. Biological Activity of the Isolated Active Substances

Two isolated substances were prepared in methanol to produce concentrations of 0.1, 0.3, 1, 3, and 10 mM. The prepared concentrations of each isolated substance (0.6 mL) were applied on *L. sativum*, *M. sativa*, *P. pratense*, and *E. crus-galli* to perform bioassays as the same protocol mentioned above. The growth parameters (shoot and root length) of the model plants were measured, and all results were calculated using the formula described above.

2.6. The Synergistic Effects of a Mixture of the Two Isolated Substances

The synergistic effects of the two isolated substances were determined. A stock solution of each isolated substance was prepared with methanol (0.3 mL) to get a concentration of 0.3 mM. The isolated substances 1 and 2 were mixed at a ratio of 1:1. The mixture of the two substances was applied on two model plants: *L. sativum* and *P. pratense*. The experimental design, incubation conditions, and data collection were the same as in the bioassay described above.

2.7. Statistical Analysis

The data from each experiment were analyzed using Tukey's test with SPSS version 16.0 (IBM, Chicago, IL, USA). Student's *t*-test was used to analyze two-group comparisons. The concentration required for 50% inhibition of each test plant was determined using GraphPad Prism 6.0 (GraphPad Software, Inc., La Jolla, CA, USA) by nonlinear regression of the inhibition results. The relationship between the seedling length of the test plants and the concentrations was determined using a two-tailed Pearson's correlation.

3. Results and Discussion

3.1. Biological Activity of the Aqueous Methanol Extracts of A. xylocarpa Leaves

The *A. xylocarpa* leaf extracts significantly reduced the seedling length of the four model plants ($p < 0.05$) (Figure 1). At the concentration of 30 mg DW equivalent extract/mL, the shoots of *L. sativum* were completely inhibited, whereas the shoots of *M. sativa*, *P. pratense*, and *E. crus-galli* were inhibited to 1.68, 30.08, and 52.44% of control length, respectively (Figure 1A). In contrast to the shoot length, the leaf extracts inhibited the root length of *L. sativum*, *M. sativa*, *P. pratense*, and *E. crus-galli* to 10.85, 10.60, 2.65, and 29.04% of control length, respectively (Figure 1B). The phytotoxic activity of the leaf extracts against the model plants significantly correlated with the extract concentrations (r = −0.92 to −0.98, $p < 0.01$) (Figure 1A,B). The IC_{50} values of the leaf extracts for the seedling length of the model plants ranged from 6.42 to 24.21 mg DW equivalent extract/mL (Table 1). In addition, the leaf extracts inhibited the root length of *L. sativum*, *P. pratense*, and *E. crus-galli* more than the shoot length (*P. pratense* was most sensitive to the extracts; inhibition of its roots was 2.3-times greater than for its shoots).

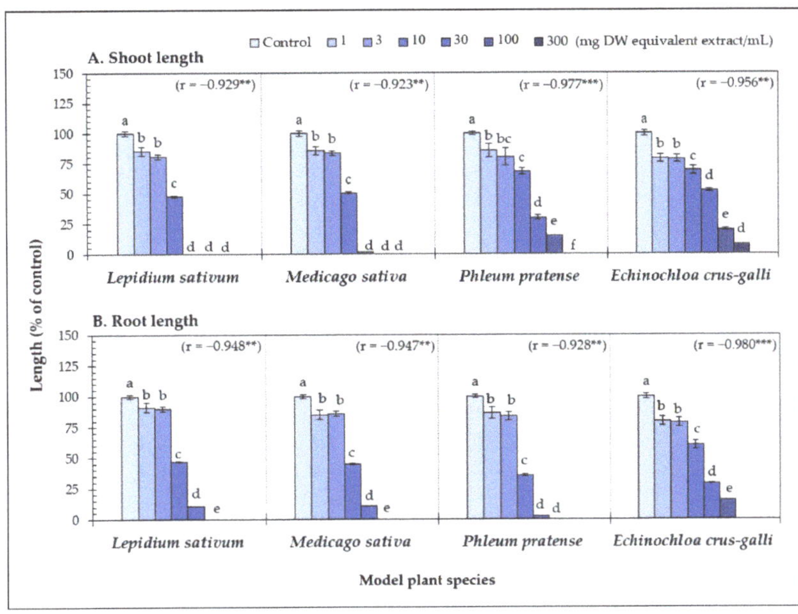

Figure 1. Effect of the *Afzelia xylocarpa* leaf extracts on the (**A**) shoot and (**B**) root length of the model plants exposed to six concentrations. Means ± SE with six replicates (10 seeds/replicates). The different letters on the columns of each model plant indicate significant differences (Tukey's HSD, $p < 0.05$). Statistical significance is represented by asterisks (two-tailed Pearson's correlation, ** $p < 0.01$ and *** $p < 0.001$). correlation coefficient (r).

Table 1. The concentration of the aqueous methanol leaf extracts and vanillic acid and *trans*-ferulic acid from *Afzelia xylocarpa* required for 50% inhibition (IC_{50} value) of the shoot and root length of the model plants.

Test plant species	IC_{50} values of *Afzelia xylocarpa* leaf extracts (mg DW equivalent extract/mL)	
	Shoot length	Root length
Lepidium sativum	7.78 [d]	7.67 [c]
Medicago sativa	8.50 [c]	8.99 [b]
Phleum pratense	15.28 [b]	6.42 [d]
Echinochloa crus-galli	24.21 [a]	12.62 [a]

Test plant species	IC_{50} values of vanillic acid (mM)	
	Shoot length	Root length
Lepidium sativum	0.81 [c]	0.93 [b]
Medicago sativa	2.72 [ab]	1.89 [a]
Phleum pratense	0.80 [c]	0.73 [c]
Echinochloa crus-galli	3.17 [a]	1.01 [b]

Test plant species	IC_{50} values of *trans*-ferulic acid (mM)	
	Shoot length	Root length
Lepidium sativum	0.65 [c]	0.76 [c]
Medicago sativa	2.43 [a]	1.40 [a]
Phleum pratense	0.61 [c]	0.42 [d]
Echinochloa crus-galli	1.10 [b]	0.97 [b]

Data with different letters in a column indicate a significant difference (Tukey's HSD, $p < 0.05$).

These results suggest that the leaves of *A. xylocarpa* may contain phytotoxic substances with phytotoxic activity. In addition, the results indicate that seedling length varied among the model plants, and the level of inhibition depended on the extract concentration (Figure 1). Similarly, several previous studies have also reported that the phytotoxic effect is concentration-dependent [48–53]. Moreover, many previous studies noted that test plant species' response varied depending on the types of phytotoxic substances [54]. Hence, it is important to understand which compounds play a major role in plant allelopathy to be used for weed control.

3.2. Isolation and Identification of the Active Substances in the A. xylocarpa Leaf Extracts

Partitioning showed that the ethyl acetate fraction had higher inhibitory activity than the aqueous fraction (Figure 2). At 300 mg DW equivalent extract/mL, the ethyl acetate fraction inhibited the seedling length of *L. sativum* by 1.9–2.1 times greater than that of the aqueous fraction. Therefore, the growth inhibitory substances in the ethyl acetate fraction were separated through a series of bioassay-guided fractionations. The most active fraction was then purified using HPLC, and two active substances were found. The molecular formula of substance 1 was determined as $C_8H_8O_4$ using HRESIMS at m/z 167.03 [M−H] − (calcd. for $C_8H_7O_4$, 167.03). The ^{1}H NMR spectrum (400 MHz, CD_3OD) of the substance showed δ_H 3.89 (3H, s), 7.56 (1H, d, *J* = 1.7), 7.55 (1H, dd, *J* = 8.7, 1.7), and 6.84 (1H, d, *J* = 8.7). The substance was identified as 4-hydroxy-3-methoxy benzoic acid (vanillic acid). Substance 2 was determined as $C_{10}H_{10}O_4$ using HRESIMS at m/z 193.05 [M−H] − (calcd. for $C_{10}H_9O_4$, 193.05). The ^{1}H NMR spectrum (400 MHz, CD_3OD) of the substance showed δ_H 3.90 (3H, s), 7.60 (1H, d, *J* = 15.9), 6.31 (1H, d, *J* =15.9), 7.18 (1H, d, *J* = 1.9), 7.06 (1H, dd, *J* = 8.1, 1.9), and 6.81 (1H, d, *J* = 8.1). The substance was identified as *trans*-4-hydroxy-3-methoxy cinnamic acid (*trans*-ferulic acid).

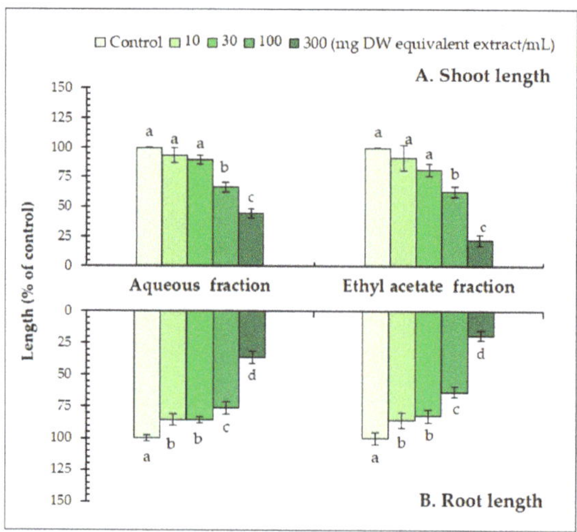

Figure 2. The effect of the aqueous and ethyl acetate fractions on the (**A**) shoot and (**B**) root length of *Lepidium sativum* exposed to four concentrations. Means ± SE with six replicates (10 seeds/replicates). The different letters on each column indicate significant differences (Tukey's HSD, $p < 0.05$).

Vanillic acid and *trans*-ferulic acid are phenolic compounds, biosynthesized from phenylalanine via the shikimic acid pathway in plants. Vanillic acid is a hydroxybenzoic acid derivative found in several natural sources [55–57]. This compound has been reported to have multiple activities, including anti-inflammatory, antimicrobial, anticancer, and antioxidant [58,59]. On the other hand, *trans*-ferulic acid, a hydroxycinnamic acid derivative, is present in many edible plants [60,61]. Many studies have found that *trans*-ferulic acid possesses different beneficial biological activities, including antioxidants [62,63]. To our best knowledge, *A. xylocarpa* leaves are a source of vanillic acid and *trans* -ferulic acid with phytotoxic effects.

3.3. Biological Activity of Vanillic Acid and Trans-Ferulic Acid

Vanillic acid and *trans*-ferulic acid showed phytotoxicity against the four model plants' seedling length (Figures 3 and 4). At the concentration 1 mM of vanillic acid, the shoot length of *L. sativum*, *M. sativa*, *P. pratense*, and *E. crus-galli* was inhibited to 47.42, 66.32, 74.34, and 66.87% of control length, respectively, whereas that of root length was inhibited to 47.79, 66.01, 63.22, and 54.75% of control length, respectively (Figure 3). On the other hand, 1 mM of *trans*-ferulic acid inhibited the shoot length of *L. sativum*, *M. sativa*, *P. pratense*, and *E. crus-galli* to 45.87, 65.81, 57.00, and 53.68% of control length, respectively, whereas root length was inhibited to 41.91, 56.25, 57.00, and 29.28% of control length, respectively (Figure 4). The correlation coefficient between vanillic acid and *trans*-ferulic acid and the seedling length of the model plants was very high, with values ranging from −0.962 to −0.996 ($p < 0.01$). The IC$_{50}$ values of vanillic acid and *trans*-ferulic acid for the seedling length of the model plants were 0.73 to 3.17 mM and 0.42 to 2.43 mM, respectively (Table 1).

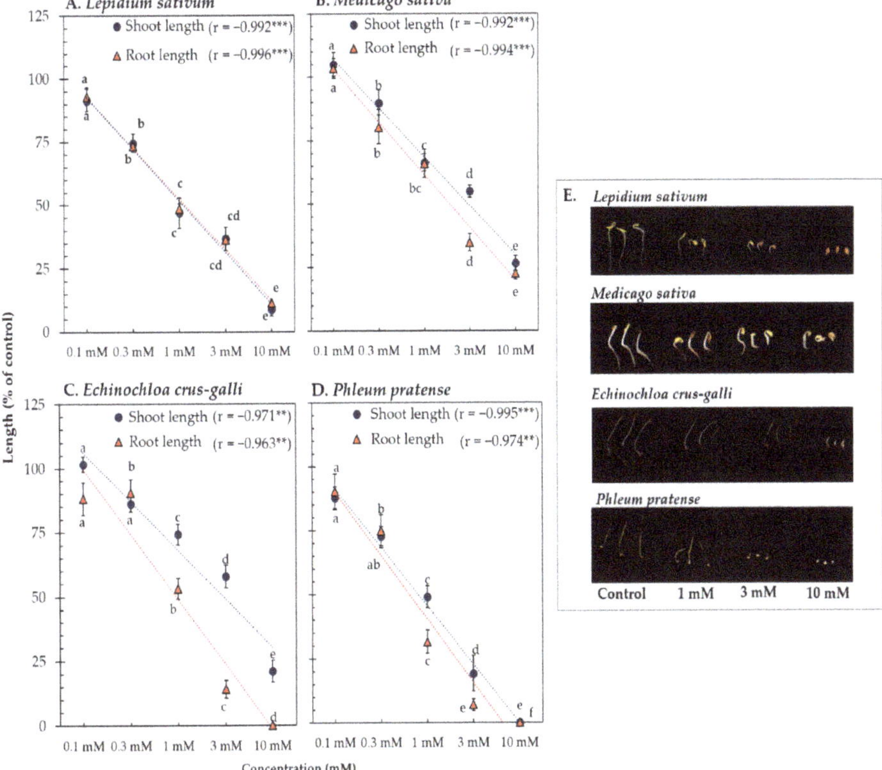

Figure 3. Effect of vanillic acid isolated from the *Afzelia xylocarpa* leaf extracts on the model plants (**A**) *Lepidium sativum*, (**B**) *Medicago sativa*, (**C**) *Phleum pratense*, (**D**) *Echinochloa crus-galli*, and (**E**) effect of vanillic acid on the seedling length of the model plants after treatment for 48 h. Means ± SE with three replicates (10 seeds/replicates). The different letters along a curve indicate significant differences (Tukey's HSD, $p < 0.05$). Statistical significance is represented by asterisks (two-tailed Pearson's correlation, ** $p < 0.01$ and *** $p < 0.001$). correlation coefficient (r).

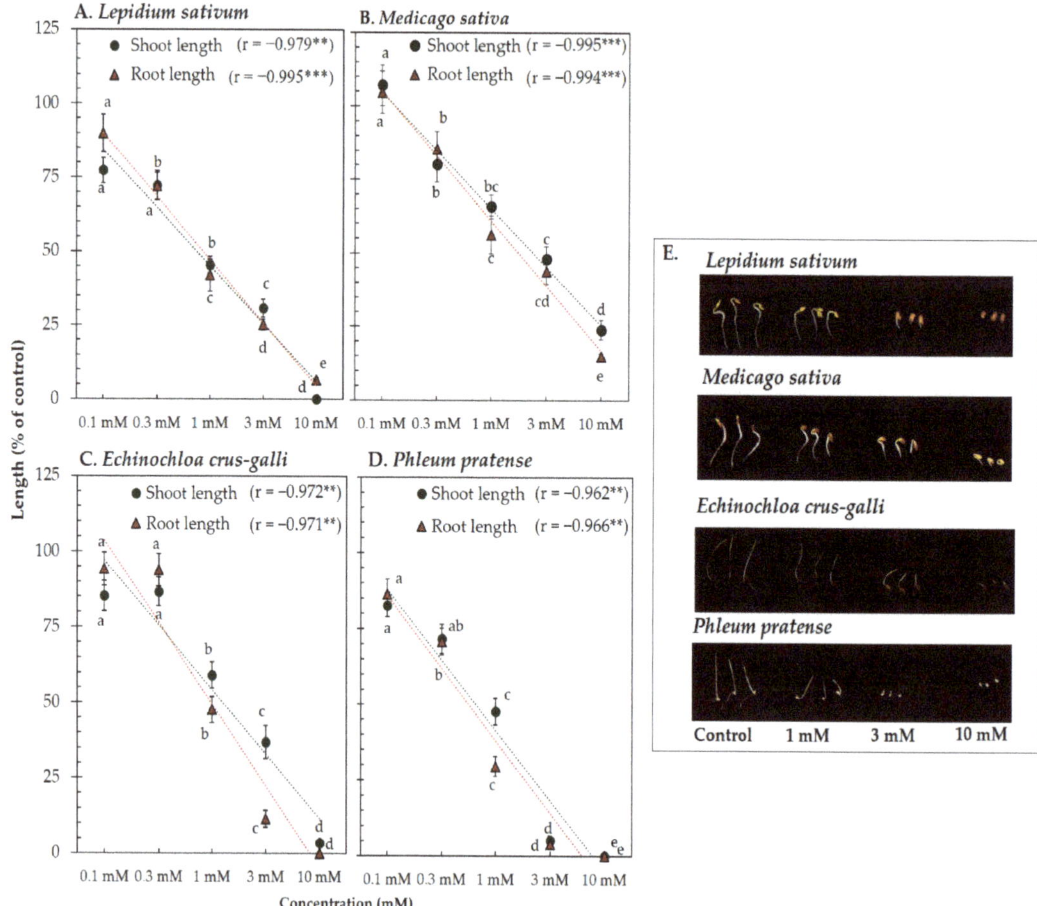

Figure 4. Effect of *trans*-ferulic acid isolated from the *Afzelia xylocarpa* leaf extracts on the model plants (**A**) *Lepidium sativum*, (**B**) *Medicago sativa*, (**C**) *Phleum pratense*, (**D**) *Echinochloa crus-galli*, and (**E**) effect of *trans*-ferulic acid on the seedling length of the model plants after treatment for 48 h. Means ± SE with three replicates (10 seeds/replicates). The different letters along a curve indicate significant differences (Tukey's HSD, $p < 0.05$). Statistical significance is represented by asterisks (two-tailed Pearson's correlation, ** $p < 0.01$ and *** $p < 0.001$). correlation coefficient (r).

These results indicate that vanillic acid and *trans*-ferulic acid have phytotoxic activity against the seedling length of the model plants in a concentration-dependent manner. According to the results of the bioassay-directed fractionation, both compounds may be the main compounds responsible for the phytotoxic activity of *A. xylocarpa* leaf. These results are supported by previous studies, which reported that vanillic acid and *trans*-ferulic acid isolated from *Triticum estivum* and *Saccharum officinarum* act as potential phytotoxic chemicals [64,65]. In addition, the IC$_{50}$ values showed the different responses of both compounds against the four model plants, indicating inhibition was species-specific (Table 1). Differences in phytotoxic effects of compounds against dicotyledonous and monocotyledonous plants have been found in many previous studies [66,67]. The different sensitivities of the model plant species may be related to the genetic, biochemical, and physiological characteristics of each specific species [68,69].

In the present study, the *P. pratense* seedlings (monocotyledonous weed) were found to be the most sensitive to the phytotoxic effects of vanillic acid and *trans*-ferulic acid. Of the model plants used in this study, the seeds of *P. pratense* were the smallest; hence, seed size may be one factor that contributed to the sensitivity to the phytotoxic substances. This supposition is consistent with the research of Leishman et al. [70], who noted that small-seeded species have greater root length per unit of root mass, which provides a greater surface area for the uptake of phytotoxic substances.

In addition, both compounds inducing the generation of reactive oxygen species (ROS) and oxidative stress may be responsible for the phytotoxic activity against test plant species [61,71,72]. Pezzani et al. [73] and Huang et al. [74] found that the accumulation of reactive oxygen species during seedling growth caused cellular damage, protein, nucleic acid, leading to irreversible damage and cell death. Nevertheless, the relationship between structure–activity and target plant species of both compounds is little understood. Both compounds contain a methoxy group at the C-3 position and a hydroxyl group at the C-4 position on the benzene ring skeleton, which may be important for their phytotoxic activity because of their radical-scavenging activity [62]. Moreover, the IC_{50} values indicate that the effectiveness of *trans*-ferulic acid against the seedling length of the model plants was greater than that of vanillic acid. The structural difference between the two compounds is present in the functional group at the C-1 position. *Trans*-ferulic acid contains the carboxylic acid group with the adjacent unsaturated carbon–carbon double bond in the functional group. It has strong electron-donating substituents that enhance its antioxidant properties compared with vanillic acid with only the carboxylic acid group [75].

3.4. The Synergistic Effects of the Mixture of Vanillic Acid and Trans-Ferulic Acid

The results of using the compounds individually at a low concentration (0.3 mM) showed a <25% decrease in the seedling length of the model plants (Figures 3 and 4). Therefore, the synergistic effects of the mixture of vanillic acid and *trans*-ferulic acid were determined at low concentrations. The results showed that the mixture of the two compounds inhibited the shoot and root length of *L. sativum* to 68.19 and 60.61% of control length, respectively (Figure 5A), whereas the *P. pratense* seedlings were inhibited to 63.22 and 52.97% of control length, respectively (Figure 5B). The effectiveness of the mixture of the compounds against the *L. sativum* roots was 1.21- and 1.15-times higher than that of vanillic acid and *trans*-ferulic acid alone, respectively. On the other hand, the effectiveness of the mixture of these compounds against the *P. pratense* shoots and roots was 1.15- and 1.41-times higher than that of vanillic acid, respectively, and 1.12- and 1.33-times higher than that of trans-ferulic acid alone, respectively.

Such inhibitory effects indicate that these two compounds work synergistically to inhibit the seedling growth of the model plants. These results support the findings of Blum [76], who reported that some phenolic compounds at low concentrations could have pronounced phytotoxic effects through synergistic action. The level of those synergistic actions depends on various factors, such as the chemical composition of each pair of compounds and the receptor species [77,78]. From our results, the mixture of vanillic acid and *trans*-ferulic acid inhibited the seedling growth of *P. pratense* more than *L. sativum*. Therefore, these results confirmed the synergistic effects of the phenolic compound mixture related to the target plant species. Consequently, the synergistic effects of the mixture of vanillic acid and *trans*-ferulic acid may lead to developing more effective activity against selected weeds and the application of smaller amounts of compounds in a mixture to achieve efficacy.

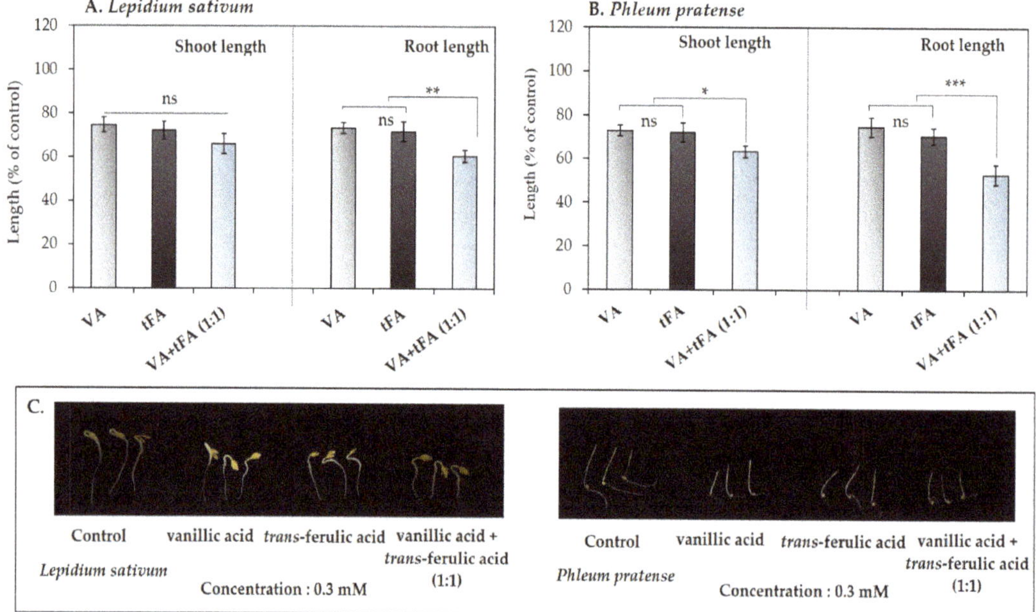

Figure 5. Effects of the mixture of vanillic acid (VA) and *trans*-ferulic acid (tFA) on the model plants (**A**) *Lepidium sativum* and (**B**) *Phleum pratense*. (**C**) effect of the mixture compound on the model plants after treatment for 48 h. The reported values are expressed as mean ± SE with three replicates (10 seeds/replicates). The asterisks represent a statistically significant comparison between treatments according to Student's *t*-test: * $p < 0.05$, ** $p < 0.01$, and *** $p < 0.001$.

4. Conclusions

This study demonstrated the phytotoxic effect of *A. xylocarpa* leaf extracts on the model plants and identified vanillic acid and *trans* ferulic acid as the major bioactive compounds. Both compounds exhibited concentration- and species-dependent inhibition. Consequently, the concentration of those compounds and differences in the model plant's sensitivity are keys to their phytotoxic action. Additionally, a mixture of both compounds at low concentrations had synergistic effects on phytotoxic activity. Our findings suggest that vanillic acid and *trans*-ferulic acid are responsible for the phytotoxic activity of *A. xylocarpa* leaves and may be useful for weed management. However, further experiments under field conditions are required to confirm our research findings.

Author Contributions: Conceptualization, R.K. and H.K.-N.; methodology, R.K., K.O., T.T., and H.K.-N.; software R.K.; validation, K.O., T.T., and H.K.-N.; formal analysis, R.K.; investigation, R.K.; data curation, H.K.-N.; writing (original draft preparation), R.K. and H.K.-N.; writing (review and editing), R.K. and H.K.-N; visualization, R.K.; supervision, H.K.-N. All authors have read and agreed to the published version of the manuscript.

Funding: This research was supported by a Ministry of Education, Culture, Sports, Science and Technology (MEXT), Japan scholarship.

Institutional Review Board Statement: Not applicable.

Informed Consent Statement: Not applicable.

Acknowledgments: We are grateful to Dennis Murphy, The United Graduate School of Agricultural Sciences, Ehime University, Japan, for editing the English of our manuscript.

Conflicts of Interest: The authors declare no conflict of interest.

References

1. Qasem, J.R.; Foy, C.L. Weed allelopathy, its ecological impacts and future prospects. *J. Crop. Prod.* **2001**, *4*, 43–119. [CrossRef]
2. Oerke, E.C. Crop losses to pests. *J. Agric. Sci.* **2006**, *144*, 31–43. [CrossRef]
3. Jabran, K.; Mahajan, G.; Sardana, V.; Chauhan, B.S. Allelopathy for weed control in agricultural systems. *J. Crop. Prot.* **2015**, *72*, 57–65. [CrossRef]
4. Aktar, M.W.; Sengupta, D.; Chowdhury, A. Impact of pesticides use in agriculture: Their benefits and hazards. *Interdisc. Toxicol.* **2009**, *2*, 569–596. [CrossRef] [PubMed]
5. Bhardwaj, T.; Sharma, J.P. Impact of pesticides application in agricultural industry: An Indian scenario. *Int. J. Agric. Food Sci. Technol.* **2013**, *4*, 817–822.
6. Awan, T.H.; Cruz, P.C.S.; Chauhan, B.S. Agronomic indices, growth, yield-contributing traits, and yield of dry-seeded rice under varying herbicides. *Field Crop. Res.* **2015**, *177*, 15–25. [CrossRef]
7. Délye, C. Unraveling the genetics bases of non-target-site-based resistance (NTSR) to herbicides: A major challenge for weed science in the forthcoming decade. *Pest Manag. Sci.* **2013**, *69*, 176–187. [CrossRef] [PubMed]
8. Bari, I.N.; Kato-Noguchi, H. Phytotoxic effects of *Cerbera manghas* L. leaf extracts on seedling elongation of four monocot and four dicot test species. *Acta Agrobot.* **2017**, *70*, 1720. [CrossRef]
9. Poonpaiboonpipat, T.; Poolkum, S. Utilization of *Bidens pilosa* var. radiata (Sch. Bip.) Sherff integrated with water irrigation for paddy weed control and rice yield production. *Weed Biol. Manag.* **2019**, *19*, 31–38. [CrossRef]
10. Kyaw, E.H.; Kato-Noguchi, H. Allelopathic potential of *Acacia pennata* (L.) Willd. leaf extracts against the seedling growth of six test plants. *Not. Bot. Horti Agrobot. Cluj-Napoca* **2020**, *48*, 1534–1542. [CrossRef]
11. Singh, N.B.; Pandey, B.N.; Singh, A. Allelopathic effects of *Cyperus rotundus* extract in vitro and ex vitro on banana. *Acta Physiol. Plant* **2009**, *31*, 633–638. [CrossRef]
12. Abbas, T.; Tanveer, A.; Khaliq, A.; Safdar, M.E. Comparative allelopathic potential of native and invasive weeds in rice ecosystem. *Pak. J. Weed. Sci. Res.* **2016**, *22*, 269–283.
13. Rice, E.L. *Allelopathy*, 2nd ed.; Academic Press: Orlando, FL, USA, 1984; pp. 67–68.
14. Chon, S.U.; Nelson, C.J. Allelopathy in compositae plants. *Agron. Sustain. Dev.* **2010**, *30*, 349–358. [CrossRef]
15. Teerarak, M.; Charoenying, P.; Laosinwattana, C. Physiological and cellular mechanisms of natural herbicide resource from *Aglaia odorata* Lour. on bioassay plants. *Acta Physiol. Plant* **2012**, *34*, 1277–1285. [CrossRef]
16. Cheng, F.; Cheng, Z. Research progress on the use of plant allelopathy in agriculture and the physiological and ecological mechanisms of allelopathy. *Front. Plant Sci.* **2015**, *6*, 1020. [CrossRef]
17. Vyvyan, J.R. Allelochemicals as leads for new herbicides and agrochemicals. *Tetrahedron* **2002**, *58*, 1631–1646. [CrossRef]
18. Dayan, F.E.; Cantrell, C.L.; Duke, S.O. Natural products in crop protection. *Bioorg. Ned. Chem.* **2009**, *17*, 4022–4034. [CrossRef]
19. Kato-Noguchi, H.; Suwitchayanon, P.; Boonmee, S.; Iwasaki, A.; Suenaga, K. Plant growth inhibitory activity of the extracts of *Acmella oleracea* and its growth inhibitory substances. *Nat. Prod. Commun.* **2019**, *14*, 1934578X19858805. [CrossRef]
20. Islam, M.S.; Zaman, F.; Iwasaki, A.; Suenaga, K.; Kato-Noguchi, H. Phytotoxic potential of *Chrysopogon aciculatus* (Retz.) Trin. (Poaceae). *Weed Biol. Manag.* **2019**, *19*, 51–58. [CrossRef]
21. Bari, I.N.; Kato-Noguchi, H.; Iwasaki, A.; Suenaga, K. Allelopathic potency and an active substance from *Anredera cordifolia* (Tenore) Steenis. *Plants* **2019**, *8*, 134. [CrossRef]
22. Rob, M.; Iwasaki, A.; Suzuki, R.; Suenaga, K.; Kato-Noguchi, H. Garcienone, a novel compound involved in allelopathic activity of *Garcinia Xanthochymus* hook. *Plants* **2019**, *8*, 301. [CrossRef] [PubMed]
23. El-Derek, M.H.; Hess, F.D. Inhibited mitotic entry is the cause of growth inhibition by cinmethylin. *Weed Sci.* **1986**, *34*, 684–688. [CrossRef]
24. Grossmann, K.; Hutzler, J.; Tresch, S.; Christiansen, N.; Looser, R.; Ehrhardt, T. On the mode of action of the herbicides cinmethylin and 5-benzyloxymethyl-1, 2-isoxazolines: Putative inhibitors of plant tyrosine aminotransferase. *Pest Manag. Sci.* **2012**, *68*, 482–492. [CrossRef] [PubMed]
25. Dayan, F.E.; Romagni, J.G.; Duke, S.O. Investigating the mode of action of natural phytotoxins. *J. Chem. Ecol.* **2000**, *26*, 2079–2094. [CrossRef]
26. Boonmee, S.; Iwasaki, A.; Suenaga, K.; Kato-Noguchi, H. Evaluation of phytotoxic activity of leaf and stem extracts and identification of a phytotoxic substance from *Caesalpinia mimosoides* Lamk. *Theor. Exp. Plant Physiol.* **2018**, *30*, 129–139. [CrossRef]
27. Krumsri, R.; Iwasaki, A.; Suenaga, K.; Kato-Noguchi, H. Assessment of allelopathic potential of *Dalbergia cochinchinensis* Pierre and its growth inhibitory substance. *Emir. J. Food. Agric.* **2020**, *32*, 513–521. [CrossRef]
28. Wink, M. Evolution of secondary metabolites in legumes (Fabaceae). *S. Afr. J. Bot.* **2013**, *89*, 164–175. [CrossRef]
29. El-Gawad, A.M.A.; El-Amier, Y.A.; Bonanomi, G. Allelopathic activity and chemical composition of *Rhynchosia minima* (L.) DC. essential oil from Egypt. *Chem. Biodivers.* **2018**, *15*, e1700438. [CrossRef]
30. Quddus, M.S.; Bellairs, S.M.; Wurm, P.A. *Acacia holosericea* (Fabaceae) litter has allelopathic and physical effects on mission grass (*Cenchrus pedicellatus* and *C. polystachios*) (Poaceae) seedling establishment. *Aust. J. Bot.* **2014**, *62*, 189–195. [CrossRef]
31. Hussain, M.I.; El-Sheikh, M.A.; Reigosa, M.J. Allelopathic Potential of Aqueous Extract from *Acacia melanoxylon* R. Br. on *Lactuca sativa*. *Plants* **2020**, *9*, 1228. [CrossRef]
32. Pakkad, G.; Ueno, S.; Yoshimaru, H. Isolation and characterization of microsatellite loci in an endangered tree species, *Afzelia xylocarpa* (Kurz) Craib (Caesalpinioideae). *Mol. Ecol. Resour.* **2009**, *9*, 880–882. [CrossRef]

33. Cai, X.; Liu, J.X.; Song, Q.S.; Chen, K.L.; Lu, Z.Y.; Zhang, Y.M. Chemical constituents of *Afzelia xylocarpa*. *Chem. Nat. Compd.* **2018**, *54*, 764–765. [CrossRef]

34. Akah, P.A.; Okpi, O.; Okoli, C.O. Evaluation of the anti-inflammatory, analgesic and antimicrobial activities of bark of *Afzelia africana*. *Niger. J. Nat. Prod. Med.* **2007**, *11*, 48–52. [CrossRef]

35. Oyedemi, S.O.; Adewusi, E.A.; Aiyegoro, O.A.; Akinpelu, D.A. Antidiabetic and haematological effect of aqueous extract of stem bark of *Afzelia africana* (Smith) on streptozotocin-induced diabetic Wistar rats. *Asian. Pac. J. Trop. Med.* **2011**, *1*, 353–358. [CrossRef]

36. Akinpelu, D.A.; Aiyegoro, O.A.; Okoh, A.I. The in vitro antioxidant property of methanolic extract of *Afzelia africana* (Smith.). *J. Med. Plant Res.* **2010**, *4*, 2022–2027. [CrossRef]

37. Pellissier, F.; Gallet, C.; Souto, X.C. Allelopathic interactions in forest ecosystems. In *Allelopathy: From Molecules to Ecosystems*; Reigosa, M.J., Nuria, P., Sanchez-Moreiras, A.M., Gonzales, L., Eds.; Science Publishers: Enfield, NH, USA, 2002; pp. 257–269.

38. Kimura, F.; Sato, M.; Kato-Noguchi, H. Allelopathy of pine litter: Delivery of allelopathic substances into forest floor. *J. Plant Biol.* **2015**, *58*, 61–67. [CrossRef]

39. Phuong, D.L.; Thuy, N.T.; Long, P.Q.; Kuo, P.C.; Thang, T.D. Composition of fatty acids, tocopherols, sterols, total phenolics, and antioxidant activity of seed oils of *Afzelia xylocarpa* and *Cassia fistula*. *Chem. Nat. Compd.* **2019**, *55*, 242–246. [CrossRef]

40. Thanh, N.D.; Nghia, N.H. Genetic diversity of *Afzelia xylocarpa* (Kurz) Craib in Vietnam based on analyses of chloroplast markers and random amplified polymorphic DNA (RAPD). *Afr. J. Biotechnol.* **2012**, *11*, 14529–14535. [CrossRef]

41. Yu, J.Q.; Shou, S.Y.; Qian, Y.R.; Zhu, Z.Z.; Hu, W.H. Autotoxic potential of cucurbit crops. *Plant Soil.* **2000**, *223*, 149–153. [CrossRef]

42. Zhou, X.; Zhang, J.; Pan, D.; Ge, X.; Jin, X.; Chen, S.; Wu, F. p-Coumaric can alter the composition of cucumber rhizosphere microbial communities and induce negative plant-microbial interactions. *Biol. Fertil. Soils.* **2018**, *54*, 683. [CrossRef]

43. Noguchi, K.; Ohno, O.; Suenaga, K.; Laosinwattana, C. A potent phytotoxic substance in *Aglaia odorata* Lour. *Chem. Biodivers.* **2016**, *13*, 549–554. [CrossRef] [PubMed]

44. Islam, M.S.; Iwasaki, A.; Suenaga, K.; Kato-Noguchi, H. 2-Methoxystypandrone, a potent phytotoxic substance in *Rumex maritimus* L. *Theor. Exp. Plant. Physiol.* **2017**, *29*, 195–202. [CrossRef]

45. Xuan, T.D.; Shinkichia, T.; Khanh, T.D.; Chung, M. Biological control of weeds and plant pathogens in paddy rice by exploiting plant allelopathy: An overview. *Crop Prot.* **2005**, *24*, 197–206. [CrossRef]

46. Krumsri, R.; Boonmee, S.; Kato-Noguchi, H. Evaluation of the allelopathic potential of leaf extracts from *Dischidia imbricata* (Blume) Steud. on the seedling growth of six test plants. *Not. Bot. Horti Agrobot. Cluj-Napoca* **2019**, *47*, 1019–1024. [CrossRef]

47. Boonmee, S.; Iwasaki, A.; Suenaga, K.; Kato-Noguchi, H. Identification of 6, 7-dimethoxychromone as a potent allelochemical from *Jatropha podagrica*. *Nat. Prod. Commun.* **2018**, *13*, 1934578X1801301126. [CrossRef]

48. Zaman, F.; Islam, M.S.; Kato-Noguchi, H. Allelopathic activity of the Oxalis europea L. extracts on the growth of eight test plant species. *Res Crop.* **2018**, *19*, 304–309. [CrossRef]

49. Poonpaiboonpipat, T.; Jumpathong, J. Evaluating herbicidal potential of aqueous–ethanol extracts of local plant species against *Echinochloa crus-galli* and *Raphanus sativus*. *Int. J. Agric. Biol.* **2019**, *21*, 648–652. [CrossRef]

50. Rob, M.M.; Kato-Noguchi, H. Study of the allelopathic activity of *Garcinia pedunculata* Roxb. *Plant Omics* **2019**, *12*, 31–36. [CrossRef]

51. Boonmee, S.; Suwitchayanon, P.; Krumsri, R.; Kato-noguchi, H. Investigation of the allelopathic potential of *Nephrolepis cordifolia* (L.) C. Presl against dicotyledonous and monocotyledonous plant species. *Environ. Control. Biol.* **2020**, *58*, 71–78. [CrossRef]

52. Krumsri, R.; Kato-Noguchi, H.; Poonpaiboonpipat, T. Allelopathic effect of *Sphenoclea zeylanica* Gaertn. on rice (*Oryza sativa* L.) germination and seedling growth. *Aust. J. Crop Sci.* **2020**, *14*, 1450–1455. [CrossRef]

53. Hossen, K.; Iwasaki, A.; Suenaga, K.; Kato-Noguchi, H. Phytotoxic Activity and Growth Inhibitory Substances from *Albizia richardiana* (Voigt.) King & Prain. *Appl. Sci.* **2021**, *11*, 1455. [CrossRef]

54. Suwitchayanon, P.; Ohno, O.; Suenaga, K.; Kato-Noguchi, H. Phytotoxic property of *Piper retrofractum* fruit extracts and compounds against the germination and seedling growth of weeds. *Acta Physiol. Plant* **2019**, *41*, 33. [CrossRef]

55. Namkeleja, H.S.; Tarimo, M.T.; Ndakidemi, P.A. Allelopathic effects of *Argemone mexicana* to growth of native plant species. *Am. J. Plant Sci.* **2014**, *5*, 1336–1344. [CrossRef]

56. Pu, W.; Wang, D.; Zhou, D. Structural characterization and evaluation of the antioxidant activity of phenolic compounds from Astragalus taipaishanensis and their structure-activity relationship. *Sci. Rep.* **2015**, *5*, 13914. [CrossRef]

57. Sethupathy, S.; Ananthi, S.; Selvaraj, A.; Shanmuganathan, B.; Vigneshwari, L.; Balamurugan, K.; Pandian, S.K. Vanillic acid from *Actinidia deliciosa* impedes virulence in Serratia marcescens by affecting S-layer, flagellin and fatty acid biosynthesis proteins. *Sci. Rep.* **2017**, *7*, 1–17. [CrossRef] [PubMed]

58. Ghareib, H.R.A.; Abdelhamed, M.S.; Ibrahim, O.H. Antioxidative effects of the acetone fraction and vanillic acid from *Chenopodium muraleon* tomato plants. *Weed Biol. Manag.* **2010**, *10*, 64–72. [CrossRef]

59. Domínguez, C.R.; Dominguez Avila, J.A.; Pareek, S.; Villegas Ochoa, M.A.; Ayala Zavala, J.F.; Yahia, E. Content of bioactive compounds and their contribution to antioxidant capacity during ripening of pineapple (*Ananas comosus* L.) cv. Esmeralda. *J. Appl. Bot. Food Qual.* **2018**, *91*, 61–68. [CrossRef]

60. Archivio, D.M.; Filesi, C.; Di Benedetto, R.; Gargiulo, R.; Giovannini, C.; Masella, R. Polyphenols, dietary sources and bioavailability. *Ann. Ist. Super. Sanita* **2007**, *43*, 348–361.

61. Zavala-López, M.; Flint-García, S.; García-Lara, S. Compositional variation in trans-ferulic, p-coumaric, and diferulic acids levels among kernels of modern and traditional maize (*Zea mays* L.) hybrids. *Front. Nutr.* **2020**, *7*, 600747. [CrossRef] [PubMed]

62. Trombino, S.; Cassano, R.; Ferrarelli, T.; Barone, E.; Picci, N.; Mancuso, C. Trans-ferulic acid-based solid lipid nanoparticles and their antioxidant effect in rat brain microsomes. *Colloids Surf. B Biointerfaces* **2013**, *109*, 273–279. [CrossRef]

63. Granata, G.; Consoli, G.M.; Nigro, R.L.; Geraci, C. Hydroxycinnamic acids loaded in lipid-core nanocapsules. *Food Chem.* **2018**, *245*, 551–556. [CrossRef] [PubMed]

64. Wu, H.; Haig, T.; Pratley, J.; Lemerle, D.; An, M. Allelochemicals in wheat (*Triticum aestivum* L.): Variation of phenolic acids in shoot tissues. *J. Chem. Ecol.* **2001**, *27*, 125–135. [CrossRef]

65. Sampietro, D.A.; Vattuone, M.A.; Isla, M.I. Plant growth inhibitors isolated from sugarcane (*Saccharum officinarum*) straw. *J. Plant Physiol.* **2006**, *163*, 837–846. [CrossRef]

66. Tuyen, P.T.; Xuan, T.D.; Tu Anh, T.T.; Mai Van, T.; Ahmad, A.; Elzaawely, A.A.; Khanh, T.D. Weed suppressing potential and isolation of potent plant growth inhibitors from *Castanea crenata* Sieb. et Zucc. *Molecules* **2018**, *23*, 345. [CrossRef] [PubMed]

67. Jmii, G.; Molinillo, J.M.; Zorrilla, J.G.; Haouala, R. Allelopathic activity of *Thapsia garganica* L. leaves on lettuce and weeds, and identification of the active principles. *S. Afr. J. Bot.* **2020**, *131*, 188–194. [CrossRef]

68. Zakaria, W.; Razak, A.R. Effects of groundnut plant residues on germination and radicle elongation of four crop species. *Pertanika* **1990**, *13*, 297–302.

69. Kobayashi, K. Factors affecting phytotoxic activity of allelochemicals in soil. *Weed Biol. Manag.* **2004**, *4*, 1–7. [CrossRef]

70. Leishman, M.R.; Wright, I.J.; Moles, A.T.; Westoby, M. The evolutionary ecology of seed size. In *Seeds: The Ecology of Regeneration in Plant Communities*; Fenner, M., Ed.; CAB International: Wallingford, UK, 2000; pp. 31–57. [CrossRef]

71. Gan, X.; Hu, D.; Wang, Y.; Yu, L.; Song, B. Novel trans-ferulic acid derivatives containing a chalcone moiety as potential activator for plant resistance induction. *J. Agric. Food Chem.* **2017**, *65*, 4367–4377. [CrossRef] [PubMed]

72. Amin, F.U.; Shah, S.A.; Kim, M.O. Vanillic acid attenuates Aβ 1-42-induced oxidative stress and cognitive impairment in mice. *Sci. Rep.* **2017**, *7*, 40753. [CrossRef] [PubMed]

73. Pezzani, R.; Vitalini, S.; Iriti, M. Bioactivities of *Origanum vulgare* L.: An update. *Phytochem. Rev.* **2017**, *16*, 1253–1268. [CrossRef]

74. Huang, H.; Ullah, F.; Zhou, D.X.; Yi, M.; Zhao, Y. Mechanisms of ROS regulation of plant development and stress responses. *Front. Plant Sci.* **2019**, *10*, 800. [CrossRef] [PubMed]

75. Jinxiang, C.; Yang, J.; Lanlan, M.; Li, J.; Nasir, S.; Kyung, K.C. Structure-antioxidant activity relationship of methoxy, phenolic hydroxyl, and carboxylic acid groups of phenolic acids. *Sci. Rep.* **2020**, *10*, 2611. [CrossRef]

76. Blum, U. Allelopathic interactions involving phenolic acids. *J. Nematol.* **1996**, *28*, 259–267. [PubMed]

77. Inderjit; Streibig, J.C.; Olofsdotter, M. Joint action of phenolic acid mixtures and its significance in allelopathy research. *Physiol. Plant* **2002**, *114*, 422–428. [CrossRef] [PubMed]

78. Jia, C.; Kudsk, P.; Mathiassen, S.K. Joint action of benzoxazinone derivatives and phenolic acids. *J. Agric. Food Chem.* **2006**, *54*, 1049–1057. [CrossRef] [PubMed]

applied
sciences

MDPI

Article

A Fast Ubiquitination of UHRF1 Oncogene Is a Unique Feature and a Common Mechanism of Thymoquinone in Cancer Cells

Mahmoud Alhosin [1,2,*], Omeima Abdullah [3], Asaad Kayali [1] and Ziad Omran [4,*]

[1] Biochemistry Department, Faculty of Science, King Abdulaziz University, Jeddah 21589, Saudi Arabia; akayali0001@stu.kau.edu.sa
[2] Centre for Artificial intelligence in Precision Medicines, King Abdulaziz University, Jeddah 21589, Saudi Arabia
[3] College of Pharmacy, Umm Al-Qura University, Makkah 21955, Saudi Arabia; oaabdullah@uqu.edu.sa
[4] Pharmacy Program, Department of Pharmaceutical Sciences, Batterjee Medical College, Jeddah 21442, Saudi Arabia
* Correspondence: malhaseen@kau.edu.sa (M.A.); ziad.omran@bmc.edu.sa (Z.O.)

Abstract: Downregulation of the ubiquitin-like containing PHD and ring finger 1 (UHRF1) oncogene in cancer cells in response to natural anticancer drugs, including thymoquinone (TQ), is a key event that induces apoptosis. TQ can induce UHRF1 autoubiquitination via the E3 ligase activity of its RING domain, most likely through the downregulation of herpes virus-associated ubiquitin-specific protease (HAUSP). In this study, we evaluated whether HAUSP downregulation and fast ubiquitination of UHRF1 are prerequisites for UHRF1 degradation in response to TQ in cancer cells and whether doxorubicin can mimic the effects of TQ on UHRF1 ubiquitination. RNA sequencing was performed to investigate differentially expressed genes in TQ-treated Jurkat cells. The protein expression of UHRF1, HAUSP and Bcl-2 was detected by means of Western blot analysis. The proliferation of human colon cancer (HCT-116) and Jurkat cells was analyzed via the WST-1 assay. RNA sequencing data revealed that TQ significantly decreased HAUSP expression. TQ triggered UHRF1 to undergo rapid ubiquitination as the first step in its degradation and the inhibition of its cell proliferation. TQ-induced UHRF1 ubiquitination is associated with HAUSP downregulation. Like TQ, doxorubicin induced a similar dose- and time-dependent downregulation of UHRF1 in cancer cells, but UHRF1 did not undergo ubiquitination as detected in response to TQ. Furthermore, TQ decreased Bcl-2 expression without triggering its ubiquitination. A fast UHRF1 ubiquitination is an indispensable event for its degradation in response to TQ but not for its responses to doxorubicin. TQ appears to trigger ubiquitination of UHRF1 but not of the Bcl-2 oncogene, thereby identifying UHRF1 as a specific target of TQ for cancer therapy.

Keywords: thymoquinone; UHRF1; ubiquitination; HAUSP; tumor suppressor genes

Citation: Alhosin, M.; Abdullah, O.; Kayali, A.; Omran, Z. A Fast Ubiquitination of UHRF1 Oncogene Is a Unique Feature and a Common Mechanism of Thymoquinone in Cancer Cells. *Appl. Sci.* **2021**, *11*, 7633. https://doi.org/10.3390/app11167633

Academic Editor: Przemysław M. Płonka

Received: 15 July 2021
Accepted: 16 August 2021
Published: 19 August 2021

1. Introduction

The ubiquitin-like containing PHD and ring finger 1 (UHRF1) oncogene is overexpressed in many human cancers and serves as a master regulator of the epigenome (DNA methylation and histone modifications) through its five functional domains [1–7], including the RING domain, which is the only domain of UHRF1 that exhibits enzymatic activity [2,8]. Indeed, UHRF1 belongs to the ring finger-type E3-ubiquitin ligases, which have in vitro autoubiquitination activity [8–10] and are currently viewed as promising anticancer drug targets due to their roles in the regulation of many tumor suppressor proteins [8,11]. UHRF1 was shown to exert E3 ubiquitin ligase activity through its RING domain, whereas the overexpression of UHRF1 RING domain mutants enhanced the sensitivity of cancer cells to several genotoxic and cytotoxic agents [8].

The herpes virus-associated ubiquitin-specific protease (HAUSP) known as USP7 was shown to protect several RING-finger E3-ubiquitin ligases, such as Mdm2 [12], ICP0 [13],

Chfr [14] and UHRF1, from autoubiquitination [15]. Indeed, HAUSP physically interacts with UHRF1 and an active HAUSP completely abolished the autoubiquitination of UHRF1 through the removal of ubiquitin adducts [15]. These findings indicate that HAUSP positively regulates the expression levels of UHRF1 through targeting the RING domain-mediated autoubiquitination activity of UHRF1, preventing proteasomal degradation of UHRF1 [15]. These data, taken together, suggest that an intact function of the UHRF1 RING domain is essential for cellular survival and its disruption through the dissociation of HAUSP from the UHRF1 complex significantly enhances the sensitivity of cancer cells to chemotherapeutics, making UHRF1 a promising target in the search for natural compound to target in cancer therapy.

Several in vitro and in vivo studies have shown that UHRF1 downregulation in cancer is sufficient to induce apoptosis through the reactivation of various tumor suppressor genes (TSGs) [7,16–18]. Indeed, UHRF1 expression decreases in response to several natural products, including naphthazarin [19], shikonin [20], curcumin [5], epigallocatechin-3-gallate (EGCG) [4], resveratrol [21], anisomycin [21], luteolin [22], doxorubicin [23] and thymoquinone (TQ) [1,11,24,25]. TQ, the most abundant biologically active component of black seed oil, shows inhibitory effects on many human cancer cells by targeting several signaling pathways, including the UHRF1/TSGs axis [1,7,11]. TQ was shown to induce the rapid autoubiquitination of UHRF1 in cancer cells through its RING domain, and this event was correlated with HAUSP downregulation [11], which is known to control the ubiquitination status of UHRF1 and protects it from degradation by the proteasome [15]. These findings suggest that HAUSP downregulation-mediated rapid autoubiquitination of UHRF1 in response to TQ is an essential first step for subsequent UHRF1 degradation [11]. In the present study, we evaluated whether TQ specifically triggers HUASP downregulation-mediated fast ubiquitination of UHRF1 in cancer cells and whether the anticancer drug doxorubicin, which is known to decrease the expression of the UHRF1 protein through an unknown mechanism [23], can mimic the effects of TQ on UHRF1.

2. Materials and Methods

2.1. Cell Culture and Treatment

The human T lymphocyte Jurkat cell line and human colon cancer cells (HCT-116) were obtained from the American Type Culture Collection (Manassas, VA, USA) and maintained in a humidified incubator at 37 °C in 5% CO_2. Jurkat cells were grown in RPMI 1640 culture medium (Sigma-Aldrich, St-Louis, MO, USA) and HCT-116 cells in Dulbecco's modified Eagle medium (DMEM) (Sigma-Aldrich, St-Louis, MO, USA). All media were supplemented with 15% (*v/v*) fetal calf serum (FCS; Biowhitaker, Lonza, Belgium), 2 mM glutamine, 100 U/mL penicillin and 50 µg/mL streptomycin (Sigma, St. Louis, MO, USA). For all treatments, a 10 mM solution of TQ (Sigma-Aldrich, St. Louis, MO, USA) was prepared in 10% DMSO (dimethylsulfoxide; Millipore, Molsheim, France) and appropriate working concentrations were prepared with cell culture medium. The final concentration of DMSO was always less than 0.1% in both control and treated conditions. Doxorubicin was obtained from Sigma-Aldrich.

2.2. RNA-Seq, Identification of Differentially Expressed Genes and Bioinformatics Analysis

Triplicate samples of Jurkat cells were treated with 20 µM TQ for 24 h, and three independent experiments were performed. The total RNA was extracted using the RNeasy kit Qiagen (Valencia, CA, USA). RNA-Seq and the determination of differentially expressed genes were carried out as described elsewhere [6].

2.3. Western Blot Analysis

Cells were treated with different concentrations of TQ or doxorubicin for different times. The cells were then harvested, centrifuged to discard the medium, washed with cold PBS (phosphate-buffered saline), and resuspended in RIPA buffer (25 mM Tris, pH 7.6, 150 mM NaCl, 1% NP-40, 1% sodium deoxycholate and 0.1% SDS; Sigma-Aldrich, USA) containing protease inhibitors. Equal amounts of total protein were separated on 10-12% polyacrylamide gels and electrophoretically transferred to nitrocellulose membranes. After blocking with 5% non-fat dry milk or 3% bovine serum albumin (BSA) and Tween 20 in PBS, the nitrocellulose membranes were incubated at 4 °C overnight with either a mouse monoclonal anti-UHRF1 antibody [26], a mouse monoclonal anti-HAUSP antibody (Santa Cruz Biotechnology, Dallas, TX, USA, Cat. No. sc-137008), a mouse monoclonal anti-Bcl-2 (Cat. No. ab59348), or a mouse monoclonal anti-GAPDH (Cat. No. ab8245) or an anti-β-actin antibody (Cat. No. ab8226) (Abcam, Paris, France) or β-tubulin (Sigma-Aldrich, Cat. No. T8328), according to the manufacturer's instructions. The membranes were then washed three times with PBS for 10 min, followed by incubation with an appropriate horseradish peroxidase-conjugated secondary antibody (diluted to 1:10,000 for anti-mouse antibodies) at room temperature for 90 min. The membranes were washed with PBS five times, and the chemiluminescence signals were detected using Amersham ECL Plus detection system (Amersham, GE Healthcare, Buckinghamshire, UK).

2.4. Cell Proliferation Assay

A colorimetric cell proliferation assay using a WST-1 Cell Proliferation Reagent Kit (Sigma-Aldrich, USA) was used to analyze the effect of TQ on cell proliferation in HCT-116 and Jurkat cells. Briefly, the cells were seeded in 96-multiwell plates at a density of 4×10^4/well for Jurkat cells and 10^4/well for HCT-116 cells. After 24 h of incubation, the cells were exposed to different concentrations of TQ for different times. The cell proliferation rate then was evaluated through a rapid WST-1 reagent. After incubation for the different times, 10 μL of the WST-1 solution was added and incubated for an additional 3 h at 37 °C. Finally, the absorbance was read at 450 nm with a microplate ELISA reader (ELx800™ Biotek, Winooski, Vermont, USA) and the results were analyzed using the Gen5 software (Biotek, Winooski, VT, USA). The reaction was based on the cleavage of the tetrazolium salt WST-1 to formazan by means of cellular mitochondrial dehydrogenases. The quantity of formazan dye in the medium is directly proportional to the number of viable metabolically active cells. The percentage of cell viability was calculated by assuming control (untreated) samples to be 100% viable.

2.5. Statistical Analysis

All the data were presented as the means $\pm$ S.E.M of triplicates carried out in the same experiment or an average of at least three separate experiments. One-way ANOVA, followed by Tukey's post hoc test using GraphPad Prism 6 (Graph Pad Software, SanDiego, CA, USA), were used for statistical analysis. Significant differences are indicated as * $p < 0.05$.

3. Results

3.1. TQ Induced UHRF1 Ubiquitination Related to HAUSP Downregulation in Human Cancer Cells

The HAUSP deubiquitinase interacts with the UHRF1 protein and controls its ubiquitination status, thereby protecting UHRF1 from degradation by the proteasome [15,27]. TQ was shown to induce a sharp decrease in HAUSP in Jurkat cells [11]. We therefore investigated whether HAUSP is a main target of TQ in its mechanism of action leading to UHRF1 downregulation by analyzing gene expression after a 24-h exposure to TQ. The RNA-seq data showed that several deubiquitinases, including HAUSP, with known oncogenic proprieties were significantly downregulated in response to TQ treatment (Table 1). By contrast, two deubiquitinases, USP27X and USP35, which are known tumor suppressors, were insignificantly upregulated (Table 2).

We then investigated whether UHRF1 ubiquitination is a common feature of TQ in cancer cells and whether this process involves the downregulation of HAUSP by evaluating the effects of TQ in two human cancer cell lines that highly express UHRF1: human colon cancer cells (HCT-116) [28,29] and Jurkat cells [1,30,31]. TQ induced a sharp decrease in the expression of UHRF1 and HAUSP at the protein level when administered at concentrations of at least 200 μM in HCT-116 cells (Figure 1A) and 30 μM in Jurkat cells (Figure 1B). The anti-proliferative effect of TQ in HCT-116 cells (Figure 1C) and in Jurkat cells (Figure 1D) was then analyzed under the same experimental conditions. Cell proliferation in response to TQ was decreased in a concentration-dependent manner in both cancer cell lines. TQ significantly inhibited cell proliferation starting from 30 μM in HCT-116 cells and the inhibition percentage reached approximately 50% at 100 μM, suggesting that TQ exerts its inhibitory effects on HCT-116 cells with an estimated half-maximal effect (IC50) value of approximately 100 μM (Figure 1C). In Jurkat cells, TQ significantly inhibited cell proliferation from 5 μM, with an IC50 value of approximately 30 μM (Figure 1D).

Table 1. Downregulation of deubiquitinases in TQ-treated Jurkat cells as compared with untreated cells.

Gene	logFc*	*p*-Value
USP4	−1.065	0.026
USP7	−0.907	0.05
USP8	−1.003	0.037
USP12	−1.067	0.0283
USP15	−1.19	0.012
USP24	−1.566	0.001
USP32	−1.239	0.01
USP34	−1.9	0.0000398

Table 2. Upregulation of deubiquitinases in TQ-treated Jurkat cells as compared with untreated cells.

Gene	logFc*	*p*-Value
USP27	0.207	0.84
USP35	0.49	0.34

A kinetic analysis of the effects of TQ on UHRF1 expression, using concentrations at which UHRF1 became undetectable, 200 μM of TQ in HCT-116 cells (Figure 2A) and 30 μM in Jurkat cells (Figure 2B), revealed time-dependent effects of TQ on UHRF1 ubiquitination, as reflected by the appearance of high-molecular-weight bands of UHRF1 at 115 and 250 kDa, rather than the expected one at 97 kDa. The TQ-induced UHRF1 ubiquitination was associated with a decrease in the expression of HAUSP, detectable at 10 min in HCT-116 cells (Figure 2A) and Jurkat cells (Figure 2B) in parallel with the appearance of high-molecular-weight bands on UHRF1. The TQ-induced downregulation of UHRF1 and HAUSP was also associated with an upregulation of the p73 tumor suppressor protein in both cancer cell lines (data not shown). The anti-proliferative effect of TQ in HCT-116 cells (Figure 2C) and Jurkat cells (Figure 2D) was also analyzed under the same experimental conditions. TQ induced a time-dependent cell proliferation inhibition in HCT-116 cells (Figure 2C) and Jurkat cells (Figure 2D) and this effect was associated with the TQ-induced UHRF1 ubiquitination. Indeed, TQ caused a significant decrease in cell proliferation after 1 h and 30 min in HCT-116 cells (Figure 2C) and Jurkat cells (Figure 2D), respectively.

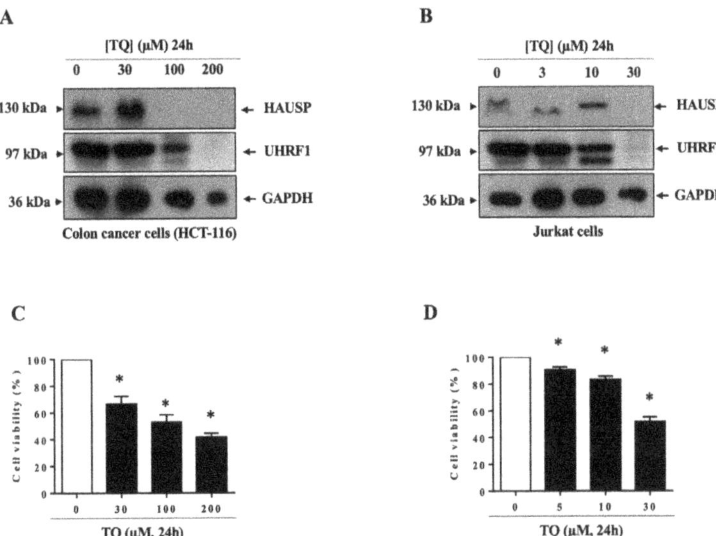

Figure 1. Dose-dependent effect of TQ on the protein expression of HAUSP and UHRF1 and cell proliferation. Human colon cancer (HCT-116) (**A,C**) and Jurkat cells (**B,D**) were exposed to increasing concentrations of TQ for 24 h. The expression of HAUSP and UHRF1 was analyzed by means of Western blotting, as described in the Materials and Methods. The cell proliferation rate was assessed via the WST-1 assay. Values are shown as means ± S.E.M of three different experiments. * $p < 0.05$ versus control.

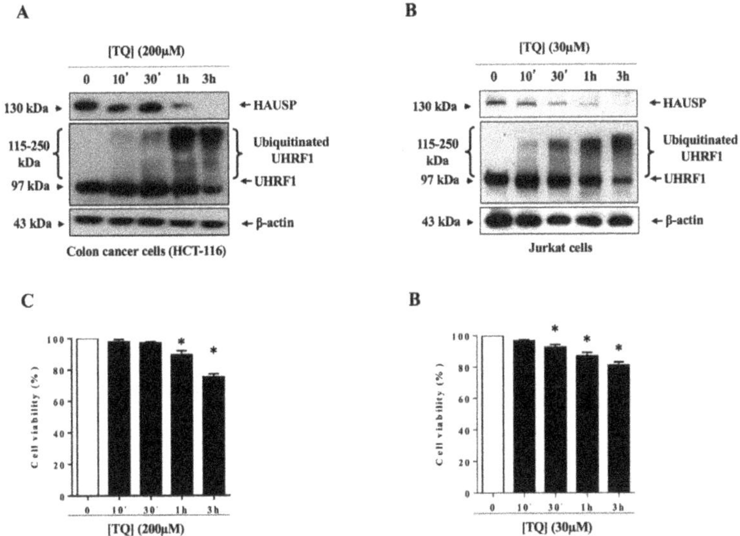

Figure 2. Time course effect of TQ on HAUSP, UHRF1 and ubiquitinated UHRF1 and cell proliferation. Human colon cancer cells (HCT-116) (**A,C**) and Jurkat cells (**B,D**) were exposed to 200 μM and 30 μM TQ, respectively, for different times. The expression of HAUSP and UHRF1 was analyzed by means of Western blotting, as described in the Materials and Methods. The cell proliferation rate was assessed via the WST-1 assay. Values are shown as means ± S.E.M of three different experiments. * $p < 0.05$ versus control.

Taken together, these findings indicate that TQ induces a fast ubiquitination of UHRF1 in cancer cells that is related to HAUSP downregulation. This process could be an indispensable event in the subsequent degradation of UHRF1 and the upregulation of the p73 tumor suppressor and the inhibition of cell proliferation in response to TQ.

3.2. Doxorubicin Induced Ubiquitination Independently of UHRF1 Downregulation in Cancer Cells

The anticancer drug doxorubicin is commonly considered to exert its anti-tumor activity through targeting several signaling pathways, including the induction of the polyubiquitination of various proteins [32–34]. Doxorubicin was shown to decrease the expression of the UHRF1 protein in HCT-116 cells through an unknown mechanism [23]. Thus, we wanted to know whether doxorubicin induces a rapid ubiquitination of UHRF1 in cancer cells, thereby mimicking the effects of TQ on the ubiquitination status of UHRF1. As shown in Figure 3A, treatment of Jurkat cells with doxorubicin resulted in a dose-dependent decrease in the expression protein level of UHRF1 starting at doxorubicin concentrations of 0.25 µg/mL. A kinetic analysis of UHRF1 expression in Jurkat cells was then conducted using doxorubicin at 2 µg/mL, a concentration at which UHRF1 became undetectable after 24 h of treatment (Figure 3A). A time-course analysis showed that doxorubicin induced a sharp decrease in the expression of UHRF1 at 6 h (Figure 3B). However, although TQ at 10 min induced UHRF1 ubiquitination, as reflected by the rapid appearance of high-molecular-weight bands of UHRF1 (Figure 3C), Western blot analysis revealed that doxorubicin treatment did not challenge UHRF1 to perform ubiquitination before the onset of its degradation in Jurkat cells (Figure 3B).

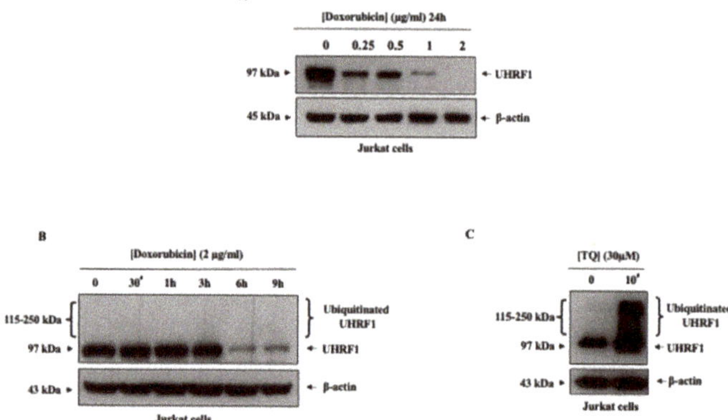

Figure 3. Dose- and time-dependent effects of doxorubicin and TQ on the protein expression of UHRF1 and ubiquitinated UHRF1 in Jurkat cells. (**A**) Jurkat cells were exposed to increasing concentrations of doxorubicin for 24 h. (**B**) Cells were exposed to 2 µg/mL doxorubicin for different times. (**C**) Cells were exposed to 30 µM TQ for 10 min. Expression of UHRF1 was analyzed via Western blotting, as described in the Materials and Methods.

The same findings were observed in HCT-116 cells treated with doxorubicin (Figure 4). Figure 4A shows that the treatment of colon cancer cells with doxorubicin resulted in a dose-dependent decrease in the protein expression level of UHRF1, with concentrations as low as 10 µg/mL. The time-course effects of 60 µg/mL of doxorubicin on UHRF1 expression showed a decrease in the expression of UHRF1 starting at 6 h (Figure 4B). Again, the doxorubicin-induced UHRF1 downregulation was not correlated with the ubiquitination process, as reflected by the lack of high-molecular-weight bands of UHRF1 (Figure 4B), as was observed in response to TQ (Figure 4C).

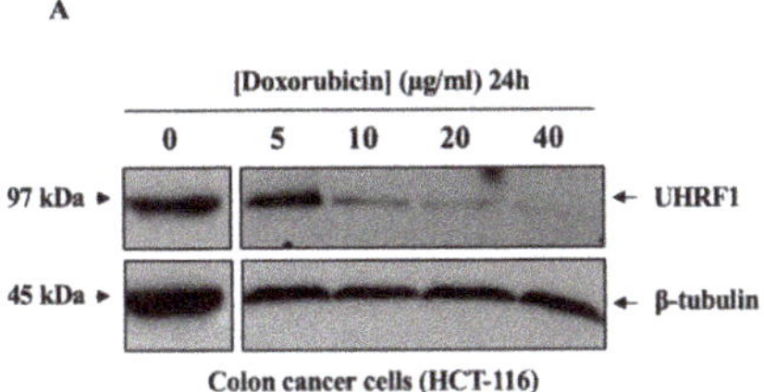

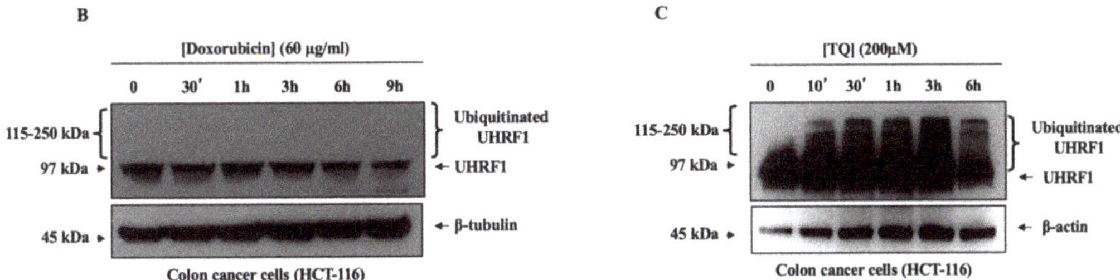

Figure 4. Dose- and time-dependent effects of doxorubicin or TQ on the protein expression of UHRF1 and ubiquitinated UHRF1 in colon cancer cells. (**A**) HCT-116 cells were exposed to increasing concentrations of doxorubicin for 24 h. (**B**) Cells were exposed to 60 μg/mL doxorubicin for different times. (**C**) Cells were exposed to 200 μM TQ for different times. Expression of UHRF1 was analyzed via Western blotting, as described in the Materials and Methods.

Taken together, these finding suggest that TQ and doxorubicin decreased UHRF1 protein expression through different mechanisms and indicate that fast UHRF1 ubiquitination is a prerequisite for its degradation in cancer cells in response to TQ but not for the response to doxorubicin.

3.3. TQ Induced the Downregulation of the Bcl-2 Oncogene in Cancer Cells through a Ubiquitination-Independent Mechanism

The B-cell lymphoma 2 (Bcl-2) oncogene is overexpressed in several tumors, including colon cancer [35–38] and Jurkat cells [39–41]. Bcl-2 was shown to be degraded in the proteasome through a ubiquitination pathway in response to several signals [42,43]. We investigated whether Bcl-2, like UHRF1, also underwent fast ubiquitination before its degradation in response to TQ in colon cancer cells and Jurkat cells. Dosage analysis of TQ on the expression of Bcl-2 in both cancer cell lines revealed that the treatment of HCT-116 cells (Figure 5A) and Jurkat cells (Figure 5B) with TQ induced a dose-dependent decrease in the protein level expression of Bcl-2, starting at 300 μM TQ in HCT-116 cells (Figure 5A) and at 10 μM in Jurkat cells (Figure 5B). A kinetic analysis of TQ was then conducted on Bcl-2 expression in HCT-116 cells, using 400 μM TQ (Figure 5C) and 200 μM TQ in Jurkat cells (Figure 5D), concentrations at which Bcl-2 became undetectable after 24 h of treatment in HCT-116 cells (Figure 5A) and Jurkat cells (Figure 5B). The time-course effects on Bcl-2 expression showed that TQ induced a decrease in the expression of Bcl-2 beginning at 1 h in HCT-116 cells (Figure 5C) and at 30 min in Jurkat cells (Figure 5D). However, Western blot analysis revealed no high-molecular-weight bands of Bcl-2 in TQ-treated HCT-116 cells (Figure 5C) or Jurkat cells (Figure 5D), indicating that, unlike its effects on UHRF1, TQ did not induce the fast ubiquitination of Bcl-2 in cancer cells.

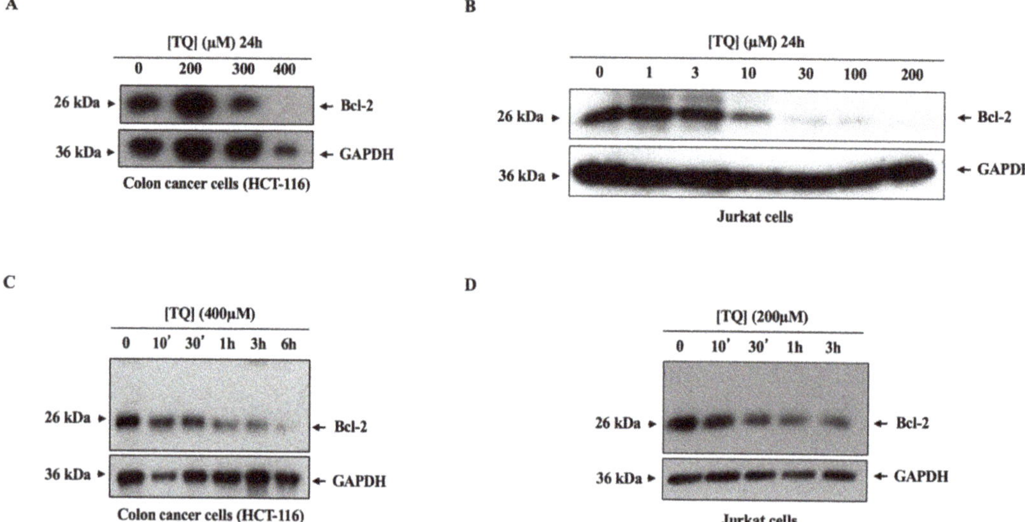

Figure 5. Dose- and time-dependent effects of TQ on the protein expression of Bcl-2 in cancer cells. HCT-116 cells were exposed to increasing concentrations of TQ for 24 h (**A**) or exposed to 400 µM TQ for different times (**C**). Jurkat cells were exposed to increasing concentrations of TQ for 24 h (**B**) or exposed to 200 µM TQ for different times (**D**). Expres-sion of Bcl-2 was analyzed via Western blotting, as described in the Materials and Methods.

4. Discussion

The overexpression of UHRF1, detected in many solid and hematological tumors, is known to inhibit the expression of various tumor suppressor genes (TSGs) and to inhibit apoptosis [4,7,17]. Several studies have shown that natural anticancer compounds, including TQ, can decrease UHRF1 expression in cancer cells and reactivate silenced TSGs with the subsequent induction of apoptosis [1,4,5,11,17,44]. Recently, TQ was shown to induce the fast auto-polyubiquitination of UHRF1 through its RING domain, and this effect was correlated with a decrease in the expression of the HAUSP deubiquitinase [11]. In the present study, two natural compounds—TQ and doxorubicin, which exhibit anticancer properties on various cancer cells through the targeting of several signaling pathways— were evaluated for their effects on UHRF1 expression in colon cancer cells and Jurkat cells. The kinetic effects of TQ on UHRF1 expression in colon cancer and Jurkat cells showed that UHRF1 was ubiquitinated, as reflected by the appearance of high-molecular-weight bands, and this effect was associated with a downregulation of HAUSP. Like the TQ effects, doxorubicin also induced a dose-dependent degradation of UHRF1 in colon cancer cells and Jurkat cells. However, the kinetic effect of doxorubicin on UHRF1 expression did not show the appearance of any high-molecular-weight bands of UHRF1 in either of the cancer cell lines, suggesting that UHRF1 ubiquitination in cancer cells is triggered only by TQ, and not by doxorubicin.

The present findings support the presence of a strong correlation between the TQ-induced ubiquitination of UHRF1 and its subsequent degradation in cancer cells [11]. In the present study, the time-course effects of doxorubicin on UHRF1 expression in colon cancer cells and Jurkat cells did not reveal any high-molecular-weight bands in either cancer cell line. This suggests that the degradation of UHRF1 in response to TQ requires auto-ubiquitination through its RING domain, but this step is not required in the response to other natural compounds that exhibit anticancer activities. This process could involve, in part, the downregulation of HAUSP to allow the auto-ubiquitination of UHRF1 and its degradation, followed by the subsequent upregulation of TSGs, the inhibition of cell proliferation and the induction of apoptosis. Indeed, several natural products, including naphthazarin [19], shikonin [20], curcumin [5], EGCG [4], resveratrol [21] and luteolin [22], as well as TQ [1,25], have been shown to decrease the expression of UHRF1 in cancer cells, but their underlying molecular mechanisms are largely unknown. Few studies have conducted kinetic analyses of the effects of these natural products on UHRF1 expression. The time-course effects of Antho 50, an anthocyanin-rich dietary bilberry extract, on UHRF1 expression in cells from patients with B-cell chronic lymphocytic leukemia (B CLL) showed a progressive decrease in UHRF1 expression, but without the appearance of high-molecular-weight bands of UHRF1 [45]. In the present study, the time-course effects of TQ on UHRF1 in colon cancer, as a model of solid tumors, and in Jurkat cells, as a hematological tumor model, revealed that TQ induced the appearance of high-molecular-weight bands of UHRF1, but doxorubicin did not. This finding suggests that TQ can induce UHRF1 ubiquitination in cancer cells as an indispensable event that triggers the onset of apoptosis, and this process could be as a result of the disassociation between HAUSP and UHRF1. Considering that TQ has no or minimal effects on normal human cells [24,46] and that UHRF1 is overexpressed in several tumors, TQ could be a promising specific regulator of the HAUSP/UHRF1 axis that functions by inducing UHRF1 ubiquitination.

Several studies in various cancer cells have shown that TQ induces a downregulation of the Bcl-2 oncogene [47–52], one of the master activators of the anti-apoptotic effects in cancer. In the present study, we found that TQ induces a dose- and time-dependent downregulation of Bcl-2 in colon cancer cells. However, it apparently does not do so by inducing ubiquitination, based on the lack of high-molecular-weight bands of Bcl-2. Therefore, TQ can induce a fast ubiquitination of UHRF1 but apparently not of other oncogenes such as Bcl-2, and this ubiquitination is indispensable for UHRF1 degradation. Bcl-2 is degraded via the activation of the proteasome ubiquitination pathway in response to the natural product pseudolaric acid B [43], whereas TQ is suggested to induce Bcl-2 degradation in cancer cells through an ubiquitination-independent mechanism. In agreement with this hypothesis, TQ induced a time-dependent downregulation of Bcl-2 in human multiple myeloma cells through the inhibition of STAT3 [52], which is a main regulator of Bcl-2 [53,54].

TQ, by virtue of its ability to induce the early ubiquitination of the oncogene UHRF1, followed by its degradation in both Jurkat cells as an experimental model of hematological malignancy as well as colon cancer as a model of solid tumors, the present study supports the chemopreventive and therapeutic potential of TQ for both human blood and solid tumors. However, due to its highly lipophilic nature, poor aqueous solubility and inadequate biological stability, TQ is characterized by poor bioavailability. However, many approaches to address TQ's pharmacokinetic limitations, such as formulations of TQ in lipid nanocarriers, are currently being developed [55,56].

In conclusion, the present study shows that, unlike the anticancer drug doxorubicin, TQ can trigger the fast ubiquitination of UHRF1, but not of another oncogene, Bcl-2. The underlying mechanism could involve the downregulation of HAUSP, which normally would protect UHRF1 from ubiquitination (see the Graphical Abstract). Nevertheless, the profound mechanisms of TQ's effects on the ubiquitination status of UHRF1 still require further investigation. Objectively, due to the underlying complexity of the mechanisms of the interaction between UHRF1 and several proteins, including HAUSP [57], further studies are needed to explore the detailed mechanisms by which UHRF1 is rapidly polyubiquitinated in cancer cells in response to TQ and how HAUSP could be involved in the process.

Author Contributions: Conceptualization, M.A., O.A. and Z.O.; methodology, M.A. and A.K.; validation, M.A and Z.O.; formal analysis, M.A.; investigation, A.K.; writing—original draft preparation, M.A. and A.K.; writing—review and editing, M.A. and O.A.; funding acquisition, O.A and Z.O. All authors have read and agreed to the published version of the manuscript.

Funding: This research was funded by the deanship of Scientific Research at Umm Al-Qura University, grant number 19-MED-1-02-0008.

Institutional Review Board Statement: Not applicable.

Informed Consent Statement: Not applicable.

Data Availability Statement: All data generated or analyzed during this study are included in this published article or are available from the corresponding authors on reasonable request.

Acknowledgments: The authors would like to thank the deanship of Scientific Research at Umm Al-Qura University for supporting this work by Grant Code: 19-MED-1-02-0008.

Conflicts of Interest: The authors declare that they have no competing interests.

References

1. Alhosin, M.; Abusnina, A.; Achour, M.; Sharif, T.; Muller, C.; Peluso, J.; Chataigneau, T.; Lugnier, C.; Schini-Kerth, V.B.; Bronner, C.; et al. Induction of apoptosis by thymoquinone in lymphoblastic leukemia Jurkat cells is mediated by a p73-dependent pathway which targets the epigenetic integrator UHRF1. *Biochem. Pharmacol.* **2010**, *79*, 1251–1260. [CrossRef]
2. Bronner, C.; Krifa, M.; Mousli, M. Increasing role of UHRF1 in the reading and inheritance of the epigenetic code as well as in tumorogenesis. *Biochem. Pharmacol.* **2013**, *86*, 1643–1649. [CrossRef]
3. Zaayter, L.; Mori, M.; Ahmad, T.; Ashraf, W.; Boudier, C.; Kilin, V.; Gavvala, K.; Richert, L.; Eiler, S.; Ruff, M.; et al. A Molecular Tool Targeting the Base-Flipping Activity of Human UHRF1. *Chemistry* **2019**, *25*, 13363–13375. [CrossRef]
4. Achour, M.; Mousli, M.; Alhosin, M.; Ibrahim, A.; Peluso, J.; Muller, C.D.; Schini-Kerth, V.B.; Hamiche, A.; Dhe-Paganon, S.; Bronner, C. Epigallocatechin-3-gallate up-regulates tumor suppressor gene expression via a reactive oxygen species-dependent down-regulation of UHRF1. *Biochem. Biophys. Res. Commun.* **2013**, *430*, 208–212. [CrossRef]
5. Abusnina, A.; Keravis, T.; Yougbaré, I.; Bronner, C.; Lugnier, C. Anti-proliferative effect of curcumin on melanoma cells is mediated by PDE1A inhibition that regulates the epigenetic integrator UHRF1. *Mol. Nutr. Food Res.* **2011**, *55*, 1677–1689. [CrossRef]
6. Qadi, S.A.; Hassan, M.A.; Sheikh, R.A.; Baothman, O.A.; Zamzami, M.A.; Choudhry, H.; Al-Malki, A.L.; Albukhari, A.; Alhosin, M. Thymoquinone-Induced Reactivation of Tumor Suppressor Genes in Cancer Cells Involves Epigenetic Mechanisms. *Epigenetics Insights* **2019**, *12*. [CrossRef]
7. Alhosin, M.; Sharif, T.; Mousli, M.; Etienne-Selloum, N.; Fuhrmann, G.; Schini-Kerth, V.B.; Bronner, C. Down-regulation of UHRF1, associated with re-expression of tumor suppressor genes, is a common feature of natural compounds exhibiting anti-cancer properties. *J. Exp. Clin. Cancer Res. CR* **2011**, *30*, 41. [CrossRef] [PubMed]
8. Jenkins, Y.; Markovtsov, V.; Lang, W.; Sharma, P.; Pearsall, D.; Warner, J.; Franci, C.; Huang, B.; Huang, J.; Yam, G.C.; et al. Critical role of the ubiquitin ligase activity of UHRF1, a nuclear RING finger protein, in tumor cell growth. *Mol. Biol. Cell* **2005**, *16*, 5621–5629. [CrossRef] [PubMed]
9. Citterio, E.; Papait, R.; Nicassio, F.; Vecchi, M.; Gomiero, P.; Mantovani, R.; Di Fiore, P.P.; Bonapace, I.M. Np95 is a histone-binding protein endowed with ubiquitin ligase activity. *Mol. Cell. Biol.* **2004**, *24*, 2526–2535. [CrossRef]
10. Karagianni, P.; Amazit, L.; Qin, J.; Wong, J. ICBP90, a novel methyl K9 H3 binding protein linking protein ubiquitination with heterochromatin formation. *Mol. Cell. Biol.* **2008**, *28*, 705–717. [CrossRef] [PubMed]
11. Ibrahim, A.; Alhosin, M.; Papin, C.; Ouararhni, K.; Omran, Z.; Zamzami, M.A.; Al-Malki, A.L.; Choudhry, H.; Mély, Y.; Hamiche, A.; et al. Thymoquinone challenges UHRF1 to commit auto-ubiquitination: A key event for apoptosis induction in cancer cells. *Oncotarget* **2018**, *9*, 28599–28611. [CrossRef] [PubMed]
12. Li, M.; Brooks, C.L.; Kon, N.; Gu, W. A dynamic role of HAUSP in the p53-Mdm2 pathway. *Mol. Cell* **2004**, *13*, 879–886. [CrossRef]
13. Canning, M.; Boutell, C.; Parkinson, J.; Everett, R.D. A RING finger ubiquitin ligase is protected from autocatalyzed ubiquitination and degradation by binding to ubiquitin-specific protease USP7. *J. Biol. Chem.* **2004**, *279*, 38160–38168. [CrossRef] [PubMed]
14. Oh, Y.M.; Yoo, S.J.; Seol, J.H. Deubiquitination of Chfr, a checkpoint protein, by USP7/HAUSP regulates its stability and activity. *Biochem. Biophys. Res. Commun.* **2007**, *357*, 615–619. [CrossRef] [PubMed]
15. Felle, M.; Joppien, S.; Németh, A.; Diermeier, S.; Thalhammer, V.; Dobner, T.; Kremmer, E.; Kappler, R.; Längst, G. The USP7/Dnmt1 complex stimulates the DNA methylation activity of Dnmt1 and regulates the stability of UHRF1. *Nucleic Acids Res.* **2011**, *39*, 8355–8365. [CrossRef] [PubMed]
16. Xia, T.; Liu, S.; Xu, G.; Zhou, S.; Luo, Z. Dihydroartemisinin induces cell apoptosis through repression of UHRF1 in prostate cancer cells. *Anticancer Drugs* **2021**. [CrossRef]
17. Alhosin, M.; Omran, Z.; Zamzami, M.A.; Al-Malki, A.L.; Choudhry, H.; Mousli, M.; Bronner, C. Signalling pathways in UHRF1-dependent regulation of tumor suppressor genes in cancer. *J. Exp. Clin. Cancer Res. CR* **2016**, *35*, 174. [CrossRef]
18. Unoki, M. Current and potential anticancer drugs targeting members of the UHRF1 complex including epigenetic modifiers. *Recent Pat. Anti-Cancer Drug Discov.* **2011**, *6*, 116–130. [CrossRef]
19. Kim, M.Y.; Park, S.J.; Shim, J.W.; Yang, K.; Kang, H.S.; Heo, K. Naphthazarin enhances ionizing radiation-induced cell cycle arrest and apoptosis in human breast cancer cells. *Int. J. Oncol.* **2015**, *46*, 1659–1666. [CrossRef]

20. Jang, S.Y.; Hong, D.; Jeong, S.Y.; Kim, J.H. Shikonin causes apoptosis by up-regulating p73 and down-regulating ICBP90 in human cancer cells. *Biochem. Biophys. Res. Commun.* **2015**, *465*, 71–76. [CrossRef]
21. Parashar, G.; Capalash, N. Promoter methylation-independent reactivation of PAX1 by curcumin and resveratrol is mediated by UHRF1. *Clin. Exp. Med.* **2016**, *16*, 471–478. [CrossRef]
22. Krifa, M.; Leloup, L.; Ghedira, K.; Mousli, M.; Chekir-Ghedira, L. Luteolin induces apoptosis in BE colorectal cancer cells by downregulating calpain, UHRF1, and DNMT1 expressions. *Nutr. Cancer* **2014**, *66*, 1220–1227. [CrossRef]
23. Arima, Y.; Hirota, T.; Bronner, C.; Mousli, M.; Fujiwara, T.; Niwa, S.; Ishikawa, H.; Saya, H. Down-regulation of nuclear protein ICBP90 by p53/p21Cip1/WAF1-dependent DNA-damage checkpoint signals contributes to cell cycle arrest at G1/S transition. *Genes Cells Devoted Mol. Cell. Mech.* **2004**, *9*, 131–142. [CrossRef]
24. Alhosin, M.; Ibrahim, A.; Boukhari, A.; Sharif, T.; Gies, J.P.; Auger, C.; Schini-Kerth, V.B. Anti-neoplastic agent thymoquinone induces degradation of alpha and beta tubulin proteins in human cancer cells without affecting their level in normal human fibroblasts. *Investig. New Drugs* **2012**, *30*, 1813–1819. [CrossRef]
25. Abusnina, A.; Alhosin, M.; Keravis, T.; Muller, C.D.; Fuhrmann, G.; Bronner, C.; Lugnier, C. Down-regulation of cyclic nucleotide phosphodiesterase PDE1A is the key event of p73 and UHRF1 deregulation in thymoquinone-induced acute lymphoblastic leukemia cell apoptosis. *Cell. Signal.* **2011**, *23*, 152–160. [CrossRef]
26. Hopfner, R.; Mousli, M.; Jeltsch, J.M.; Voulgaris, A.; Lutz, Y.; Marin, C.; Bellocq, J.P.; Oudet, P.; Bronner, C. ICBP90, a novel human CCAAT binding protein, involved in the regulation of topoisomerase IIalpha expression. *Cancer Res.* **2000**, *60*, 121–128. [PubMed]
27. Bronner, C. Control of DNMT1 abundance in epigenetic inheritance by acetylation, ubiquitylation, and the histone code. *Sci. Signal.* **2011**, *4*, pe3. [CrossRef] [PubMed]
28. Kofunato, Y.; Kumamoto, K.; Saitou, K.; Hayase, S.; Okayama, H.; Miyamoto, K.; Sato, Y.; Katakura, K.; Nakamura, I.; Ohki, S.; et al. UHRF1 expression is upregulated and associated with cellular proliferation in colorectal cancer. *Oncol. Rep.* **2012**, *28*, 1997–2002. [CrossRef] [PubMed]
29. Wang, F.; Yang, Y.Z.; Shi, C.Z.; Zhang, P.; Moyer, M.P.; Zhang, H.Z.; Zou, Y.; Qin, H.L. UHRF1 promotes cell growth and metastasis through repression of p16(ink(4)a) in colorectal cancer. *Ann. Surg. Oncol.* **2012**, *19*, 2753–2762. [CrossRef] [PubMed]
30. Abbady, A.Q.; Bronner, C.; Trotzier, M.A.; Hopfner, R.; Bathami, K.; Muller, C.D.; Jeanblanc, M.; Mousli, M. ICBP90 expression is downregulated in apoptosis-induced Jurkat cells. *Ann. N. Y. Acad. Sci.* **2003**, *1010*, 300–303. [CrossRef] [PubMed]
31. Yu, C.; Xing, F.; Tang, Z.; Bronner, C.; Lu, X.; Di, J.; Zeng, S.; Liu, J. Anisomycin suppresses Jurkat T cell growth by the cell cycle-regulating proteins. *Pharmacol. Rep. PR* **2013**, *65*, 435–444. [CrossRef]
32. Mandili, G.; Khadjavi, A.; Gallo, V.; Minero, V.G.; Bessone, L.; Carta, F.; Giribaldi, G.; Turrini, F. Characterization of the protein ubiquitination response induced by Doxorubicin. *FEBS J.* **2012**, *279*, 2182–2191. [CrossRef] [PubMed]
33. Halim, V.A.; García-Santisteban, I.; Warmerdam, D.O.; van den Broek, B.; Heck, A.J.R.; Mohammed, S.; Medema, R.H. Doxorubicin-induced DNA Damage Causes Extensive Ubiquitination of Ribosomal Proteins Associated with a Decrease in Protein Translation. *Mol. Cell. Proteom. MCP* **2018**, *17*, 2297–2308. [CrossRef]
34. Lang, V.; Aillet, F.; Xolalpa, W.; Serna, S.; Ceccato, L.; Lopez-Reyes, R.G.; Lopez-Mato, M.P.; Januchowski, R.; Reichardt, N.C.; Rodriguez, M.S. Analysis of defective protein ubiquitylation associated to adriamycin resistant cells. *Cell Cycle* **2017**, *16*, 2337–2344. [CrossRef]
35. Yuan, L.; Tian, J. LIN28B promotes the progression of colon cancer by increasing B-cell lymphoma 2 expression. *Biomed. Pharmacother. Biomed. Pharmacother.* **2018**, *103*, 355–361. [CrossRef] [PubMed]
36. Sun, N.; Meng, Q.; Tian, A. Expressions of the anti-apoptotic genes Bag-1 and Bcl-2 in colon cancer and their relationship. *Am. J. Surg.* **2010**, *200*, 341–345. [CrossRef] [PubMed]
37. Meterissian, S.H.; Kontogiannea, M.; Al-Sowaidi, M.; Linjawi, A.; Halwani, F.; Jamison, B.; Edwardes, M. Bcl-2 is a useful prognostic marker in Dukes' B colon cancer. *Ann. Surg. Oncol.* **2001**, *8*, 533–537. [CrossRef] [PubMed]
38. Paul-Samojedny, M.; Kokocińska, D.; Samojedny, A.; Mazurek, U.; Partyka, R.; Lorenz, Z.; Wilczok, T. Expression of cell survival/death genes: Bcl-2 and Bax at the rate of colon cancer prognosis. *Biochim. Et Biophys. Acta* **2005**, *1741*, 25–29. [CrossRef]
39. Kleschyov, A.L.; Strand, S.; Schmitt, S.; Gottfried, D.; Skatchkov, M.; Sjakste, N.; Daiber, A.; Umansky, V.; Munzel, T. Dinitrosyl-iron triggers apoptosis in Jurkat cells despite overexpression of Bcl-2. *Free Radic. Biol. Med.* **2006**, *40*, 1340–1348. [CrossRef]
40. Thomson, S.J.; Brown, K.K.; Pullar, J.M.; Hampton, M.B. Phenethyl isothiocyanate triggers apoptosis in Jurkat cells made resistant by the overexpression of Bcl-2. *Cancer Res.* **2006**, *66*, 6772–6777. [CrossRef]
41. Molto, L.; Rayman, P.; Paszkiewicz-Kozik, E.; Thornton, M.; Reese, L.; Thomas, J.C.; Das, T.; Kudo, D.; Bukowski, R.; Finke, J.; et al. The Bcl-2 transgene protects T cells from renal cell carcinoma-mediated apoptosis. *Clin. Cancer Res.* **2003**, *9*, 4060–4068.
42. Breitschopf, K.; Haendeler, J.; Malchow, P.; Zeiher, A.M.; Dimmeler, S. Posttranslational modification of Bcl-2 facilitates its proteasome-dependent degradation: Molecular characterization of the involved signaling pathway. *Mol. Cell. Biol.* **2000**, *20*, 1886–1896. [CrossRef] [PubMed]
43. Zhao, D.; Lin, F.; Wu, X.; Zhao, Q.; Zhao, B.; Lin, P.; Zhang, Y.; Yu, X. Pseudolaric acid B induces apoptosis via proteasome-mediated Bcl-2 degradation in hormone-refractory prostate cancer DU145 cells. *Toxicol. Vitr.* **2012**, *26*, 595–602. [CrossRef] [PubMed]
44. Sidhu, H.; Capalash, N. UHRF1: The key regulator of epigenetics and molecular target for cancer therapeutics. *Tumour Biol.* **2017**, *39*. [CrossRef] [PubMed]

45. Alhosin, M.; Leon-Gonzalez, A.J.; Dandache, I.; Lelay, A.; Rashid, S.K.; Kevers, C.; Pincemail, J.; Fornecker, L.M.; Mauvieux, L.; Herbrecht, R.; et al. Bilberry extract (Antho 50) selectively induces redox-sensitive caspase 3-related apoptosis in chronic lymphocytic leukemia cells by targeting the Bcl-2/Bad pathway. *Sci. Rep.* **2015**, *5*, 8996. [CrossRef]
46. Sutton, K.M.; Greenshields, A.L.; Hoskin, D.W. Thymoquinone, a bioactive component of black caraway seeds, causes G1 phase cell cycle arrest and apoptosis in triple-negative breast cancer cells with mutant p53. *Nutr. Cancer* **2014**, *66*, 408–418. [CrossRef] [PubMed]
47. Liu, X.; Dong, J.; Cai, W.; Pan, Y.; Li, R.; Li, B. The Effect of Thymoquinone on Apoptosis of SK-OV-3 Ovarian Cancer Cell by Regulation of Bcl-2 and Bax. *Int. J. Gynecol. Cancer* **2017**, *27*, 1596–1601. [CrossRef]
48. Park, E.J.; Chauhan, A.K.; Min, K.J.; Park, D.C.; Kwon, T.K. Thymoquinone induces apoptosis through downregulation of c-FLIP and Bcl-2 in renal carcinoma Caki cells. *Oncol. Rep.* **2016**, *36*, 2261–2267. [CrossRef]
49. Badr, G.; Mohany, M.; Abu-Tarboush, F. Thymoquinone decreases F-actin polymerization and the proliferation of human multiple myeloma cells by suppressing STAT3 phosphorylation and Bcl2/Bcl-XL expression. *Lipids Health Dis.* **2011**, *10*, 236. [CrossRef]
50. Samarghandian, S.; Azimi-Nezhad, M.; Farkhondeh, T. Thymoquinone-induced antitumor and apoptosis in human lung adenocarcinoma cells. *J. Cell. Physiol.* **2019**, *234*, 10421–10431. [CrossRef]
51. Ng, W.K.; Yazan, L.S.; Ismail, M. Thymoquinone from Nigella sativa was more potent than cisplatin in eliminating of SiHa cells via apoptosis with down-regulation of Bcl-2 protein. *Toxicol. Vitr.* **2011**, *25*, 1392–1398. [CrossRef] [PubMed]
52. Li, F.; Rajendran, P.; Sethi, G. Thymoquinone inhibits proliferation, induces apoptosis and chemosensitizes human multiple myeloma cells through suppression of signal transducer and activator of transcription 3 activation pathway. *Br. J. Pharmacol.* **2010**, *161*, 541–554. [CrossRef] [PubMed]
53. Costantino, L.; Barlocco, D. STAT 3 as a target for cancer drug discovery. *Curr. Med. Chem.* **2008**, *15*, 834–843. [CrossRef]
54. Zhao, X.; Guo, X.; Shen, J.; Hua, D. Alpinetin inhibits proliferation and migration of ovarian cancer cells via suppression of STAT3 signaling. *Mol. Med. Rep.* **2018**, *18*, 4030–4036. [CrossRef] [PubMed]
55. Mohammadabadi, M.R.; Mozafari, M.R. Enhanced efficacy and bioavailability of thymoquinone using nanoliposomal dosage form. *J. Drug Deliv. Sci. Technol.* **2018**, *47*, 445–453. [CrossRef]
56. Pal, R.R.; Rajpal, V.; Singh, P.; Saraf, S.A. Recent Findings on Thymoquinone and Its Applications as a Nanocarrier for the Treatment of Cancer and Rheumatoid Arthritis. *Pharmaceutics* **2021**, *13*, 775. [CrossRef]
57. Abdullah, O.; Omran, Z.; Hosawi, S.; Hamiche, A.; Bronner, C.; Alhosin, M. Thymoquinone Is a Multitarget Single Epidrug That Inhibits the UHRF1 Protein Complex. *Genes* **2021**, *12*, 622. [CrossRef]

Communication

Synthesis of Proposed Structure of Aaptoline B via Transition Metal-Catalyzed Cycloisomerization and Evaluation of Its Neuroprotective Properties in *C. Elegans*

Soobin Kim [1], Wooin Yang [2], Dong-Seok Cha [2,*] and Young-Taek Han [1,*]

[1] College of Pharmacy, Dankook University, 119 Dandae-ro, Dongnam-gu, Cheonan-si 31116, Chungnam, Korea; rue42351@naver.com

[2] College of Pharmacy, Woosuk University, 443 Samnye-ro, Wanju-gun 55338, Jeonbuk, Korea; benefinn@naver.com

* Correspondence: hanyt@dankook.ac.kr (Y.-T.H.); cha@woosuk.ac.kr (D.-S.C.); Tel.: +82-41-550-1431 (Y.-T.H.)

Abstract: A concise synthesis of the proposed structure of aaptoline B, a pyrroloquinoline derived from a marine sponge, was accomplished. A key feature of this synthesis is the versatile transition metal-catalyzed cycloisomerization of *N*-propargylaniline to construct a quinoline skeleton. However, the spectral data of the synthesized aaptoline B did not agree with those of previous studies. The structure of the synthesized aaptoline B was confirmed using a combined 2D NMR analysis. Furthermore, we assessed the possible neuroprotective potential of aaptoline B using the *C. elegans* model system. In this study, aaptoline B significantly improved the viability and the morphology of dopaminergic neurons of nematodes under MPP$^+$ exposure conditions. We also found that MPP$^+$-induced motor deficits in nematodes were efficiently restored by aaptoline B treatment. Our findings demonstrate the neuroprotective effects of aaptoline B against MPP$^+$-induced dopaminergic neuronal damage. Further studies are underway to explain its pharmacological mechanism.

Keywords: aaptoline B; pyrroloquinoline; Ag(I)-catalyzed cycloisomerization; dopaminergic neuroprotection; Parkinson's disease

Citation: Kim, S.; Yang, W.; Cha, D.-S.; Han, Y.-T. Synthesis of Proposed Structure of Aaptoline B via Transition Metal-Catalyzed Cycloisomerization and Evaluation of Its Neuroprotective Properties in C. Elegans. *Appl. Sci.* **2021**, *11*, 9125. https://doi.org/10.3390/app11199125

Academic Editors: Ana M. L. Seca and Monica Gallo

Received: 13 September 2021
Accepted: 28 September 2021
Published: 30 September 2021

Publisher's Note: MDPI stays neutral with regard to jurisdictional claims in published maps and institutional affiliations.

1. Introduction

Parkinson's disease (PD) is a progressive neurodegenerative disorder of the central nervous system that affects millions of people worldwide. The clinical features of PD include impaired motor functions and nonmotor symptoms, which are mainly due to the selective loss of dopaminergic neurons in the substantia nigra pars compacta (SNpc) [1]. Although the underlying pathological mechanism of PD remains unknown, overproduction of reactive oxygen species (ROS) is considered a major trigger of neuronal death in the SNpc [2]. In addition, neurotoxins such as 1-methyl-4-phenylpyridinium (MPP$^+$) and 6-hydroxydopamine (6-OHDA) can induce dopaminergic neurodegeneration via mitochondrial dysfunction and consequent ROS generation, and the treated animals exhibit symptoms similar to humans with PD [3].

Aaptoline B (**1**) has been reported as a pyrroloquinoline alkaloid isolated from the marine sponge *Aaptos aaptos* by a Chinese research group [4]. Structurally, as shown in Figure 1, **1** is characterized by a unique 7*H*-pyrrolo [2,3-h]quinoline skeleton embedded with privileged scaffolds, such as indole and quinoline. Indole and quinoline derivatives exhibit diverse biological activities, including anticancer and anti-neurodegenerative diseases. Hence, these two scaffolds, including their heterocycle-linked and fused analogs, have been considered important privileged scaffolds in drug discovery [5–7].

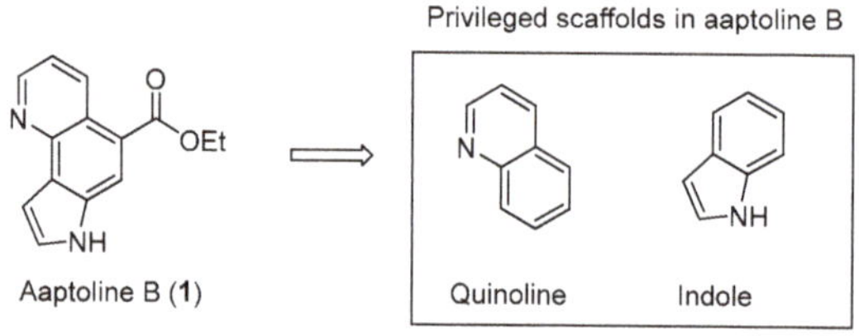

Aaptoline B (**1**)   Quinoline   Indole

Figure 1. Proposed structure of pyrroloquinoline alkaloid aaptoline B (**1**) and embedded privileged scaffolds.

Recently, we have intensively worked on the synthesis and biological applications of pyridine-fused privileged scaffolds, such as quinoline and pyridocoumarin, using Ag(I)-catalyzed cycloisomerization [8–10]. Taking advantage of its high yield and regioselectivity, Ag(I)-catalyzed cycloisomerization has also been used for the synthesis of natural products, including goniothalines [8] and polynemoraline C [10]. These successful results of the construction of pyridine-fused skeletons via Ag(I)-catalyzed cycloisomerization as well as interesting structural features of the pyrroloquinoline scaffold of **1** prompted us to attempt the total synthesis of **1**. This paper describes the total synthesis of the proposed structure of **1** and its protective properties against MPP^+-induced dopaminergic neurodegeneration using the *Caenorhabditis elegans* model, which offers an excellent environment for PD research.

2. Materials and Methods

2.1. Chemistry

2.1.1. General Experimental

Unless otherwise noted, all reactions were performed under an argon atmosphere in oven-dried glassware. The starting materials and solvents were used as received from commercial suppliers without further purification. Thin layer chromatography was carried out using Merck silica gel 60 F254 plates, and visualized with a combination of UV, *p*-anisaldehyde, and potassium permanganate staining. Flash chromatography was performed using Merck silica gel 60 (0.040–0.063 mm, 230–400 mesh). Mass spectra were obtained using an Agilent 6530 Q-TOF unit. ^1H and ^{13}C spectra were recorded on a Bruker 500′54 Ascend (500 MHz) spectrometer at the Center for Bio-medical Engineering Core Facility (Dankook University, Yongin-si, Korea), Jeol RESONANCE ECZ 400S (400 MHz), or Bruker Ascend III (700 MHz) in deuterated solvents. ^1H and ^{13}C NMR chemical shifts are reported in parts per million (ppm) relative to TMS, with the residual solvent peak used as an internal reference. Signals are reported as m (multiplet), s (singlet), d (doublet), t (triplet), q (quartet), bs (broad singlet), bd (broad doublet), dd (doublet of doublets), dt (doublet of triplets), or dq (doublet of quartets); the coupling constants are reported in hertz (Hz)

2.1.2. Ethyl 4-methyl-3,5-dinitrobenzoate (**3**)

Ethyl benzoic acid **3** was obtained from **4** according to literature procedures with some modifications [11]. Briefly, to a solution of 3,5-dinitro-4-methylbenzoic acid (**4**; 3.00 g, 13.27 mmol) in 20 mL of anhydrous EtOH was slowly added 3 mL of H_2SO_4 at ambient temperature. After being refluxed overnight, the reaction mixture was cooled to ambient temperature, and concentrated in vacuo. Purification of residue via column chromatography on silica gel with CH_2Cl_2 afforded **3** (3.14 g, 93%) as a yellowish solid. The NMR spectra were consistent with the previously reported data. Melting point: 74~76 °C; ^1H-NMR (500 MHz, $CDCl_3$) δ 8.59 (s, 2H), 4.46 (q, 2H, *J* = 7.1 Hz), 2.63 (s, 3H), 1.43 (t, 3H, *J* = 7.1 Hz); ^{13}C-NMR (125 MHz, $CDCl_3$) δ 162.5, 151.6, 131.3, 130.6, 127.8, 62.7, 15.2, 14.2.

2.1.3. Ethyl 4-amino-1*H*-indole-6-carboxylate (**6**)

To a solution of **3** (300 mg, 1.18 mmol) in 0.3 mL of DMF was added *N,N*-dimethylformamide diethyl acetal (0.45 mL, 2.6 mmol) at ambient temperature. The reaction mixture was heated at 50 °C for 1 h, and concentrated in vacuo to give crude enamine **5**. To a solution of the residue in EtOH (60 mL) was added catalytic amount of 10% palladium on carbon. The reaction mixture was stirred under a hydrogen atmosphere until TLC analysis showed the complete disappearance of **5**, and filtered using a Celite pad. The filtrate was concentrated in vacuo. Purification of the residue via flash column chromatography (EtOAc:*n*-hexane = 1:1) afforded amino-1*H*-indole **6** (191 mg, 78% in 2 steps) as a dark-red gum. ^{1}H-NMR (500 MHz, CD$_3$OD) δ 7.57 (s, 1H), 7.28 (s, 1H), 7.02 (s, 1H), 6.56 (s, 1H), 4.59 (s, 1H), 4.32 (d, 2H, *J* = 5.8 Hz), 1.38 (s, 3H); ^{13}C-NMR (125 MHz, CD$_3$OD) δ 170.1, 140.8, 137.7, 127.0, 125.3, 123.2, 106.1, 105.0, 99.9, 61.6, 14.7; HR-MS (Q-ToF): Calcd for C$_{11}$H$_{13}$N$_2$O$_2{}^+$ (M + H$^+$): 205.0972. Found: 205.0977.

2.1.4. Ethyl 4-(prop-2-yn-1-ylamino)-1*H*-indole-6-carboxylate (**2**)

To a solution of **6** (30 mg, 0.147 mmol), KI (2.4 mg, 14.7 μmol) and K$_2$CO$_3$ (44.6 mg, 0.323 mmol) in DMF (1.5 mL) was added 24.1 mg (0.162 mmol) of 80 w/w% solution of propargyl bromide in toluene at ambient temperature. The reaction mixture was stirred for 72 h at 65 °C, cooled to ambient temperature, and then diluted with water and EtOAc. The organic layer was washed with water and brine, dried over MgSO$_4$, and concentrated in vacuo. Purification of residue via column chromatography on silica gel (EtOAc:*n*-hexane = 1:4~1:1) afforded *N*-propargyl aminobenzoate **2** (21.8 mg, 61%) as a dark-yellow oil along with 4.2 mg of starting material **6** (71% yield brsm). ^{1}H-NMR (400 MHz, CDCl$_3$) δ 8.47 (s, 1H), 7.71 (d, 1H, *J* = 0.9 Hz), 7.25 (m, 2H), 7.09 (d, 1H, *J* = 0.9 Hz), 6.52 (m, 1H), 4.38 (q, 2H, *J* = 7.1 Hz), 4.15 (d, 2H, *J* = 2.5 Hz), 2.26 (t, 1H, *J* = 2.4 Hz), 1.40 (t, 3H, *J* = 7.1 Hz); ^{13}C-NMR (100 MHz, CDCl$_3$) δ 167.4, 135.1, 125.0, 125.0, 120.5, 105.4, 100.9, 98.7, 80.1, 71.5, 60.2, 33.6, 14.0; HR-MS (Q-ToF): Calcd for C$_{14}$H$_{15}$N$_2$O$_2{}^+$ (M + H$^+$): 243.1128. Found: 243.1131.

2.1.5. Representative Procedure of Transition Metal-Catalyzed Cycloisomerization for Ethyl 7*H*-pyrrolo [2,3-h]quinoline-5-carboxylate (Aaptoline B; **1**)

To a solution of **2** (21.8 mg, 0.09 mmol) in DMSO (2.3 mL) was added AgSbF$_6$ (3.1 mg, 9 μmol) at ambient temperature. The reaction mixture was stirred at 110 °C for 4.5 h, cooled to ambient temperature, diluted with EtOAc and quenched with saturated aq. NaHCO$_3$ solution. The organic layer was washed with water and brine, dried over Na$_2$SO$_4$, and concentrated in vacuo. Purification of residue via column chromatography on silica gel (EtOAc:*n*-hexane = 1:3~1:1) afforded aaptoline B (**1**; 14.1 mg, 73%) as a yellowish solid. Melting point: 187~190 °C; ^{1}H-NMR (500 MHz, CDCl$_3$) δ 9.56 (d, 1H, *J* = 8.4 Hz), 9.43 (s, 1H), 8.96 (d, 1H, *J* = 3.2 Hz), 8.52 (s, 1H), 7.52 (m, 3H), 4.47 (q, 2H, *J* = 7.1 Hz), 1.46 (t, 3H, *J* = 7.1 Hz); ^{13}C-NMR (125 MHz, CDCl$_3$) δ 167.3, 148.2, 143.1, 136.3, 133.1, 127.1, 126.8, 122.9, 120.4, 120.1, 119.5, 104.0, 61.2, 14.6; ^{1}H-NMR (700 MHz, CD$_3$OD) δ 9.52 (dd, 1H, *J* = 1.4, 8.4 Hz), 8.79 (dd, 1H, *J* = 1.8, 4.6 Hz), 8.51 (s, 1H), 7.59 (d, 1H, *J* = 2.8 Hz), 7.51 (dd, 1H, *J* = 4.6, 8.8 Hz), 7.34 (t, 1H, *J* = 1.4 Hz), 4.45 (q, 2H, *J* = 7.2 Hz), 1.46 (t, 3H, *J* = 7.0 Hz); ^{13}C-NMR (175 MHz, CD$_3$OD) δ 168.7, 149.2, 144.7, 137.2, 134.7, 128.6, 128.2, 124.0, 121.1, 120.9, 120.8, 103.9, 62.2, 14.9; HR-MS (Q-ToF): Calcd for C$_{14}$H$_{13}$N$_2$O$_2{}^+$ (M + H$^+$): 241.0972. Found: 241.0977.

2.2. Biology

2.2.1. *C. Elegans* Maintenance and MPP$^+$ Treatment

Wild-type Bristol N2 and transgenic strain: BZ555 (egIs1, Pdat-1::GFP) were provided from the Caenorhabditis Genetic Center (CGC; University of Minnesota, Minneapolis, MN, USA). Nematodes were maintained in the liquid medium with live *Escherichia coli* bacteria (OP50) at 20 °C. After embryo isolation, age-synchronized young-adult nematodes were treated with **1** or vehicle for 2 h. Then the nematodes were incubated another 96 h under 4 mM of MPP$^+$ exposure.

2.2.2. Fluorescence Microscopy and Visualization

On the 4th day of adulthood, 10% sodium azide was applied to nematodes and immobilized nematodes were transferred onto 2% agarose pad. The DA neuronal viability of the nematodes was evaluated by using a fluorescence microscope (Nikon Eclipse Ni-u, Tokyo, Japan). The GFP fluorescence signals of each animal were photographed and calculated the number of intact DA neurons.

2.2.3. Behavioral Assay

Age-synchronized N2 nematodes were incubated with or without **1**. On the 4th day of adulthood, their motor abilities were observed using a dissecting microscope (Nikon SMZ1500, Tokyo, Japan) and Nikon image analysis software 4.0(NIS-Elements, Tokyo, Japan). The travel distance of nematodes was automatically recorded for 20 s. For the food sensing assay, nematodes were washed with M9 buffer 3 times to remove any residual food and then transferred to a nonfood plate. The velocity of nematodes on the nonfood plate was measured and then the velocity was measured again on the food plate.

2.2.4. Statistical Analysis

All experiments were independently conducted at least thrice. The results were shown in mean $\pm$ S.D. Statistical significance of differences between groups were compared with one-way analysis of variance (ANOVA) followed by Tukey test.

3. Results and Discussion

3.1. Synthesis and Structure Elucidation of Aaptoline B

The retrosynthetic analysis for **1** is depicted in Figure 2, which includes the 6-endo selective transition metal-catalyzed cycloisomerization of 4-propargylaminoindole **2**. Precursor **2** was expected to be readily synthesized from the known dinitrobenzoate **3** by Batcho–Leimgruber indole synthesis, a powerful synthetic method for the synthesis of indoles from ortho-nitrotoluenes, which proceeds via the formation of the enamine intermediates and subsequent reductive cyclization [12], followed by regioselective *N*-propargylation.

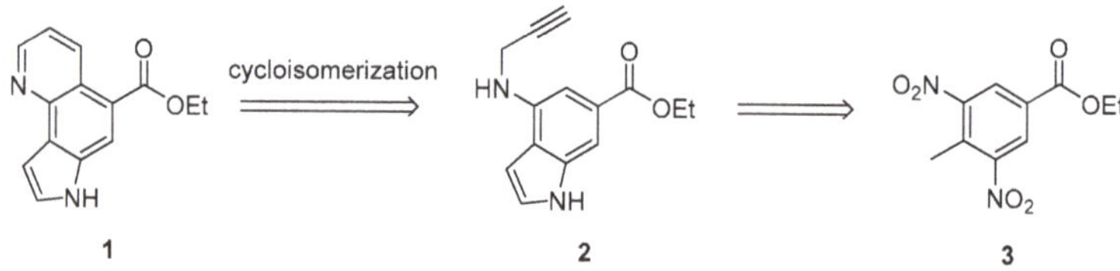

Figure 2. Retrosynthetic analysis of aaptoline B (**1**).

As shown in Scheme 1, our synthesis commenced with the preparation of previously reported ethyl 4-methyl-3,5-dinitrobenzoate **3**. Thus, benzoic acid **4** was refluxed in anhydrous ethyl alcohol in the presence of sulfuric acid to afford ethyl benzoate **3** [11]. When *N,N*-dimethylformamide dimethyl acetal was used, according to the literature [13] to obtain enamine **5** from **3**, followed by reductive cyclization, the indole ethyl carboxylate **6** was obtained as a mixture with the corresponding methyl ester **6′**. We supposed that methoxide or methanol, possibly generated from dimethylformamide dimethyl acetal, led to unexpected transesterification. Hence, we performed the Batcho–Leimgruber indole synthesis reaction of **3** using *N,N*-dimethylformamide diethyl acetal instead of dimethyl acetal, and obtained ethyl indole carboxylate **6** in 74% (two steps from **3**) without methyl ester byproducts. To minimize side reactions, such as over-alkylation and *N*-alkylation of pyrrolic nitrogen atom, we intensively investigated the reaction conditions of *N*-propargylation of **6** into 4-propargylaminoindole **2** by altering the amount of reagents, reaction time, and temperature. The optimal yield (61% isolated yield; 71% brsm yield) was achieved without *N*-alkylation of aromatic pyrrolic amine when the DMF solution of **2** and 1.1 equivalents of propargyl bromide was heated at 65 °C for 72 h with 2.2 equivalents of K_2CO_3 and a catalytic amount of KI [14].

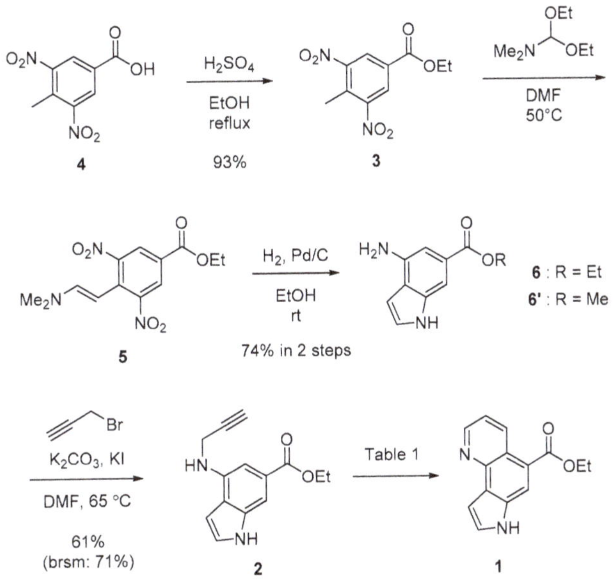

Scheme 1. Synthesis of proposed structure of aaptoline B (**1**).

With the cycloisomerization precursor **2** in hand, we tested diverse transition metal catalysts (Table 1) to find out the optimal reaction conditions. We initially carried out cycloisomerization of **2** with 0.1 equivalent of CuI (entry 1), and obtained **1** with 56% yield. When the reaction was conducted with $InCl_3$ (entry 2) or AuCl (entry 3), the yields (18% and 27%, respectively) were poorer. The use of $AgSbF_6$ as catalyst resulted in the best yield (73%, entry 4), being superior to other catalysts of Ag(I), such as $AgNO_3$ (61%; entry 5) and AgOTf (35%; entry 6).

Table 1. Catalyst screening for the completion of synthesis of proposed structure of **1**.

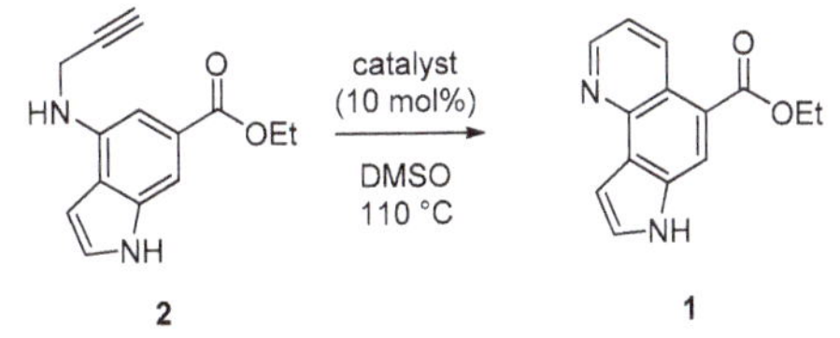

Entry [1]	Catalyst	Time (h)	Yield [2] (%)
1	CuI	2	56
2	$InCl_3$	27	18
3	AuCl	8	27
4	$AgSbF_6$	4.5	73
5	$AgNO_3$	5	61
6	AgOTf	5	35

[1] 30 mg of **7** in DMSO (0.1 M solution) was reacted; [2] Isolated yields.

However, the NMR spectra of **1** differed from those reported when using the same deuterated solvent ($CDCl_3$). Significant differences were observed in both the ^1H and ^{13}C

NMR spectra of our synthetic sample compared to those of the isolated product [4]. To precisely confirm the structure of synthetic **1**, we employed two-dimensional NMR analysis (700 MHz, CD$_3$OD; see the Supplementary Materials). Proton at C-3 was clearly revealed by homonuclear correlation spectroscopy (^{1}H-^{1}H COSY) by correlations with the neighboring aromatic protons on C-2 and C-4, which were confirmed by further investigation employing heteronuclear correlation spectroscopy. Singlet peak at 8.51 ppm (^{1}H NMR in CD$_3$OD) did not exhibit any correlation in ^{1}H-^{1}H COSY, and it was confirmed as proton at C-6. The ^{13}C-NMR peaks of all hydrogenated carbons could be assigned unambiguously based on heteronuclear single quantum coherence (^{1}H-^{13}C HSQC) analysis. The ^{13}C-NMR peaks of the quaternary carbons (5 and 7–12) were carefully assigned using the HMBC correlation. Because it was difficult to find a significant difference in HMBC correlations from their neighboring carbons, C-9~12 were hard to be confirmed using 2D NMR analysis. For instance, C-9 and C-10 at the quinoline moiety exhibited the same HMBC correlation with neighboring protons.

Taking into consideration the previous assignments of ^{1}H-NMR of the indole system [15], protons at δ 7.34 and δ 7.59 were assigned as protons on C-11 and C-12, respectively. Furthermore, based on preceding literature in which the peaks observed at about 125 and 145 ppm were designated as C-9 and C-10 carbon adjacent to nitrogen atom, respectively [16–18], the peak at δ 144.7 was assigned as carbons at C-10. Through two-dimensional NMR analysis, we were also able to assign the ^{13}C-NMR peak of C-9 (δ 120.8) that could not be assigned in the previous study [4]. Overall, our assignment of **1** seems to be more reasonable compared to the previous report. The selected HMBC correlations are shown in Figure 3.

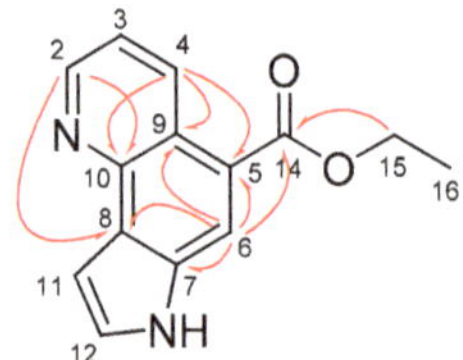

Figure 3. Selected HMBC correlations (H→C) of synthesized **1**.

3.2. Evaluation of Neuroprotective Potential of Aaptoline B

The neuroprotective potential of aaptoline B (**1**) was investigated using a *C. elegans* PD model. To test the effect of **1** on dopaminergic neuronal loss induced by MPP$^+$, we analyzed the fluorescent signals of the transgenic nematode BZ555 (Pdat-1::GFP), which expresses GFP in dopaminergic neurons. As shown in Figure 4A,B, the treatment of nematodes with MPP$^+$ for 96 h significantly decreased the viability of neurons to 46.3 ± 5.6%. Compared to untreated nematodes, **1**-fed nematodes exhibited a notable increase in GFP signals of all eight dopaminergic neurons in a dose-dependent manner (Figure 4A,B). These findings indicate that **1** might have the ability to protect dopaminergic neurons against MPP$^+$-induced neurotoxicity. To confirm the protective role of **1** in the MPP$^+$-mediated dopaminergic neuronal damage, we examined whether **1** could modulate behaviors requiring dopamine function, such as food-sensing response and locomotion. Well-fed nematodes reduce their movement in the presence of food as compared to that afforded in its absence. This basal slowing response is known to be mediated solely by dopaminergic neural circuitry. In the current study, normal nematodes demonstrated a 73.5% reduction in locomotory velocity when they came across the bacterial lawn (Figure 4C). When the nematodes were exposed to MPP$^+$, the slowing response was significantly decreased to 37.2%, and treatment with **1** efficiently restored this reduction in the slowing response induced by MPP$^+$ (Figure 4C). In addition, we observed the effect of **1** on the travel distance of nematodes under MPP$^+$ exposure conditions. Similar to the results from the food-sensing assay, the travel distance of nematodes was significantly decreased in the presence of MPP$^+$, while **1** supplementation restored these MPP$^+$-induced defects in motor activities of nematodes in a dose-dependent manner (26.0%, 35.5%, and 38.2% at 5, 10, and 20 μM, respectively) (Figure 4D). Overall, our findings indicate that **1** could contribute to ameliorating dopaminergic neurodegeneration not only morphologically but also functionally.

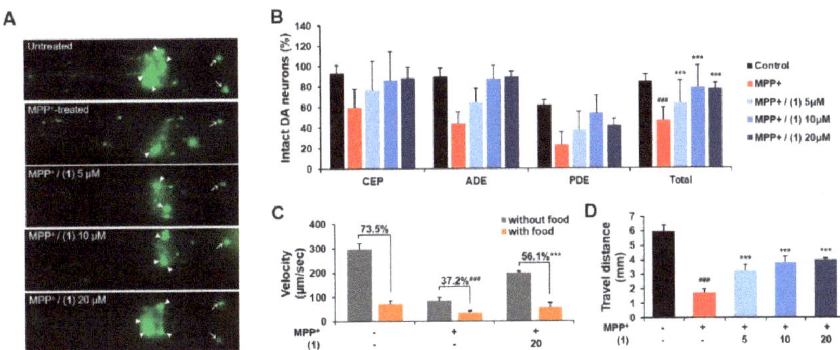

Figure 4. Neuroprotective effects of synthesized **1** on the DA neurodegeneration in the MPP$^+$-exposed *C. elegans*. (**A**) GFP fluorescence signals of CEP (arrowhead) and ADE (arrow) in the Pdat-1::GFP nematode were photographed at 400× magnification. (**B**) The viability of all eight DA neurons was scored by inspecting the GFP expressions of Pdat-1::GFP nematode. (**C**) The velocity of wild-type nematodes was observed in the presence or absence of food (*E. coli* OP50) and the basal slowing response was calculated. (**D**) The travel distance of wild-type nematodes was recorded for 20 s under a dissecting microscope. Statistical significance was determined by one-way ANOVA; $^{\#\#\#}$ $p < 0.001$ compared with vehicle alone; *** $p < 0.001$ compared with MPP$^+$-treated control.

4. Conclusions

In summary, a concise and efficient synthesis of the proposed structure of aaptoline B was accomplished. Key features of the synthesis involve Batcho–Leimgruber indole synthesis and subsequent AgSbF$_6$-catalyzed cycloisomerization for the construction of the unique pyrrolo [2,3-h]quinoline skeleton of **1**. The spectral data of synthesized **1** did not agree with those in a previous report. However, the structure of the synthesized ethyl 7*H*-pyrrolo [2,3-h]quinoline-5-carboxylate (**1**), originally proposed as aaptoline B, was confirmed using combined two-dimensional NMR analysis. Our findings from a biological study revealed that **1** could protect *C. elegans* from MPP$^+$-induced dopaminergic neuronal damage and restore dopamine-related behavioral functions, suggesting its therapeutic potential in PD. Further studies, such as pharmacophore identification and structure modification of **1**, were currently underway.

Supplementary Materials: The following are available online at https://www.mdpi.com/article/10.3390/app11199125/s1, Table S1: ^{1}H- and ^{13}C-NMR assignment of aaptoline B.

Author Contributions: Y.-T.H. and D.-S.C. conceived and designed the experiments, and wrote the paper; S.K. performed the synthesis and chemical analysis; W.Y. performed biological studies. All authors have read and agreed to the published version of the manuscript.

Funding: This research was funded by National Research Foundation of Korea (NRF-2020R1F1A1058295).

Institutional Review Board Statement: Not applicable.

Informed Consent Statement: Not applicable.

Data Availability Statement: Not applicable.

Acknowledgments: The authors gratefully acknowledge Center for Bio-Medical Engineering Core Facility for providing analytical equipment of Dankook University.

Conflicts of Interest: The authors declare no conflict of interest.

References

1. Sveinbjornsdottir, S. The clinical symptoms of Parkinson's disease. *J. Neurochem.* **2016**, *139*, 318–324. [CrossRef]
2. Mosley, R.L.; Benner, E.J.; Kadiu, I.; Thomas, M.; Boska, M.D.; Hasan, K.; Laurie, C.; Gendelman, H.E. Neuroinflammation, oxidative stress and the pathogenesis of Parkinson's disease. *Clin. Neurosci. Res.* **2006**, *6*, 261–281. [CrossRef] [PubMed]
3. Nass, R.; Hall, D.H.; Miller, D.M.; Blakely, R.D. Neurotoxin-induced degeneration of dopamine neurons in *Caenorhabditis elegans*. *Proc. Natl. Acad. Sci. USA* **2002**, *99*, 3264–3269. [CrossRef] [PubMed]
4. Tang, W.Z.; Yu, H.B.; Lu, J.R.; Lin, H.W.; Sun, F.; Wang, S.P.; Yang, F. Aaptolines A and B, two new quinoline alkaloids from the marine sponge *Aaptos aaptos*. *Chem. Biodivers.* **2020**, *17*, e2000074. [CrossRef] [PubMed]
5. de Sá Alves, F.R.; Barreiro, E.J.; Fraga, C.A. From nature to drug discovery: The indole scaffold as a 'privileged structure'. *Mini Rev. Med. Chem.* **2009**, *9*, 782–793. [CrossRef] [PubMed]
6. Bongarzone, S.; Bolognesi, M.L. The concept of privileged structures in rational drug design: Focus on acridine and quinoline scaffolds in neurodegenerative and protozoan diseases. *Expert Opin. Drug Discov.* **2011**, *6*, 251–268. [CrossRef] [PubMed]
7. Musiol, R. An overview of quinoline as a privileged scaffold in cancer drug discovery. *Expert Opin. Drug Discov.* **2017**, *12*, 583–597. [CrossRef] [PubMed]
8. Ahn, S.; Yoon, J.A.; Han, Y.T. Total synthesis of the natural pyridocoumarins goniothaline A and B. *Synthesis* **2019**, *51*, 552–556.
9. Yoon, J.A.; Lim, C.; Han, Y.T. Preliminary study on novel expedient synthesis of 5-azaisocoumarins by transition metal-catalyzed cycloisomerization. *Front. Chem.* **2020**, *8*, 772. [CrossRef] [PubMed]
10. Yoon, J.A.; Han, Y.T. Efficient synthesis of pyrido [3, 2-c] coumarins via silver nitrate catalyzed cycloisomerization and application to the first synthesis of polyneomarline C. *Synthesis* **2019**, *51*, 4611–4618. [CrossRef]
11. Pickaert, G.; Ziessel, R. Synthesis of oligopyridinic scaffolds from amido substituted phenyl rings for extended hydrogen bonding. *Synthesis* **2004**, *2004*, 2716–2726.
12. Li, J.J. *Name Reactions in Heterocyclic Chemistry*; John Wiley & Sons: Hoboken, NJ, USA, 2005; pp. 104–109.
13. Demont, H.E.; Faller, A.; MacPherson, D.T.; Milner, P.H.; Naylor, A.; Redshaw, S.; Stanway, S.J.; Vesey, R.D.; Walter, D.S. Preparation of Hydroxyethylamine Derivatives for the Treatment of Alzheimer's Disease. WO 2004050619A1, 17 June 2004.
14. Zhao, Y.; Cai, L.; Huang, T.; Meng, S.; Chan, A.S.; Zhao, J. Solvent-mediated C3/C7 regioselective switch in chiral phosphoric acid-catalyzed enantioselective Friedel-Crafts alkylation of indoles with α-ketiminoesters. *Adv. Synth. Catal.* **2020**, *362*, 1309–1316. [CrossRef]
15. Joule, J.A. *Product Class 13: Indole and Its Derivatives*; Thieme: Stuttgart, Germany, 2001; Volume 10, pp. 361–652.
16. da Rosa Monte Machado, G.; Diedrich, D.; Ruaro, T.C.; Zimmer, A.R.; Lettieri Teixeira, M.; de Oliveira, L.F.; Jean, M.; Van de Weghe, P.; de Andrade, S.F.; Baggio Gnoatto, S.C.; et al. Quinolines derivatives as promising new antifungal candidates for the treatment of candidiasis and dermatophytosis. *Braz. J. Microbiol.* **2020**, *51*, 1691–1701. [CrossRef]
17. Lloyd, L.S.; Adams, R.W.; Bernstein, M.; Coombes, S.; Duckett, S.B.; Green, G.G.; Lewis, R.J.; Mewis, R.E.; Sleigh, C.J. Utilization of SABRE-derived hyperpolarization to detect low-concentration analytes via 1D and 2D NMR methods. *J. Am. Chem. Soc.* **2012**, *134*, 12904–12907. [CrossRef]
18. Krzymiński, K.; Malecha, P.; Zadykowicz, B.; Wróblewska, A.; Błażejowski, J. ^{1}H and ^{13}C NMR spectra, structure and physico-chemical features of phenyl acridine-9-carboxylates and 10-methyl-9-(phenoxycarbonyl)acridinium trifluoromethanesulphonates–alkyl substituted in the phenyl fragment. *Spectrochim. Acta A Mol. Biomol. Spectrosc.* **2012**, *78*, 401–409. [CrossRef]

applied sciences

MDPI

Review

Recent Advances in the Heterologous Biosynthesis of Natural Products from *Streptomyces*

Van Thuy Thi Pham [1], Chung Thanh Nguyen [1], Dipesh Dhakal [1], Hue Thi Nguyen [1], Tae-Su Kim [1] and Jae Kyung Sohng [1,2,*]

[1] Department of Life Science and Biochemical Engineering, SunMoon University, 70 Sunmoon-ro 221, Tangjeong-myeon, Asan-si, Chungnam 31460, Korea; pham@sunmoon.ac.kr (V.T.T.P.); chungnguyen@sunmoon.ac.kr (C.T.N.); dipeshdhakal@sunmoon.ac.kr (D.D.); hue219@snu.ac.kr (H.T.N.); taesuda@sunmoon.ac.kr (T.-S.K.)

[2] Department of Pharmaceutical Engineering and Biotechnology, SunMoon University, 70 Sunmoon-ro 221, Tangjeong-myeon, Asan-si, Chungnam 31460, Korea

* Correspondence: sohng@sunmoon.ac.kr

Abstract: *Streptomyces* is a significant source of natural products that are used as therapeutic antibiotics, anticancer and antitumor agents, pesticides, and dyes. Recently, with the advances in metabolite analysis, many new secondary metabolites have been characterized. Moreover, genome mining approaches demonstrate that many silent and cryptic biosynthetic gene clusters (BGCs) and many secondary metabolites are produced in very low amounts under laboratory conditions. One strain many compounds (OSMAC), overexpression/deletion of regulatory genes, ribosome engineering, and promoter replacement have been utilized to activate or enhance the production titer of target compounds. Hence, the heterologous expression of BGCs by transferring to a suitable production platform has been successfully employed for the detection, characterization, and yield quantity production of many secondary metabolites. In this review, we introduce the systematic approach for the heterologous production of secondary metabolites from *Streptomyces* in *Streptomyces* and other hosts, the genome analysis tools, the host selection, and the development of genetic control elements for heterologous expression and the production of secondary metabolites.

Keywords: heterologous expression; *Streptomyces*; secondary metabolite

Citation: Pham, V.T.T.; Nguyen, C.T.; Dhakal, D.; Nguyen, H.T.; Kim, T.-S.; Sohng, J.K. Recent Advances in the Heterologous Biosynthesis of Natural Products from *Streptomyces*. *Appl. Sci.* **2021**, *11*, 1851. https://doi.org/10.3390/app11041851

Academic Editor: Ana M. L. Seca

Received: 21 December 2020
Accepted: 12 February 2021
Published: 19 February 2021 .

Publisher's Note: MDPI stays neutral with regard to jurisdictional claims in published maps and institutional affiliations.

1. Introduction

Streptomyces is a group of filamentous, Gram-positive bacteria with a high GC (guanine-cytosine) content in their genomes. The secondary metabolites or natural products of *Streptomyces*, such as antibiotic, anticancer, antitumor, antiviral, and immune suppressive compounds, are some of the industrially important compounds that are used for biotechnological applications [1]. *Streptomyces* is the main source of novel secondary metabolites, including polyketides, peptides, terpenoids, alkaloids, and saccharides [2–6]. Some natural secondary metabolites from *Streptomyces* are shown in Figure 1.

99

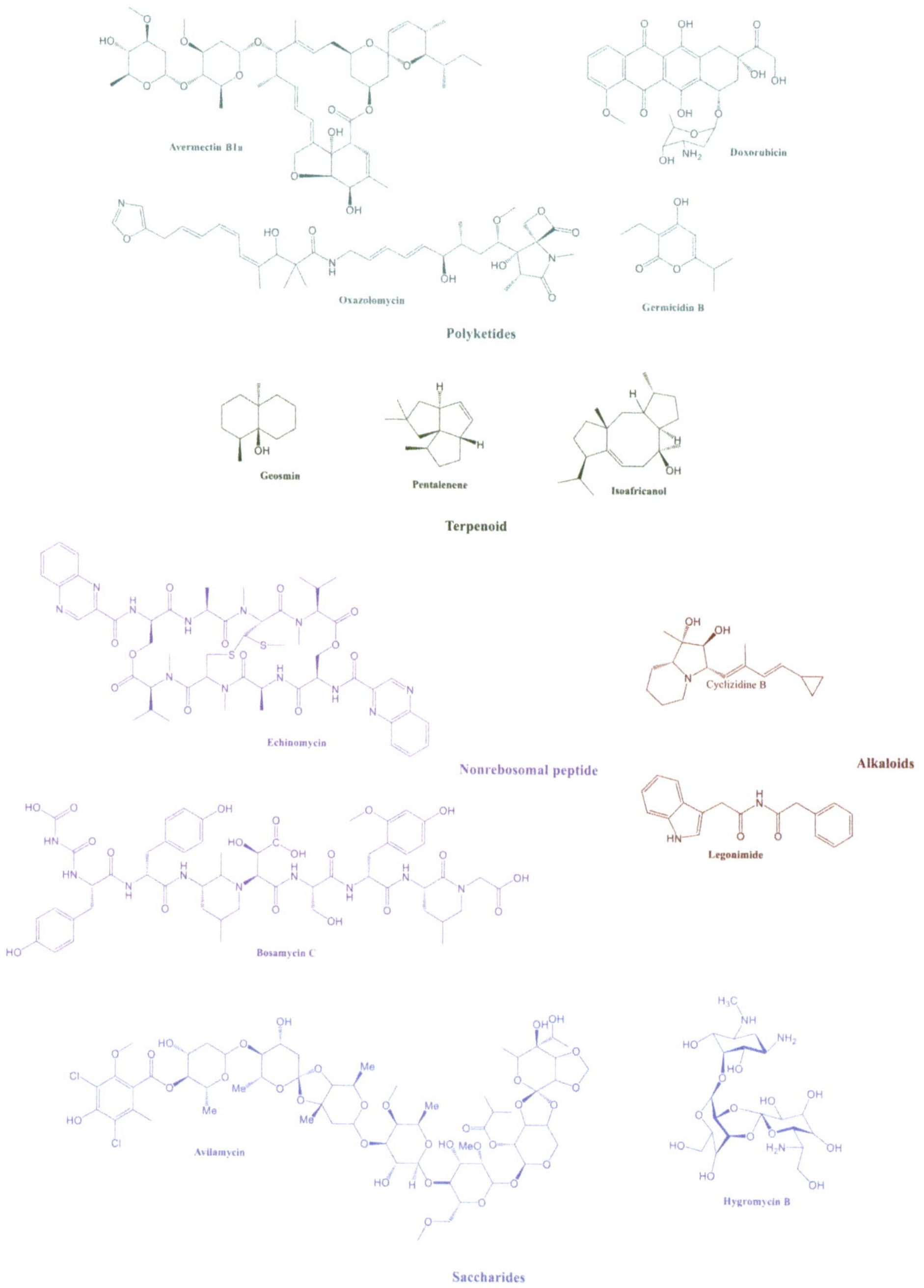

Figure 1. Natural secondary metabolites synthesized in *Streptomyces*: avermectin, pentalenene from *Streptomyces avermitilis* [2,7], doxorubicin and geosmin from *Streptomyces peucetius* ATCC 27952 [8,9], oxazolomycin from *Streptomyces albus* JA3453 [10], germicidin from *Streptomyces lividans* TK24 [11], isoafricanol from *Streptomyces malaysiensis* [4], echinomycin from *Streptomyces lasaliensis* ATCC 35851 [12], bosamycin from *Streptomyces* sp. 120454 [3], cyclizidine from *Streptomyces* sp. HNA39 [13], legonimide from *Streptomyces* sp. CT37 [6], avilamycin from *Streptomyces viridochromogenes* Tü57 [5,14], and hygromycin from *Streptomyces hygroscopicus* [15].

Previous studies showed that natural products have greater chemical diversity and a more extensive chemical space than synthetic compounds [16]. The structure of natural products is also more complex than the structure of synthetic compounds. Therefore, natural products are useful templates for developing relevant bioactive compound classes [17]. Many chemical synthetic drugs cause several side effects when curing disease. In contrast, natural products are produced by living cells and they may avoid the side effects caused by synthetic drugs [18]. Furthermore, bacterial resistance emerges as a current problem for treatment; therefore, discovering novel drugs is required and the exploration of natural products is one solution [19].

Genome sequences and genome mining revealed that *Streptomyces* contains many secondary metabolite biosynthetic gene clusters (BGCs); however, some BGCs of native strains are silent or cryptic under laboratory conditions [20]. The heterologous expression is a powerful strategy for discovering uncharacterized BGCs or increasing the production of many secondary metabolites BGCs. For example, hybrubins were successfully explored via the heterologous expression of the *hbn* BGC from *Streptomyces variabilis* Snt24 in *S. lividans* SBT5 [21]. The yield of pikromycin from the heterologous expression in *S. lividans* TK21 showed a 2.1-fold increase compared to the production of pikromycin from the parental *Streptomyces venezuelae* ATCC 15439 [22].

The BGCs vary in size from a small size to large sizes, such as 12 kb for the kocurin BGC [23], 60 kb for the pikromycin BGC [22], 65 kb for the daptomycin BGC [24], 90 kb for the meridamycin BGC [25], and 141 kb for the vancoresmycin BGC [26]. Depending on the size of the BGC, several cloning methods and vector systems have been developed for the heterologous expression of BGCs. Many techniques are applied for cloning target BGCs, such as the Gibson assembly [27], site-specific recombination-based tandem assembly (SSRTA) [28], the bacterial artificial chromosomal (BAC) system [2], and the transformation-associated recombination (TAR) system [29]. The selection of hosts and other elements for heterologous expression depends on the targeted products.

In this review, we provide insight into the systematic approach for the heterologous production of secondary metabolites from *Streptomyces* in *Streptomyces* and other hosts, such as *Escherichia coli*, *Pseudomonas putida*, *Saccharomyces cerevisiae*, *Bacillus subtilis*, *Corynebacterium glutamicum*, and *Rhodococcus erythropolis*. The tools for genome analysis assist with precisely identifying and cloning BGCs. This review also presents the process for selecting hosts and developing genetic control elements for heterologous expression.

2. In Silico Prediction of BGCs

Sequence analysis provides opportunities to produce novel secondary metabolites based on their genetic potential. The BGCs are chosen for their heterologous expression of the natural products by using bioinformatics tools. Nowadays, various bioinformatics tools are developed for analyzing the BGCs, such as antiSMASH [30], NaPDoS [31,32], NRPSpredictor [33], ClusterFinder [34], PRISM [35], ClustScan [36], NP.searcher [37], EvoMining [38], ARTS [39], SBSPKS [40], BAGEL4 [41], RiPPER [42], NeuRiPP [43], RippMiner [44], and RODEO [45].

AntiSMASH, PRISM, and Clusterfinder are widely used to detect the secondary metabolite gene clusters. They are powerful tools for identifying the number of putative BGCs. These tools determine the BGCs of polyketide synthase (PKS), nonribosomal peptide synthase (NRPS), hybrid, ribosomally synthesized and post-translationally modified peptide (RiPP), terpene, melanin, and siderophore [30,34,35]. For example, antiSMASH was used to identify the genome of *Streptomyces actuosus* ATCC 25421, and the avermipeptin BGC achieved from that analysis was successfully expressed in *S. lividans* TK24 [46].

SBSPKS, NaPDoS, ClustScan, and NP.searcher focus on nonribosomal peptide and polyketide biosynthetic pathways. SBSPKSv2 is a powerful web server. It facilitates the analysis and identification of not only NRPS/PKS domains but also other clusters with similar open reading frames (ORFs); three-dimensional modeling; chemical structure similarities; a simplified molecular-input line-entry system (SMILES) for starter, extender, intermediates, and final secondary metabolites; specificity prediction [40]. NaPDoS predicts and analyzes ketosynthase and the condensation domain from deoxyribonucleic acid (DNA) and protein sequences based on the domain phylogeny. This tool contributes to assessing the secondary metabolite gene diversity [31]. ClustScan and NP.searcher also predict NRPSs, PKSs, and hybrid NRPSs/PKSs. ClustScan can predict linear and cyclical structures of polyketides, while NP.searcher can predict the putative structure of both non-ribosomal peptides and polyketides. NRPSpredictor2 only works on NRPS to predict specific substrates of adenylation domains based on transductive support vector machines [33].

In contrast to PKS and NRPS BGCs, some RiPPs do not exhibit common genetic features for identification, where the gene encoding precursors are small in size and easily ignored [42]. Many bioinformatics tools were developed for predicting the precursor peptide sequences of RiPPs, such as BAGEL, RODEO, RiPPMiner, RiPPER, and NeuRiPP. BAGEL also enables the analysis of ORF predictions and facilitates discovering novel classes of peptides, and it is used to analyze small gene encoding for RiPPs [41]. RODEO mainly focuses on exploring certain RiPP classes [45]. RiPPMiner is an informatics tool for identifying the correct crosslink sequence in a core peptide of RiPPs [44]. RiPPER and NeuRiPP are new tools for identifying an RiPP precursor peptide family. The genes encoding for RiPP precursors are often missed in genome analysis. These tools are advantageous for predicting the novel RiPP classes because they identify precursor peptide events of unknown homology or novel RiPP structural classes [47].

Minimum Information about a BGC (MIBiG) is the global repository of microbial BGCs, which provides another useful resource for the accurate prediction and analysis of *Streptomyces* BGCs. MIBiG was established in 2015 with 1170 known BGCs; up to now, MIBiG contains 1923 known secondary metabolite BGCs [48,49]. These resources provide new opportunities to discover *Streptomyces* natural products using the bottom-up approach [1].

In addition to the exploration of unidentified or less studied resources, the metabolomics-based approach is based on reference compounds as potential biomarkers or the identification of molecular fingerprints that determine bioactivities has recently been used to identify new chemical diversities with versatile activities. The top-down approach focuses on the production of natural products without the knowledge of BGCs. In contrast, the bottom-up approach first focuses on BGCs and then uses genetic techniques to produce natural products from target BGCs [50]. Molecular networking was introduced in 2012 [51] and was built using simplified tandem mass spectrometry (MS/MS) data for molecular network analysis. The mass spectrometry data of putative compounds can reveal similarities in the fragmentation patterns of a single compound using the Global Natural Products Search (GNPS) website. The putative structure of molecules is closely linked to the structure of compounds found in the GNPS. The GNPS is a powerful tool for the discovery of new drugs and metabolites [52].

Molecular networking entails the computational analysis of metabolic fragments. It enables the rapid identification of compounds in a broad diversity of unknown natural products related to potential drugs, where even the compounds are in the low titer or difficult to separate. Based on the MS/MS data, bioactive natural products were explored by visualizing and organizing them using untargeted mass spectrometry. The advantage of molecular networking is the possibility of detecting analog compounds and identifying the biosynthetic pathways. Molecular networking permits finding unknown compounds and connecting "standard networks" for the annotation of new compounds [52]. For example, a saponin glycoside from the British Bluebell, a map of matlystatin congeners in various media, and the proposal of bottromycin pathway were explored via molecular networking [53–55].

3. Selecting a Suitable Host for Heterologous *Streptomyces* Products

3.1. Streptomyces

Heterologous expression is a promising tool for exploring compounds from silent BGCs or increasing the titers of natural products. *Streptomyces* produces numerous biologically active compounds with various effects, including agrobiological and pharmacologic agents. Many *Streptomyces* strains grow slowly, are difficult to genetically manipulate, and can produce many types of compounds. Thus, to select strains for the activation or enhancement of target products, host strains should contain some attractive features, such as fast growth, a well-studied secondary metabolism, and being easily genetically manipulated. Based on this, some *Streptomyces* hosts have been used for heterologous expression, such as *S. albus*, *Streptomyces ambofaciens*, *S. avermitilis*, *Streptomyces coelicolor*, *Streptomyces fradiae*, *Streptomyces roseosporus*, *Streptomyces toyocaensis*, *S. venezuelae*, and *S. lividans* [56]. Therefore, heterologous expression in *Streptomyces* enables the activation or enhancement of the production of secondary metabolic products. For example, albucidin was produced in *S. albus* Del14 from *S. albus* subsp. *chlorinus* NRRL B-24108 [57], kinamycin from *Streptomyces galtieri* Sgt26 was successfully biosynthesized in *S. albus* J1074 [58], and thaxtomins were successfully produced at a 10-fold higher titer in *S. albus* J1074 than in native *Streptomyces scabiei* [59].

Streptomyces host can also be used for the production of metabolites from rare *Actinomycetes*. Some rare *Actinomycetes* exhibit slow growth and carry a smaller population than *Streptomyces*, which is also considered a bioactive compound source. Some compounds were discovered from rare *Actinomycetes* via heterologous expression in a *Streptomyces* host, such as chuangxinmycin (from *Actinoplanes*) [60], tunicamycin (from *Actinosynnema*) [61], huimycin (from *Kutzneria*) [62], thiocoraline (from *Micromonospora*) [63], nargenicin (from *Nocardia*) [64], kocurin (from *Nonomuraea*) [65], GE2270 (from *Planobispora*) [66], shinorine (from *Pseudonocardia*) [67], taromycin B (from *Saccharomonospora*) [68], and A201A (from *Saccharothrix*) [69].

3.2. Other Hosts

Naturally, *Streptomyces* exhibits a slow growth rate (about 0.16–0.25 h^{-1}) [70], and their species are many times more difficult to genetically engineer. In most cases, the BGCs corresponding to potent compounds are silent/cryptic or the production titers are only produced in a detectable range. The BGCs derived from the native organism can be introduced into surrogate hosts to overcome the undesirable biological features of native producers. *Streptomyces* compounds were expressed heterologously in other hosts, such as *E. coli*, *P. putida*, *S. cerevisiae*, *B. subtilis*, *C. glutamicum*, and *R. erythropolis*.

S. cerevisiae is a generally recognized as safe (GRAS) organism with DNA recombinant stability, easy genome engineering, well-known genomics and proteomics, and a cell factory for producing natural products. Engineered *S. cerevisiae* was successfully used to produce a high level of natural products from other sources, for example, artemisinic acid up to 100 mg/L [71], tetrahydrocannabinolic acid up to 8.0 mg/L, cannabidiolic acid, and tetrahydrocannabivarinic acid up to 4.8 mg/L [72]. Engineered *S. cerevisiae* also produced some natural products from *Streptomyces* with high titers, such as gamma aminobutyric acid up to 62.6 g/L [73] and cannabigerolic acid up to 299.8 mg/L [74].

B. subtilis exhibits a rapid growth rate of 0.5 h^{-1} [75], which facilitates its use for heterologous expression. In previous work, *B. subtilis* was successfully used for the heterologous expression of secondary metabolites from *Streptomyces*, such as 6-deoxyerythronolide B from *Saccharopolyspora erythraea* [76] and enniatin from *Fusarium oxysporum* [77]. Therefore, *B. subtilis* is a promising host for the heterologous expression of BGCs from *Streptomyces*.

E. coli also exhibits fast growth with a growth rate of 2–3 h^{-1} [78] and it is an easy target for genetic manipulation. Its native biochemistry, physiology, and metabolomics are extensively understood [79]. Natural products, such as isoprenoids, non-ribosomal peptides, alkaloids, and polyketides, were successfully biosynthesized in *E. coli* [79]. However, the disadvantage of using *E. coli* for the heterologous expression of natural products from *Streptomyces* is the lack of building blocks and phosphopantetheinyl moiety in *E. coli*. Many complex enzymes were expressed in insoluble form in *E. coli*. *E. coli* lacks some specific building blocks, such as propionyl-CoA, methylmalonyl-CoA, and benzoyl-CoA, which can be supplied to the culture [80] or an engineered strain can be used that could provide the necessary precursors [81]. Many factors should be considered when doing a heterologous expression in *E. coli*, such as cofactors, the functional generation of the holoenzymes, the copy number of BGCs, promoters, native regulatory elements, and transcription factors [82,83]. Therefore, for the heterologous expression of BGCs, the specific genes must be introduced or re-engineered to produce attractive products. On the other hand, the BGCs were re-engineered for expression in *E. coli*. For example, type II PKS systems from *Streptomyces* contains an insoluble ketosynthase chain length factor in *E. coli*. Cummings and co-workers developed a plug-and-play production system to express type II PKSs for generating aromatic polyketides. An iterative monomodular type I was used for minimal PKS (mPKS) components expressing type II PKS clusters in *E. coli* using mPKS. They successfully used *Photorhabdus luminescens* mPKS combined with tailored enzymes from other organisms to produce anthraquinones, dianthrones, and benzoisochromanequinones [84]. Liu et al. also successfully used mPKS to generate TW95C and dehydrorabelomycin in *E. coli* [85].

C. glutamicum is a GRAS strain, generates amino acids at a large scale, and highly resists aromatic compounds. *C. glutamicum* belongs to *Actinomycetes* and is closely related to *Streptomyces*. Several investigators reported that engineered *Corynebacterium* could biosynthesize target compounds, such as alcohols, aromatic compounds, and other secondary metabolites. *Corynebacterium* possesses endogenous 4′-phosphopantetheinyl transferase (PPTase), PptA$_{Cg}$, which can be utilized in heterologous expression of type I PKS, and NPRS from *Streptomyces* [86]. For example, roseoflavin from *Streptomyces davaonensis* was produced in *C. glutamicum* via the heterologous expression of its BGCs [87].

P. putida is also a GRAS strain with a rapid growth rate of 0.3–0.6 h^{-1} [88]. It has a high GC content in its genome and a wide range of PPTases, similar to *Streptomyces* [89]. *P. putida* is a useful host for natural products because many natural products have been produced in this strain using heterologous expression [90]. For instance, flaviolin, monorhamnolipid, zeaxanthin, and prodigiosin were produced via the heterologous expression of their BGCs from *Streptomyces griseus*, *Pseudomonas aeruginosa*, *Pantoea ananatis*, and *Serratia marcescens* in *P. putida*, respectively [90,91]. However, *P. putida* requires introducing intracellular substrates and other features for the expression of target products, such as methylmalonyl-CoA [92] and coumaroyl-CoA [93].

R. erythropolis is an aerobic, non-motile, and Gram-positive bacteria with a high GC content in its genome, similar to the *Streptomyces* genome. Kasuga et al. demonstrated that *Rhodococcus* is a candidate for the expression of BGCs from *Streptomyces* [94]. For example, kasugamycin from *Streptomyces kasugaensis* was produced in *R. erythropolis* [94]. The processing of heterologous in different hosts is shown in Figure 2.

From previous experiments, *Streptomyces* spp., *S. cerevisiae*, *B. subtilis*, *E. coli*, *C. glutamicum*, *P. putida*, and *R. erythropolis* were shown to be hosts for the heterologous expression of BGCs from *Streptomyces*. Some natural products, which were produced by heterologous expression in *Streptomyces* and other hosts, are listed in Table 1. The successful heterologous expression of BGCs depends on many elements, such as regulators, promoters, terminators, ribosome binding sites (RBSs), riboswitches, and transfer ribonucleic acids (tRNAs). However, to achieve this, the host should be engineered by deleting unwanted BGCs, inducing transcriptional terminators, using riboswitches, engineering tRNAs, and deleting or inserting regulators.

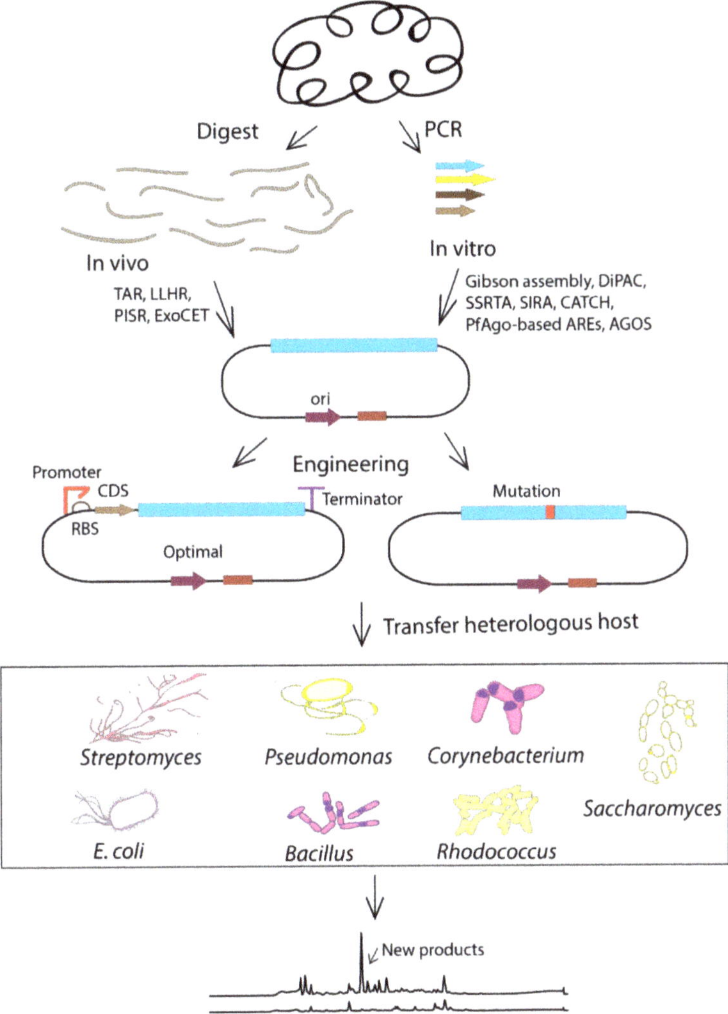

Figure 2. Processing of heterologous biosynthetic gene clusters. RBS, ribosome binding site; CDS, coding sequence; TAR, transformation-associated recombination; LLHR, linear-plus-linear homologous recombination-mediated recombineering; PISR, phage φBT1 integrase-mediated site-specific recombination; ExoCET, exonuclease combined with RecET recombination; DiPAC, direct pathway cloning; SSRTA, site-specific recombination-based tandem assembly; SIRA, serine integrase recombinational assembly; CATCH, Cas9-assisted targeting of chromosome segments; PfAgo-based AREs, *Pyrococcus furiosus* Argonaute-protein based artificial restriction enzymes; AGOS, artificial gene operon assembly system; PCR, polymerase chain reaction.

Table 1. List of compounds synthesized heterologously.

Native Strain	Heterologous Host	Method Clone	Products
Streptomyces leeuwenhoekii C34[T]	*S. coelicolor* M1152 and M1154	PCR, cloning to pIJ10257	Leepeptin [95]
S. variabilis Snt24	*S. lividans* SBT5	BAC clone	Tetramic acid [21]
Streptomyces koyangensis SCSIO 5802	*S. coelicolor* M1152	Phage-P1-derived artificial chromosome (PAC) library	Neoabyssomicin, abyssomicin [96]
Streptomyces sp. Acta1362	*S. albus* J1074	TAR system	Grecocycline [97]
S. venezuelae	*S. lividans* and *S. coelicolor*	BAC cloning	Pikromycin [22]
S. galtieri Sgt26	*S. albus* J1074	BAC cloning	Kinamycin [58]
S. albus subsp. *chlorinus* NRRL B-24108	*S. albus* Del14, *S. lividans* TK24	BAC clone	Albucidin [57]
S. erythraea DSM 40517	*S. coelicolor* M1152 and M1154	Direct pathway cloning (DiPaC)	Erythromycin [98]
Streptomyces tsusimaensis ATCC 15141	*E. coli*	Co-expression genes	Valinomycin [99]
S. erythraea	*B. subtilis*	PCR, cloning	6-deoxyerythronolide B [76]
S. davaonensis	*C. glutamicum*	PCR, cloning	Roseoflavin [87]
S. griseus	*P. putida*	PCR, cloning	Flaviolin [91]
S. kasugaensis	*R. erythropolis*	PCR, cloning	Kasugamycin [94]
Actinosynnema mirum DSM 43827	*S. avermitilis* SUK22	PCR, cloning	Mycosporine-glycine-alanine [67]
Kocuria rosea s17	*S. coelicolor* M1146, *S.* sp. s120	Gibson assembly, integration vector pSET152	Kocurin [23]
Amycolatopsis sp. DEM30355	*S. coelicolor* M1152	PAC library	Vancoresmycin [26]
Myxococcus xanthus DK1622	*S. albus* Del14	PCR and subcloned	Cittilins [100]
Actinoplanes tsinanensis CPCC 200056	*S. coelicolor* M1146	DNA assembler by yeast	Chuangxinmycin [60]
Kutzneria albida DSM 43870	*S. albus* Del14	BAC vector	Huimycin [62]
Planobisporarosea ATCC 53733	*S.coelicolor* M1146	Cosmid vector	GE2270 [66]
A. mirum DSM 43827	*S. avermitilis* SUKA22	In vivo by phage λ-Red recombining system	Shinorine, porphyra-334, mycosporine-glycine, and mycosporine-glycine-alanine [67]
Kitasatospora setae KM-6054	*Streptomyces lohii* JCM 14114, *S. griseus* DSM 2608	BAC cloning	Bafilomycins A1, C1, and B1 (setamycin) [101]

4. Host Engineering Approaches

4.1. Host Cleaning

Analysis of the *Streptomyces* genome has demonstrated that its genomes contain many BGCs; therefore, it is required that the genome be minimized by deleting unnecessary BGCs. The advantages of a minimal genome include faster growth of the host strains, a low background of native secondary metabolites, absence of biological activity, and a high chance of successful heterologous expression of BGCs [102]. Clean host strains are used for the characterization and modification of the BGCs, activation of the silent BGCs, and increasing the titer of target compounds.

Streptomyces host strains were engineered by deleting or repressing the expression of unwanted BGCs, and the resulting engineered strains were employed as hosts for the integration of heterologous BGCs, leading to the production of the target products [103]. Several methods are available to develop clean hosts, such as a PCR-targeted system [104], site-specific recombination using Cre-*loxP* [105,106], meganuclease I-SceI [107], and clustered regularly interspaced short palindromic repeat (CRISPR)/Cas genome editing tools [108]. A PCR-targeted system is a useful tool for gene knockout. This system contains a selectable gene, two *FRT* or *loxP* sites with 40–50 bp of homologous extension flank at both ends, and a λ recombinant system. Xu et al. successfully deleted 10 regulator genes in *S. coelicolor* by using a PCR-targeted method [109]. Site-specific recombination using Cre/*loxP* is applied in the knockout and/or insertion of huge BGC regions in many *Streptomyces* spp. The Cre-*loxP* system needs a double crossover to introduce the *loxP* site into the genome and the expression of Cre recombinase to delete target genes. The Cre-loxP system was successfully used to delete 1.4 Mb of *S. avermitilis*'s genome [110]. Meganuclease I-SceI is an endonucleave from *S. cerevisiae* that recognizes an 18 bp unique sequence and cleaves DNA for engineering a genome with high efficiency and multiple deletions [111]. This method was successfully applied to delete the red-pigmented undecylprodiginine BGC in *S. coelicolor* M1141 [111]. Recently, the CRISPR/Cas system of *Streptococcus pyogenes* (*Sp*Cas9) is a vital tool in genome engineering that is widely applied in different strains [112]. It has been developed and used to clean many *Streptomyces* hosts, such as *S. coelicolor*, *S. lividans*, and *S. albus* [113,114]. However, *Sp*Cas9 does not properly work in some *Streptomyces* strains [115,116]; therefore, the *Fn*Cas12a system (from *Francisella novicida* U112), *fn*Cpf1 (from *Francisella tularensis* subsp. *novicida* U112 Cpf1), *sth1*Cas9 (from *Streptococcus thermophilus* CRISPR1 Cas9), and *sa*Cas9 (from *Staphylococcus aureus* Cas9) were developed for genome editing *Streptomyces* strains [116,117]. For example, the *Fn*Cas12a system was used to disrupt 128 kb of chromosome of *S. hygroscopicus* [117].

Additionally, the specific integration ΦC31 *attB* loci sites were introduced into a clean host to facilitate the integration of exogenous biosynthetic clusters into the chromosomes [11,102]. For example, in the *S. albus* Del14 strain, fifteen BGCs were deleted, and one, two, and four ΦC31 *attB* sites were introduced to make *S. albus* B2P1, B3P1, and B4, respectively [102]. *S. lividans* ΔYA9, ΔYA10, and ΔYA11 were achieved by deleting 9, 10, and 11 BGCs and one, two, and three additional introduced *attB* sites, respectively [11].

4.2. Transcription Terminators

Besides the regulator and promoter, the transcriptional terminator is another essential element that affects the gene expression level. The transcription terminator includes protein-dependent factors, such as *rho*, *nusA*, and *tau*, and protein-independent factors structured by a stem-loop form in messenger ribonucleic acid (mRNA) [118]. Based on the heterologous expression level of glucuronidase, Horbal et al. found that the tt_{sbiB} terminator is the strongest compared with U, V, T4 lang, and T4 kurz terminators [119]. Curran et al. reported that terminators impact the mRNA half-life and can be used to modulate gene expression [120]. The terminators carry larger stem-loop structures to increase the efficiency of the mRNA released. Inverted repeat sequences play a significant role in the stability of the transcription by a stem-loop structure. The *Streptomyces* genome, with its high GC, requires a long terminator and a powerful stem-loop structure to increase the transcription termination efficiency [118]. For example, the *TD1* terminator from *B. subtilis* was successfully introduced into *S. lividans* [121]. Many transcription terminator sequences were found in several genes, such as *aph*, *vio*, *tsr*, and *hyg*. The terminator of the *aph* gene was successfully utilized for the heterologous expression of the interferon α2 gene from humans in *S. lividans* [118].

4.3. Riboswitches

A riboswitch is a non-coding RNA structure, which includes an aptamer domain and an expression platform. It can regulate the expression of the genes at both the transcription and translation levels by interacting with mRNA [122]. Rudolph et al. found several theophylline-dependent riboswitches to induce gene expression in *S. coelicolor*. Riboswitch E* was the best-activated regulator for the expression of agarase gene *dagA* [123]. The heterologous expression of *btm* BGC from *Streptomyces scabies*, together with a gene from yeast in *S. coelicolor*, lead to a 120-fold increase in bottromycin production [124]. The cumate and theophylline dual-riboswitch control system were expressed in *S. albus*. They could control the production of pamamycin [125].

4.4. tRNA Engineering

Studies on tRNA have demonstrated the presence of 30–40 different tRNAs in bacteria. The tRNAs carry amino acids to build proteins. A few specific aminoacyl-tRNAs also participate in the biosynthesis of natural products, such as non-ribosomal peptides, tRNA-dependent cyclodipeptides, diketopiperazines, and cyclodipeptides, by using aminoacyl-tRNA to form peptide bonds. Gondry et al. reported that the active form of AlbC from

Streptomyces noursei required aminoacyl-tRNA as a substrate to catalyze the cyclodipeptide formation [126]. Some BGCs required specific tRNAs to build the diketopiperazine precursors for the biosynthesis of natural compounds. For example, the BGC of bicyclomycin from *Streptomyces sapporonensis* ATCC 21532 was expressed in *E. coli*, suggesting that Leu-tRNA^{Leu} and Ile-tRNA^{Ile} are precursors to initiate the biosynthesis of bicyclomycin [127].

The biosynthesis of some compounds occurs in different ways, such as the C4 (Shemin pathway) or the C5 pathway. For example, the biosynthesis of tetrapyrroles required 5-aminolevulinic acid as the precursor, which is generated via the condensation of glycine and succinyl-CoA in the C4 pathway in animal mitochondria and alpha-proteobacteria, whereas glutamyl-tRNA is needed in the C5 pathway in all other cells (chloroplasts) [128]. The heterologous expression of the gene encoding aminoacyl-tRNA led to increasing the titer of secondary metabolites in expression hosts. For example, the overexpression of *hemA1* encoding glutamyl-tRNA in *S. roseosporus* HCCB10043 and *Streptomyces gilvosporeus* HCCB 13086 led to a 1.4-fold increase in daptomycin and a 1.64-fold increase in natamycin, respectively [129]. However, specific tRNAs also indirectly affect the morphology or the yield of secondary metabolites depending on the need for a specific tRNA as an adaptor molecule for the biosynthesis of polypeptide chains.

The *bldA* gene from *S. coelicolor* encodes specific tRNA$^{Leu}_{UUA}$, which recognizes the rare UUA codon in *Streptomyces*. It can control the expression of phenotypic *carB* gene in *S. coelicolor* [130] and the heterologous expression of *pur* BGC from *Streptomyces alboniger* in *S. lividans* [131]. The interaction between bldA-tRNA and DnrO led to increasing the daunorubicin production in *S. peucetius* [8]. Some natural products were produced by the heterologous expression of BGCs involving tRNA synthases genes; for instance, the valanimycin BGC consists of the gene *vlmL* encoding Ser-tRNA^{Ser} that produces seryl-tRNA, which may take part in the biosynthetic pathway of valanimycin [132].

4.5. Regulators

Regulators positively or negatively affect the expression of target BGCs by directly or indirectly regulating the transcription of target genes. *Streptomyces* contains many transcriptional regulator families, such as TetR, LuxR, MarR, and ArfR. Each family has different functions. The TetR family regulator is the most popular in *Streptomyces* and controls the expression of multiple genes in secondary metabolite BGCs. For example, the expression of GdmRIII led to increasing the production of geldanamycin while decreasing the production of elaiophylin in *Streptomyces autolyticus* CGMCC0516 [133]. AccR bound to a 16-nucleotide palindromic binding motif in the promoter of target genes resulted in the decreasing biosynthesis of malonyl-CoA and methylmalonyl-CoA in *S. avermitilis*. The production of avermectin was enhanced by 14.5% when *accR* gene was deleted [134]. The LuxR family regulator is a pathway-specific transcriptional regulator of BGCs, including AniF-activated anisomycin production [135]. The MarR family regulators contain winged helix-turn-helix DNA binding domains, which can work like transcriptional repressors or activators. Gou et al. indicated that SAV4189, a MarR-family regulator, acts as an activator of avermectin production in *S. avermitilis*. They also produced the heterologous expression of *sav_4189* in *S. coelicolor*, resulting in enhancing the antibiotic production of this strain [136]. Peng et al. engineered *S. lividans* STB5 by integrating global regulator genes, such as *mdfA$_{co}$*, *lmrA$_{co}$*, *nusG$_{sc}$*, and *afsRS$_{cla}$*, and deleting a transcription regulator gene, namely, *wblA$_{sl}$*, to achieve *S. lividans* LJ1018. The production of murayaquinone, hybrubins, actinorhodin, dehydrorabelomycin, and actinomycin D achieved via heterologous expression in *S. lividans* LJ1088 were increased 96-, 29-, 23.3-, 12.7-, and 3.5-fold compared with the production achieved via heterologous expression in *S. lividans* STB5, respectively [137]. Therefore, for improving the heterologous expression level of BGCs, host strains should be engineered by deleting repressor genes and integrating activator genes.

5. Strategies for the Construction of a Biosynthetic Gene Cluster

5.1. Vectors and Cloning Methods

BGCs contain many genes, which encode one or some secondary metabolites. Some cloning methods were used to clone large BGCs. Traditionally based on a library, the large genomic DNA has been digested and ligated into high-capacity vectors, such as cosmids, fosmids, or BACs. Cosmids can insert DNA fragments ranging in size between 32 kb and 45 kb based on the bacteriophage vector [138]. Fosmids, which are similar to cosmids but contain an F-factor to control the copy number of vectors, can insert DNA fragments up to 100 kb in size [139]. For example, the *mat* gene cluster used a fosmid vector to produce deshydroxymatlystatins via heterologous expression in *S. coelicolor* and *S. albus* [53]. The BAC vector contains the F-factor, multiple cloning sites, and is capable of inserting DNA that is approximately 490 kb in size [140]. PAC vector could insert DNA that was between 100 and 300 kb in size [141].

BGCs can be cloned in a directed manner by employing in vivo cloning methods, such as TAR, LLHR, ExoCET, and PISR. TAR constructs target BGC from a digested genome to an expression vector using yeast. This technique successfully cloned a DNA fragment with a size up to 250 kb [142]. As the recombinant vector was constructed in yeast, maintained in *E. coli*, and expressed in a host, it has to carry elements from each of the yeast, *E. coli*, and the host. For example, the 67 kb BGC of taromycin A was successfully expressed from *Saccharomonospora* sp. CNQ-490 in *S. coelicolor* using the TAR cloning system [143]. The LLHR used *E. coli* to clone a large gene cluster up to 100 kb to an expression vector based on the activity of RecE/RecT. Fu et al. successfully used this method to clone and express PKS-NRPS BGCs from *P. luminescens* in *E. coli* [144]. Recently, ExoCET was used to directly clone large selected regions (up to 106 kb) into a BAC vector in a single step. ExoCET combines an in vitro exonuclease of T4 polymerase with a full-length RecE/RecT to clone homology regions in *E. coli*. The efficiency of vector construction depends on the position of the homologous sequence in the digested genome. For increasing the efficiency of this method, Cas9 was applied to digest the genome. For instance, the salinomycin gene cluster was cloned to a BAC vector by using the ExoCET method [145]. PISR was used to clone and delete large BGCs directly from the *Streptomyces* genome based on the activity of the φBT1 integrase. Du et al. applied this technique to delete and clone actinorhodin, nikkomycin, and gougerotin from *S. coelicolor*, *Streptomyces ansochromogene*, and *Streptomyces graminearus*, respectively [146].

Besides in vivo cloning, some BGCs were also successfully cloned using PCR gene and DNA assembly in different ways. In vitro cloning includes DNA assembly methods, such as Gibson assembly, DiPAC, PfAgo-based AREs, SSRTA, SIRA, and CATCH. Gibson assembly was applied to clone DNA fragments with a size up to 14 kb, such as bicyclomycin [147], kocurin [23], and lasso peptides [148]. Gibson assembly needs DNA polymerases, exonucleases, and DNA ligases to assemble BGCs with vectors. DiPAC was applied to clone DNA fragments amplified from PCR based on HiFi DNA assembly. DiPAC is suitable for cloning short BGCs, with a capacity of up to 23 kb, such as *apt* [98], sodorifen [149], and hapalosin [150] BGCs. SIRA used Φ31 integrase and a pair of different orthogonal *attP/attB* recombination sites to form *attL/attR* for a recombinational assembly and cloning strategy [151]. For example, Gao et al. successfully cloned and expressed erythromycin BGC using SIRA [152]. The SSRTA method is based on the operation of *Streptomyces* φBT1 integrase and *attB/attP* recombination sites for the tandem assembly of multiple DNA fragments. Some BGCs were cloned using this method, such as epothilone [28] and lycopene BGCs [153]. PfAgo-based AREs utilize PfAgo to create artificial restriction enzyme sites, which possess the ability to recognize and cleave a specific DNA sequence. Therefore, PfAgo-based AREs can become a robust tool for cloning BGCs because PfAgo can induce the unique restriction sites, which are often limited in the target DNA [154]. The AGOS is a plug-and-play method based on a SuperCos1 and Red/ET-mediated assembly that introduces unique restriction sites to target BGCs. From this target, BGCs can be cloned by using the restriction enzymes. AGOS is a power tool for refactoring and cloning target

BGCs [155]. CATCH assembles DNA target sequences that are digested using Cas9 from the bacterial chromosomes with vectors via Gibson assembly [156]. In vivo cloning methods are powerful tools for cloning large-size BGCs (up to 250 kb when using the TAR cloning method). However, the efficiency of these tools is not so high, from 8.3 to 58% [143,145]. In vitro cloning methods are useful for cloning multiple DNA fragments with very high cloning efficiencies from 87% up to almost 100%; however, these methods are usually used for cloning the small and medium DNA sizes (up to 100 kb) [98,147,152,153]. The cloning and expression protocol for the BGCs is shown in Figure 3.

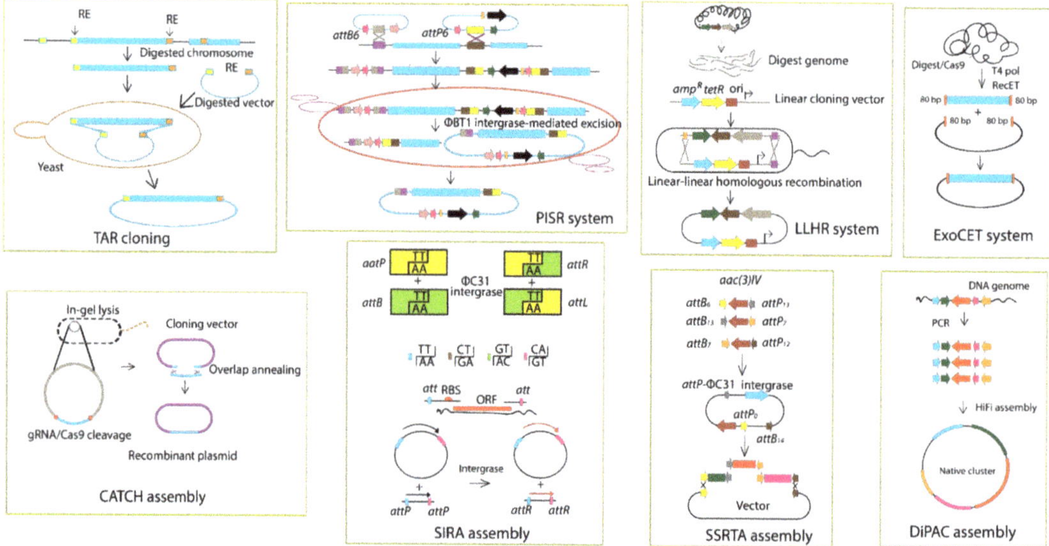

Figure 3. Schematic diagram of the BGC cloning methods. The in vivo cloning methods include TAR, PIRS, LLHR, and ExoCET systems, while the in vitro cloning methods include CATCH, SIRA, and SSRTA assembly. RE, restriction enzymes; *kanR*, kanamycin restriction gene gene; *ampR*, ampicillin restriction gene; *tetR*, tetracycline restriction enzyme; gRNA/Cas9, guide RNA/Cas9; T4 pol, T4 polymerase.

5.2. Promoter Engineering

The promoter is one of the most important factors for the heterologous expression of BGCs at the transcription level. Each host has different promoter systems. Therefore, after the engineering host, powerful and suitable promoters should be selected for developing heterologous expression systems. When the BGCs are cloned to vectors, the powerful and suitable promoter can replace the native promoter. This helps to enhance or activate the heterologous expression of BGCs.

The most suitable part of the promoter activity relies on the -10 and -35 regions. These regions determine the strength of the RNA polymerase binding. There are three types of promoters in *Streptomyces*. A type I promoter includes the -10 element (5'-TANNNT-3') and -35 element (5'-TTGNC-3') with an 18–19 nucleotide spacer length [157,158], such as the *hrdB* promoter from *S. coelicolor* [159]. In contrast, a type II promoter contains the -10 element, similar to type I, and other elements (5'-TGTC-3') with a different spacer length [160], and a type III does not display the typical -10 and -35 sequences that are present in different types [157,161].

Native promoters were used in studies for a long time, but they do not work at a wide range of transcription levels. Based on the properties of native promoters, powerful synthetic promoters were developed and selected for the expression of the target gene at a high level. Promoter engineering is performed by changing the sequence between the -10 and

−35 regions of a native promoter. Several native promoters have been studied and used to drive the expression of genes in *Streptomyces*, such as inducible promoters *tipAp* from *S. lividans* 66 [162] and *nitAp* from *Rhodococcus rhodochrous* J1 [163], and the constitutive promoters *ermEp* from *Streptomyces erythraeus* [164], *SF14p* from *Streptomyces ghanaensis* phage I19 [165], and *kasOp* from *S. coelicolor* [166]. The activity of promoter *thlM4p* from *Streptomyces chattanoogensis* L10, which was selected via a transcriptome-based screening method, was seven times stronger than the promoter *ermE*p* [167]. Many synthetic promoters were developed; among them, *ermE*p* and *kasO*p* were widely used in the gene expression in *Streptomyces*. Wang et al. demonstrated that the *kasO*p* promoter is stronger than the *SF14p* promoter, which is stronger than the *erm*p* promoter, as seen from the production level of actinorhodin when used in *S. coelicolor* [166].

5.3. RBS Tuning

Engineering the expression of genes is focused not only on the transcriptional level but also on the translation level. The protein expression strongly depends on the RBS, which controls the gene expression at the translation level. The position and sequence of the RBS affect the translational efficiency and protein expression. The experiment showed that an unsuitable RBS could reduce the gene expression to zero [119], which then affects the biosynthesis of secondary metabolite products. For identifying a suitable RBS, Farasat et al. developed a Library Calculator [168] and Na et al. introduced the RBSDesigner [169] for designing a powerful synthetic RBS sequence [168] to enhance the protein expression.

6. Conclusions

The development of genome sequencing and bioinformatics tools has greatly facilitated the discovery of BGCs. Advances in heterologous expression have resulted in the development of novel secondary metabolites. The heterologous expression of natural products depends on two crucial factors: host engineering and BGC construction. Host engineering involves choosing hosts, deleting unwanted BGCs, inducing transcriptional terminators, riboswitches, engineering tRNAs, and deleting or inserting regulators. In comparison, BGC construction involves vectors, cloning methods, promoters, and RBSs.

Many hosts were used for the heterologous expression of natural products from *Streptomyces*, such as *Streptomyces* spp., *S. cerevisiae*, *B. subtilis*, *E. coli*, *C. glutamicum*, *P. putida*, and *R. erythropolis*. For easily detecting and purifying heterologous products, unwanted secondary metabolite BGCs from host strains should be deleted. Many genetic systems were developed for genome editing, such as CRISPR/Cas, meganuclease I-SceI, site-specific recombination using Cre-loxP, and PCR-targeted systems. The transcriptional regulator and terminator are two factors that affect the transcription level. Negative regulator genes of heterologous BGCs should be deleted. Moreover, positive regulator genes and transcriptional terminator genes should be introduced in hosts to improve the heterologous expression level of natural products. Riboswitches could regulate the heterologous expression of BGCs in both transcription and translation levels. They are also crucial factors that can be used to control the heterologous expression. Furthermore, some active tRNAs are precursors for the biosynthesis of secondary metabolites. Therefore, inducing specific tRNA genes into hosts could also enhance the production achieved from the heterologous expression of BGCs.

Two cloning method classes have been applied to clone the target BGCs, namely, in vitro and in vivo cloning methods. The vectors and the cloning methods were chosen depending on the heterologous hosts and the size of the target BGCs. In vitro cloning methods are often used for cloning the small- or medium-sized attractive BGCs, while in vivo cloning methods are usually applied for cloning the large-sized target BGCs. For the activation or enhancement of the expression products, recombinant vectors are also needed for the reconstruction. Robust and suitable promoters and RBSs were used to replace the native promoters and RBSs.

In conclusion, the advances of synthetic biology and powerful genome mining techniques led to rapidly discovering and enhancing the production of natural products. Besides OSMAC, the overexpression/deletion of regulatory genes, ribosome engineering, and promoter replacement strategies that are utilized in heterologous expression provide robust strategies.

Author Contributions: Conceptualization, V.T.T.P., C.T.N. and J.K.S.; writing-original draft preparation, V.T.T.P. and C.T.N.; writing-review and editing, V.T.T.P., C.T.N., D.D., H.T.N., and T.-S.K.; supervision, J.K.S.; funding acquisition, J.K.S. All authors have read and agreed to the published version of the manuscript.

Funding: This work was supported by the National Research Foundation of Korea (NRF) grant funded by the Korea government (MEST) (NRF-2017R1A2A2A05000939.

Institutional Review Board Statement: Not applicable.

Informed Consent Statement: Not applicable.

Data Availability Statement: Not applicable.

Conflicts of Interest: The authors declare no conflict of interest.

Abbreviations

AGOS, artificial gene operon assembly system; ARE, artificial restriction enzyme; BAC, bacterial artificial chromosomal; BGC, biosynthetic gene cluster; CATCH, Cas9_Assisted Targeting of Chromosome segments; CLF, chain length factor; CRISPR, clustered regularly interspaced short palindromic repeat; DiPAC, direct pathway cloning; DNA, deoxyribonucleic acid; ExoCET, exonuclease combined with RecET recombination; GC, guanine-cytosine; GNPS, Global Natural Products Search; GRAS, generally recognized as safe; LLHR, linear-plus-linear molecule homologous recombination; MIBiG, Minimum Information about a Biosynthetic Gene Cluster; mPKS, minimal PKS; mRNA, messenger ribonucleic acid; MS/MS, tandem mass spectrometry; NRPS, nonribosomal peptide synthase; ORF, open reading flame; OSMAC, One Strain Many Compounds; PCR, polymerase chain reaction; PfAgo-based ARE, *Pyrococcus furiosus* Argonaute protein-based artificial restriction enzyme; PISR, phage φBT1 integrase-mediated site-specific recombination; PKS, polyketides synthase; PPTase, 4′-phosphopantetheinyl transferase; RBS(s), ribosome binding site(s); RiPP, ribosomally synthesized and post-translationally modified peptide; SIRA, serine integrase recombinational assembly; SMILES, simplified molecular-input line-entry system; SSRTA, site-specific recombination-based tandem assembly; TAR, transformation associated recombination; tRNA(s), transfer ribonucleic acid(s).

References

1. Nguyen, C.T.; Dhakal, D.; Pham, V.T.T.; Nguyen, H.T.; Sohng, J.K. Recent Advances in Strategies for Activation and Discovery/Characterization of Cryptic Biosynthetic Gene Clusters in *Streptomyces*. *Microorganisms* **2020**, *8*, 616. [CrossRef]
2. Deng, Q.; Zhou, L.; Luo, M.; Deng, Z.; Zhao, C. Heterologous expression of Avermectins biosynthetic gene cluster by construction of a Bacterial Artificial Chromosome library of the producers. *Synth. Syst. Biotechnol.* **2017**, *2*, 59–64. [CrossRef]
3. Xu, Z.F.; Bo, S.T.; Wang, M.J.; Shi, J.; Jiao, R.H.; Sun, Y.; Xu, Q.; Tan, R.X.; Ge, H.M. Discovery and biosynthesis of bosamycins from *Streptomyces* sp. 120454. *Chem. Sci.* **2020**, *11*, 9237–9245. [CrossRef]
4. Martín-Sánchez, L.; Singh, K.S.; Avalos, M.; Van Wezel, G.P.; Dickschat, J.S.; Garbeva, P. Phylogenomic analyses and distribution of terpene synthases among *Streptomyces*. *Beilstein J. Org. Chem.* **2019**, *15*, 1181–1193. [CrossRef]
5. McCulloch, K.M.; McCranie, E.K.; Smith, J.A.; Sarwar, M.; Mathieu, J.L.; Gitschlag, B.L.; Du, Y.; Bachmannb, B.O.; Iversona, T.M. Oxidative cyclizations in orthosomycin biosynthesis expand the known chemistry of an oxygenase superfamily. *Proc. Natl. Acad. Sci. USA* **2015**, *112*, 11547–11552. [CrossRef]
6. Fang, Q.; Maglangit, F.; Mugat, M.; Urwald, C.; Kyeremeh, K.; Deng, H. Targeted isolation of indole alkaloids from *Streptomyces* sp. CT37. *Molecules* **2020**, *25*, 1108. [CrossRef]
7. Tetzlaff, C.N.; You, Z.; Cane, D.E.; Takamatsu, S.; Omura, S.; Ikeda, H. A gene cluster for biosynthesis of the sesquiterpenoid antibiotic pentalenolactone in *Streptomyces avermitilis*. *Biochemistry* **2006**, *45*, 6179–6186. [CrossRef]
8. Pokhrel, A.R.; Chaudhary, A.K.; Nguyen, H.T.; Dhakal, D.; Le, T.T.; Shrestha, A.; Liou, K.; Sohng, J.K. Overexpression of a pathway specific negative regulator enhances production of daunorubicin in *bldA* deficient *Streptomyces peucetius* ATCC 27952. *Microbiol. Res.* **2016**, *192*, 96–102. [CrossRef]

9.   Singh, B.; Oh, T.J.; Sohng, J.K. Exploration of geosmin synthase from *Streptomyces peucetius* ATCC 27952 by deletion of doxorubicin biosynthetic gene cluster. *J. Ind. Microbiol. Biotechnol.* **2009**, *36*, 1257–1265. [CrossRef]

10. Helfrich, E.J.N.; Piel, J. Biosynthesis of polyketides by trans-AT polyketide synthases. *Nat. Prod. Rep.* **2016**, *33*, 231–316. [CrossRef]

11. Ahmed, Y.; Rebets, Y.; Estévez, M.R.; Zapp, J.; Myronovskyi, M.; Luzhetskyy, A. Engineering of *Streptomyces lividans* for heterologous expression of secondary metabolite gene clusters. *Microb. Cell Factories* **2020**, *19*, 5. [CrossRef]

12. Migita, A.; Watanabe, M.; Hirose, Y.; Watanabe, K.; Tokiwano, T.; Kinashi, H.; Oikawa, H. Identification of a gene cluster of polyether antibiotic lasalocid from *Streptomyces lasaliensis*. *Biosci. Biotechnol. Biochem.* **2009**, *73*, 169–176. [CrossRef]

13. Jiang, Y.J.; Li, J.Q.; Zhang, H.J.; Ding, W.J.; Ma, Z.J. Cyclizidine-Type Alkaloids from *Streptomyces* sp. HNA39. *J. Nat. Prod.* **2018**, *81*, 394–399. [CrossRef]

14. Arenz, S.; Juette, M.F.; Graf, M.; Nguyen, F.; Huter, P.; Polikanov, Y.S.; Blanchard, S.C.; Wilson, D.N. Structures of the orthosomycin antibiotics avilamycin and evernimicin in complex with the bacterial 70S ribosome. *Proc. Natl. Acad. Sci. USA* **2016**, *113*, 7527–7532. [CrossRef]

15. Mann, R.L.; Bromer, W.W. The Isolation of a Second Antibiotic from *Streptomyces hygroscopicus*. *J. Am. Chem. Soc.* **1958**, *80*, 2714–2716. [CrossRef]

16. Myronovskyi, M.; Luzhetskyy, A. Heterologous production of small molecules in the optimized: *Streptomyces* hosts. *Nat. Prod. Rep.* **2019**, *36*, 1281–1294. [CrossRef]

17. Lachance, H.; Wetzel, S.; Kumar, K.; Waldmann, H. Charting, navigating, and populating natural product chemical space for drug discovery. *J. Med. Chem.* **2012**, *55*, 5989–6001. [CrossRef]

18. Lahlou, M. The Success of Natural Products in Drug Discovery. *Pharmacol. Pharm.* **2013**, *4*, 17–31. [CrossRef]

19. Waldetoft, K.W.; Gurney, J.; Lachance, J.; Hoskisson, P.A.; Brown, S.P. Evolving antibiotics against resistance: A potential platform for natural product development? *mBio* **2019**, *10*, e02946-19. [CrossRef]

20. Moussa, M.; Ebrahim, W.; Bonus, M.; Gohlke, H.; Mándi, A.; Kurtán, T.; Hartmann, R.; Kalscheuer, R.; Lin, W.; Liu, Z.; et al. Co-culture of the fungus *Fusarium tricinctum* with *Streptomyces lividans* induces production of cryptic naphthoquinone dimers. *RSC Adv.* **2019**, *9*, 1491–1500. [CrossRef]

21. Zhao, Z.; Shi, T.; Xu, M.; Brock, N.L.; Zhao, Y.L.; Wang, Y.; Deng, Z.; Pang, X.; Tao, M. Hybrubins: Bipyrrole tetramic acids obtained by crosstalk between a truncated undecylprodigiosin pathway and heterologous tetramic acid biosynthetic genes. *Org. Lett.* **2016**, *18*, 572–575. [CrossRef]

22. Pyeon, H.R.; Nah, H.J.; Kang, S.H.; Choi, S.S.; Kim, E.S. Heterologous expression of pikromycin biosynthetic gene cluster using *Streptomyces* artificial chromosome system. *Microb. Cell Factories* **2017**, *16*, 96. [CrossRef]

23. Linares-Otoya, L.; Linares-Otoya, V.; Armas-Mantilla, L.; Blanco-Olano, C.; Crüsemann, M.; Ganoza-Yupanqui, M.L.; Campos-Florian, J.; König, G.M.; Schäberle, T.F. Identification and heterologous expression of the kocurin biosynthetic gene cluster. *Microbiology* **2017**, *163*, 1409–1414. [CrossRef]

24. Choi, S.; Nah, H.J.; Choi, S.; Kim, E.S. Heterologous expression of daptomycin biosynthetic gene cluster via *Streptomyces* artificial chromosome vector system. *J. Microbiol. Biotechnol.* **2019**, *29*, 1931–1937. [CrossRef]

25. Liu, H.; Jiang, H.; Haltli, B.; Kulowski, K.; Muszynska, E.; Feng, X.; Summers, M.; Young, M.; Graziani, E.; Koehn, F.; et al. Rapid Cloning and Heterologous Expression of the Meridamycin Biosynthetic Gene Cluster Using a Versatile *Escherichia coli–Streptomyces* Artificial Chromosome Vector, pSBAC. *J. Nat. Prod.* **2009**, *72*, 389–395. [CrossRef]

26. Kepplinger, B.; Morton-Laing, S.; Seistrup, K.H.; Marrs, E.C.L.; Hopkins, A.P.; Perry, J.D.; Strahl, H.; Hall, M.J.; Errington, J.; Ellis Allenby, N.E. Mode of Action and Heterologous Expression of the Natural Product Antibiotic Vancoresmycin. *ACS Chem. Biol.* **2018**, *13*, 207–214. [CrossRef]

27. Gibson, D.G.; Young, L.; Chuang, R.Y.; Venter, J.C.; Hutchison, C.A.; Smith, H.O. Enzymatic assembly of DNA molecules up to several hundred kilobases. *Nat. Methods* **2009**, *6*, 343–345. [CrossRef]

28. Zhang, L.; Zhao, G.; Ding, X. Tandem assembly of the epothilone biosynthetic gene cluster by in vitro site-specific recombination. *Sci. Rep.* **2011**, *1*, 141. [CrossRef]

29. Kim, J.H.; Feng, Z.; Bauer, J.D.; Kallifidas, D.; Calle, P.Y.; Brady, S.F. Cloning large natural product gene clusters from the environment: Piecing environmental DNA gene clusters back together with TAR. *Biopolymers* **2010**, *93*, 833–844. [CrossRef]

30. Blin, K.; Shaw, S.; Steinke, K.; Villebro, R.; Ziemert, N.; Lee, S.Y.; Medema, M.H.; Weber, T. antiSMASH 5.0: Updates to the secondary metabolite genome mining pipeline. *Nucleic Acids Res.* **2019**, *47*, W81–W87. [CrossRef]

31. Ziemert, N.; Podell, S.; Penn, K.; Badger, J.H.; Allen, E.; Jensen, P.R. The natural product domain seeker NaPDoS: A phylogeny based bioinformatic tool to classify secondary metabolite gene diversity. *PLoS ONE* **2012**, *7*, e34064. [CrossRef]

32. Ziemert, N.; Lechner, A.; Wietz, M.; Millán-Aguiñaga, N.; Chavarria, K.L.; Jensen, P.R. Diversity and evolution of secondary metabolism in the marine actinomycete genus *Salinispora*. *Proc. Natl. Acad. Sci. USA* **2014**, *111*, E1130–E1139. [CrossRef]

33. Röttig, M.; Medema, M.H.; Blin, K.; Weber, T.; Rausch, C.; Kohlbacher, O. NRPSpredictor2—A web server for predicting NRPS adenylation domain specificity. *Nucleic Acids Res.* **2011**, *39*, 362–367. [CrossRef]

34. Cimermancic, P.; Medema, M.H.; Claesen, J.; Kurita, K.; Wieland Brown, L.C.; Mavrommatis, K.; Pati, A.; Godfrey, P.A.; Koehrsen, M.; Clardy, J.; et al. Insights into secondary metabolism from a global analysis of prokaryotic biosynthetic gene clusters. *Cell* **2014**, *158*, 412–421. [CrossRef]

35. Skinnider, M.A.; Johnston, C.W.; Gunabalasingam, M.; Merwin, N.J.; Kieliszek, A.M.; MacLellan, R.J.; Li, H.; Ranieri, M.R.M.; Webster, A.L.H.; Cao, M.P.T.; et al. Comprehensive prediction of secondary metabolite structure and biological activity from microbial genome sequences. *Nat. Commun.* **2020**, *11*, 6058. [CrossRef]
36. Starcevic, A.; Zucko, J.; Simunkovic, J.; Long, P.F.; Cullum, J.; Hranueli, D. ClustScan: An integrated program package for the semi-automatic annotation of modular biosynthetic gene clusters and in silico prediction of novel chemical structures. *Nucleic Acids Res.* **2008**, *36*, 6882–6892. [CrossRef]
37. Li, M.H.T.; Ung, P.M.U.; Zajkowski, J.; Garneau-Tsodikova, S.; Sherman, D.H. Automated genome mining for natural products. *BMC Bioinform.* **2009**, *10*, 185. [CrossRef]
38. Cruz-Morales, P.; Kopp, J.F.; Martínez-Guerrero, C.; Yáñez-Guerra, L.A.; Selem-Mojica, N.; Ramos-Aboites, H.; Feldmann, J.; Barona-Gómez, F. Phylogenomic analysis of natural products biosynthetic gene clusters allows discovery of arseno-organic metabolites in model *Streptomycetes*. *Genome Biol. Evol.* **2016**, *8*, 1906–1916. [CrossRef]
39. Alanjary, M.; Kronmiller, B.; Adamek, M.; Blin, K.; Weber, T.; Huson, D.; Philmus, B.; Ziemert, N. The Antibiotic Resistant Target Seeker (ARTS), an exploration engine for antibiotic cluster prioritization and novel drug target discovery. *Nucleic Acids Res.* **2017**, *45*, W42–W48. [CrossRef]
40. Khater, S.; Gupta, M.; Agrawal, P.; Sain, N.; Prava, J.; Gupta, P.; Grover, M.; Kumar, N.; Mohanty, D. SBSPKSv2: Structure-based sequence analysis of polyketide synthases and non-ribosomal peptide synthetases. *Nucleic Acids Res.* **2017**, *45*, W72–W79. [CrossRef]
41. Van Heel, A.J.; De Jong, A.; Song, C.; Viel, J.H.; Kok, J.; Kuipers, O.P. BAGEL4: A user-friendly web server to thoroughly mine RiPPs and bacteriocins. *Nucleic Acids Res.* **2018**, *46*, W278–W281. [CrossRef]
42. Santos-Aberturas, J.; Chandra, G.; Frattaruolo, L.; Lacret, R.; Pham, T.H.; Vior, N.M.; Eyles, T.H.; Truman, A.W. Uncovering the unexplored diversity of thioamidated ribosomal peptides in *Actinobacteria* using the RiPPER genome mining tool. *Nucleic Acids Res.* **2019**, *47*, 4624–4637. [CrossRef]
43. De Los Santos, E.L.C. NeuRiPP: Neural network identification of RiPP precursor peptides. *Sci. Rep.* **2019**, *9*, 13406. [CrossRef]
44. Agrawal, P.; Khater, S.; Gupta, M.; Sain, N.; Mohanty, D. RiPPMiner: A bioinformatics resource for deciphering chemical structures of RiPPs based on prediction of cleavage and cross-links. *Nucleic Acids Res.* **2017**, *45*, W80–W88. [CrossRef]
45. Tietz, J.I.; Schwalen, C.J.; Patel, P.S.; Maxson, T.; Blair, P.M.; Tai, H.C.; Zakai, U.I.; Mitchell, D.A. A new genome-mining tool redefines the lasso peptide biosynthetic landscape. *Nat. Chem. Biol.* **2017**, *13*, 470–478. [CrossRef]
46. Liu, W.; Sun, F.; Hu, Y. Genome Mining-Mediated Discovery of a New Avermipeptin Analogue in *Streptomyces actuosus* ATCC 25421. *ChemistryOpen* **2018**, *7*, 558–561. [CrossRef]
47. Kloosterman, A.M.; Medema, M.H.; van Wezel, G.P. Omics-based strategies to discover novel classes of RiPP natural products. *Curr. Opin. Biotechnol.* **2021**, *69*, 60–67. [CrossRef]
48. Kautsar, S.A.; Blin, K.; Shaw, S.; Navarro-Muñoz, J.C.; Terlouw, B.R.; Van Der Hooft, J.J.J.; Van Santen, J.A.; Tracanna, V.; Suarez Duran, H.G.; Pascal Andreu, V.; et al. MIBiG 2.0: A repository for biosynthetic gene clusters of known function. *Nucleic Acids Res.* **2020**, *48*, D454–D458. [CrossRef]
49. MIBiG Logo Minimum Information about a Biosynthetic Gene Cluster. Available online: https://mibig.secondarymetabolites.org/ (accessed on 18 March 2020).
50. Luo, Y.; Cobb, R.E.; Zhao, H. Recent Advances in Natural Product Discovery Yunzi. *Curr. Opin. Biotechnol.* **2014**, *30*, 230–237. [CrossRef]
51. Watrous, J.; Roach, P.; Alexandrov, T.; Heath, B.S.; Yang, J.Y.; Kersten, R.D.; Van Der Voort, M.; Pogliano, K.; Gross, H.; Raaijmakers, J.M.; et al. Mass spectral molecular networking of living microbial colonies. *Proc. Natl. Acad. Sci. USA* **2012**, *109*, 1743–1752. [CrossRef]
52. Vincenti, F.; Montesano, C.; Di Ottavio, F.; Gregori, A.; Compagnone, D.; Sergi, M.; Dorrestein, P. Molecular Networking: A Useful Tool for the Identification of New Psychoactive Substances in Seizures by LC–HRMS. *Front. Chem.* **2020**, *8*, 572952. [CrossRef]
53. Leipoldt, F.; Santos-Aberturas, J.; Stegmann, D.P.; Wolf, F.; Kulik, A.; Lacret, R.; Popadić, D.; Keinhörster, D.; Kirchner, N.; Bekiesch, P.; et al. Warhead biosynthesis and the origin of structural diversity in hydroxamate metalloproteinase inhibitors. *Nat. Commun.* **2017**, *8*, 1965. [CrossRef]
54. Raheem, D.J.; Tawfike, A.F.; Abdelmohsen, U.R.; Edrada-Ebel, R.A.; Fitzsimmons-Thoss, V. Application of metabolomics and molecular networking in investigating the chemical profile and antitrypanosomal activity of British bluebells (Hyacinthoides non-scripta). *Sci. Rep.* **2019**, *9*, 2547. [CrossRef]
55. Crone, W.J.K.; Vior, N.M.; Santos-Aberturas, J.; Schmitz, L.G.; Leeper, F.J.; Truman, A.W. Dissecting Bottromycin Biosynthesis Using Comparative Untargeted Metabolomics. *Angew. Chem. Int. Ed.* **2016**, *55*, 9639–9643. [CrossRef]
56. Nepal, K.K.; Wang, G. *Streptomycetes*: Surrogate Hosts for the Genetic Manipulation of Biosynthetic Gene Clusters and Production of Natural Products. *Biotechnol. Adv.* **2019**, *37*, 1–20. [CrossRef]
57. Myronovskyi, M.; Rosenkränzer, B.; Stierhof, M.; Petzke, L.; Seiser, T.; Luzhetskyy, A. Identification and heterologous expression of the albucidin gene cluster from the marine strain *Streptomyces albus* subsp. *chlorinus* NRRL B-24108. *Microorganisms* **2020**, *8*, 237. [CrossRef]
58. Liu, X.; Liu, D.; Xu, M.; Tao, M.; Bai, L.; Deng, Z.; Pfeifer, B.A.; Jiang, M. Reconstitution of Kinamycin Biosynthesis within the Heterologous Host *Streptomyces albus* J1074. *J. Nat. Prod.* **2018**, *81*, 72–77. [CrossRef]

59. Jiang, G.; Zhang, Y.; Powell, M.M.; Zhang, P.; Zuo, R.; Zhang, Y.; Kallifidas, D.; Tieu, A.M.; Luesch, H.; Loria, R.; et al. High-Yield Production of Herbicidal Thaxtomins and Thaxtomin Analogs in a Nonpathogenic *Streptomyces* Strain. *Appl. Environ. Microbiol.* **2018**, *84*, e00164-18. [CrossRef]

60. Shi, Y.; Jiang, Z.; Li, X.; Zuo, L.; Lei, X.; Yu, L.; Wu, L.; Jiang, J.; Hong, B. Biosynthesis of antibiotic chuangxinmycin from *Actinoplanes tsinanensis*. *Acta Pharm. Sin. B* **2018**, *8*, 283–294. [CrossRef]

61. Chen, W.; Qu, D.; Zhai, L.; Tao, M.; Wang, Y.; Lin, S.; Price, N.P.J.; Deng, Z. Characterization of the tunicamycin gene cluster unveiling unique steps involved in its biosynthesis. *Protein Cell* **2010**, *1*, 1093–1105. [CrossRef]

62. Shuai, H.; Myronovskyi, M.; Nadmid, S.; Luzhetskyy, A. Identification of a biosynthetic gene cluster responsible for the production of a new pyrrolopyrimidine natural product—Huimycin. *Biomolecules* **2020**, *10*, 1074. [CrossRef]

63. Lombó, F.; Velasco, A.; Castro, A.; De La Calle, F.; Braña, A.F.; Sánchez-Puelles, J.M.; Méndez, C.; Salas, J.A. Deciphering the biosynthesis pathway of the antitumor thiocoraline from a marine actinomycete and its expression in two *Streptomyces* species. *ChemBioChem* **2006**, *7*, 366–376. [CrossRef]

64. Dhakal, D.; Han, J.M.; Mishra, R.; Pandey, R.P.; Kim, T.S.; Rayamajhi, V.; Jung, H.J.; Yamaguchi, T.; Sohng, J.K. Characterization of Tailoring Steps of Nargenicin A1 Biosynthesis Reveals a Novel Analogue with Anticancer Activities. *ACS Chem. Biol.* **2020**, *15*, 1370–1380. [CrossRef]

65. Alduina, R.; Giardina, A.; Gallo, G.; Renzone, G.; Ferraro, C.; Contino, A.; Scaloni, A.; Donadio, S.; Puglia, A.M. Expression in *Streptomyces lividans* of *Nonomuraea* genes cloned in an artificial chromosome. *Appl. Microbiol. Biotechnol.* **2005**, *68*, 656–662. [CrossRef]

66. Flinspach, K.; Kapitzke, C.; Tocchetti, A.; Sosio, M.; Apel, A.K. Heterologous expression of the thiopeptide antibiotic GE2270 from *Planobispora rosea* ATCC 53733 in *Streptomyces coelicolor* requires deletion of ribosomal genes from the expression construct. *PLoS ONE* **2014**, *9*, e90449.

67. Miyamoto, K.T.; Komatsu, M.; Ikeda, H. Discovery of gene cluster for mycosporine-like amino acid biosynthesis from *Actinomycetales* microorganisms and production of a novel mycosporine-like amino acid by heterologous expression. *Appl. Environ. Microbiol.* **2014**, *80*, 5028–5036. [CrossRef]

68. Reynolds, K.A.; Luhavaya, H.; Li, J.; Dahesh, S.; Nizet, V.; Yamanaka, K.; Moore, B.S. Isolation and structure elucidation of lipopeptide antibiotic taromycin B from the activated taromycin biosynthetic gene cluster. *J. Antibiot.* **2018**, *71*, 333–338. [CrossRef]

69. Saugar, I.; Molloy, B.; Sanz, E.; Blanca Sánchez, M.; Fernández-Lobato, M.; Jiménez, A. Characterization of the biosynthetic gene cluster (*ata*) for the A201A aminonucleoside antibiotic from *Saccharothrix mutabilis* subsp. capreolus. *J. Antibiot.* **2017**, *70*, 404–413. [CrossRef]

70. Shepherd, M.D.; Kharel, M.K.; Bosserman, M.A.; Rohr, J. Laboratory maintenance of *Streptomyces* species. *Curr. Protoc. Microbiol.* **2010**, *18*, 10E.1.1–10E.1.8. [CrossRef]

71. Ro, D.K.; Paradise, E.M.; Quellet, M.; Fisher, K.J.; Newman, K.L.; Ndungu, J.M.; Ho, K.A.; Eachus, R.A.; Ham, T.S.; Kirby, J.; et al. Production of the antimalarial drug precursor artemisinic acid in engineered yeast. *Nature* **2006**, *440*, 940–943. [CrossRef]

72. Luo, X.; Reiter, M.A.; d'Espaux, L.; Wong, J.; Denby, C.M.; Lechner, A.; Zhang, Y.; Grzybowski, A.T.; Harth, S.; Lin, W.; et al. Complete biosynthesis of cannabinoids and their unnatural analogues in yeast. *Nature* **2019**, *567*, 123–126. [CrossRef]

73. Yuan, H.; Zhang, W.; Xiao, G.; Zhan, J. Efficient production of gamma-aminobutyric acid by engineered *Saccharomyces cerevisiae* with glutamate decarboxylases from *Streptomyces*. *Biotechnol. Appl. Biochem.* **2020**, *67*, 240–248. [CrossRef]

74. Thomas, F.; Schmidt, C.; Kayser, O. Bioengineering studies and pathway modeling of the heterologous biosynthesis of tetrahydrocannabinolic acid in yeast. *Appl. Microbiol. Biotechnol.* **2020**, *104*, 9551–9563. [CrossRef]

75. Burdett, I.D.; Kirkwood, T.B.; Whalley, J.B. Growth Kinetics of Individual *Bacillus subtilis* Cells and Correlation with Nucleoid Extension. *J. Bacteriol* **1986**, *167*, 219–230. [CrossRef]

76. Kumpfmüller, J.; Methling, K.; Fang, L.; Pfeifer, B.A.; Lalk, M.; Schweder, T. Production of the polyketide 6-deoxyerythronolide B in the heterologous host *Bacillus subtilis*. *Appl. Microbiol. Biotechnol.* **2016**, *100*, 1209–1220. [CrossRef]

77. Zobel, S.; Kumpfmüller, J.; Süssmuth, R.D.; Schweder, T. *Bacillus subtilis* as heterologous host for the secretory production of the non-ribosomal cyclodepsipeptide enniatin. *Appl. Microbiol. Biotechnol.* **2015**, *99*, 681–691. [CrossRef]

78. Jaruszewicz-Błońska, J.; Lipniacki, T. Genetic toggle switch controlled by bacterial growth rate. *BMC Syst. Biol.* **2017**, *11*, 117. [CrossRef]

79. Pontrelli, S.; Chiu, T.Y.; Lan, E.I.; Chen, F.Y.H.; Chang, P.; Liao, J.C. *Escherichia coli* as a host for metabolic engineering. *Metab. Eng.* **2018**, *50*, 16–46. [CrossRef]

80. Boghigian, B.A.; Pfeifer, B.A. Current status, strategies, and potential for the metabolic engineering of heterologous polyketides in *Escherichia coli*. *Biotechnol. Lett.* **2008**, *30*, 1323–1330. [CrossRef]

81. Wu, J.; Boghigian, B.A.; Myint, M.; Zhang, H.; Zhang, S.; Pfeifer, B.A. Construction and performance of heterologous polyketide-producing K-12- and B-derived *Escherichia coli*. *Lett. Appl. Microbiol.* **2010**, *51*, 196–204. [CrossRef]

82. Yang, D.; Park, S.Y.; Park, Y.S.; Eun, H.; Lee, S.Y. Metabolic Engineering of *Escherichia coli* for Natural Product Biosynthesis. *Trends Biotechnol.* **2020**, *38*, 745–765. [CrossRef]

83. Huo, L.; Hug, J.J.; Fu, C.; Bian, X.; Zhang, Y.; Müller, R. Heterologous expression of bacterial natural product biosynthetic pathways. *Nat. Prod. Rep.* **2019**, *41*, 425–431. [CrossRef]

84. Cummings, M.; Peters, A.D.; Whitehead, G.F.S.; Menon, B.R.K.; Micklefield, J.; Webb, S.J.; Takano, E. Assembling a plug-and-play production line for combinatorial biosynthesis of aromatic polyketides in *Escherichia coli*. *PLoS Biol.* **2019**, *17*, e3000347. [CrossRef]

85. Liu, X.; Hua, K.; Liu, D.; Wu, Z.L.; Wang, Y.; Zhang, H.; Deng, Z.; Pfeifer, B.A.; Jiang, M. Heterologous Biosynthesis of Type II Polyketide Products Using *E. coli. ACS Chem. Biol.* **2020**, *15*, 1177–1183. [CrossRef]

86. Kallscheuer, N.; Kage, H.; Milke, L.; Nett, M.; Marienhagen, J. Microbial synthesis of the type I polyketide 6-methylsalicylate with *Corynebacterium glutamicum. Appl. Microbiol. Biotechnol.* **2019**, *103*, 9619–9631. [CrossRef]

87. Mora-Lugo, R.; Stegmüller, J.; MacK, M. Metabolic engineering of roseoflavin-overproducing microorganisms. *Microbial Cell Factories* **2019**, *18*, 146. [CrossRef]

88. Molin, G. Effect of carbon dioxide on growth of *Pseudomonas putida* ATCC 11172 on asparagine, citrate, glucose, and lactate in batch and continuous culture. *Can. J. Microbiol.* **1985**, *31*, 763–766. [CrossRef]

89. Gross, F.; Gottschalk, D.; Müller, R. Posttranslational modification of myxobacterial carrier protein domains in *Pseudomonas sp.* by an intrinsic phosphopantetheinyl transferase. *Appl. Microbiol. Biotechnol.* **2005**, *68*, 66–74. [CrossRef]

90. Loeschcke, A.; Thies, S. *Pseudomonas putida*—A versatile host for the production of natural products. *Appl. Microbiol. Biotechnol.* **2015**, *99*, 6197–6214. [CrossRef]

91. Choi, K.R.; Cho, J.S.; Cho, I.J.; Park, D.; Lee, S.Y. Markerless gene knockout and integration to express heterologous biosynthetic gene clusters in *Pseudomonas putida. Metab. Eng.* **2018**, *47*, 463–474. [CrossRef]

92. Zhang, H.; Wang, Y.; Pfeifer, B.A. Bacterial hosts for natural product production. *Mol. Pharm.* **2008**, *5*, 212–225. [CrossRef]

93. Incha, M.R.; Thompson, M.G.; Blake-Hedges, J.M.; Liu, Y.; Pearson, A.N.; Schmidt, M.; Gin, J.W.; Petzold, C.J.; Deutschbauer, A.M.; Keasling, J.D. Leveraging host metabolism for bisdemethoxycurcumin production in *Pseudomonas Putida. Metab. Eng. Commun.* **2020**, *10*, e00119. [CrossRef]

94. Kasuga, K.; Sasaki, A.; Matsuo, T.; Yamamoto, C.; Minato, Y.; Kuwahara, N.; Fujii, C.; Kobayashi, M.; Agematu, H.; Tamura, T.; et al. Heterologous production of kasugamycin, an aminoglycoside antibiotic from *Streptomyces kasugaensis*, in *Streptomyces lividans* and *Rhodococcus erythropolis* L-88 by constitutive expression of the biosynthetic gene cluster. *Appl. Microbiol. Biotechnol.* **2017**, *101*, 4259–4268. [CrossRef]

95. Gomez-Escribano, J.P.; Castro, J.F.; Razmilic, V.; Jarmusch, S.A.; Saalbach, G.; Ebel, R.; Jaspars, M.; Andrews, B.; Asenjo, J.A.; Bibb, M.J. Heterologous Expression of a Cryptic Gene Cluster from *Streptomyces leeuwenhoekii* C34^T Yields a Novel Lasso Peptide, Leepeptin. *Appl. Environ. Microbiol.* **2019**, *85*, e01752-19. [CrossRef]

96. Tu, J.; Li, S.; Chen, J.; Song, Y.; Fu, S.; Ju, J.; Li, Q. Characterization and heterologous expression of the neoabyssomicin/abyssomicin biosynthetic gene cluster from *Streptomyces koyangensis* SCSIO 5802. *Microb. Cell Factories* **2018**, *17*, 28. [CrossRef]

97. Bilyk, O.; Sekurova, O.N.; Zotchev, S.B.; Luzhetskyy, A. Cloning and heterologous expression of the grecocycline biosynthetic gene cluster. *PLoS ONE* **2016**, *11*, e0158682. [CrossRef]

98. Greunke, C.; Duell, E.R.; D'Agostino, P.M.; Glöckle, A.; Lamm, K.; Gulder, T.A.M. Direct Pathway Cloning (DiPaC) to unlock natural product biosynthetic potential. *Metab. Eng.* **2018**, *47*, 334–345. [CrossRef]

99. Li, J.; Jaitzig, J.; Theuer, L.; Legala, O.E.; Süssmuth, R.D.; Neubauer, P. Type II thioesterase improves heterologous biosynthesis of valinomycin in *Escherichia coli. J. Biotechnol.* **2015**, *193*, 16–22. [CrossRef]

100. Hug, J.J.; Dastbaz, J.; Adam, S.; Revermann, O.; Koehnke, J.; Krug, D.; Müller, R. Biosynthesis of Cittilins, Unusual Ribosomally Synthesized and Post-translationally Modified Peptides from *Myxococcus xanthus. ACS Chem. Biol.* **2020**, *15*, 2221–2231. [CrossRef]

101. Nara, A.; Hashimoto, T.; Komatsu, M.; Nishiyama, M.; Kuzuyama, T.; Ikeda, H. Characterization of bafilomycin biosynthesis in *Kitasatospora setae* KM-6054 and comparative analysis of gene clusters in *Actinomycetales* microorganisms. *J. Antibiot.* **2017**, *70*, 616–624. [CrossRef]

102. Myronovskyi, M.; Rosenkränzer, B.; Nadmid, S.; Pujic, P.; Normand, P.; Luzhetskyy, A. Generation of a cluster-free *Streptomyces albus* chassis strains for improved heterologous expression of secondary metabolite clusters. *Metab. Eng.* **2018**, *49*, 316–324. [CrossRef]

103. Wang, X.; Yin, S.; Bai, J.; Liu, Y.; Fan, K.; Wang, H.; Yuan, F.; Zhao, B.; Li, Z.; Wang, W. Heterologous production of chlortetracycline in an industrial grade *Streptomyces rimosus* host. *Appl. Microbiol. Biotechnol.* **2019**, *103*, 6645–6655. [CrossRef]

104. Xu, Z.; Li, Y. A MarR-family transcriptional factor MapR positively regulates actinorhodin production in *Streptomyces coelicolor. FEMS Microbiol. Lett.* **2020**, *367*, fnaa140. [CrossRef]

105. Herrmann, S.; Siegl, T.; Luzhetska, M.; Jilg, L.P.; Welle, E.; Erb, A.; Leadlay, P.F.; Bechthold, A.; Luzhetskyy, A. Site-specific recombination strategies for engineering *Actinomycete* genomes. *Appl. Environ. Microbiol.* **2012**, *78*, 1804–1812. [CrossRef]

106. Bu, Q.T.; Yu, P.; Wang, J.; Li, Z.Y.; Chen, X.A.; Mao, X.M.; Li, Y.Q. Rational construction of genome-reduced and high-efficient industrial *Streptomyces* chassis based on multiple comparative genomic approaches. *Microb. Cell Factories* **2019**, *18*, 16. [CrossRef]

107. Lu, Z.; Xie, P.; Qin, Z. Promotion of markerless deletion of the actinorhodin biosynthetic gene cluster in *Streptomyces coelicolor. Acta Biochim. Biophys. Sin.* **2010**, *42*, 717–721. [CrossRef]

108. Fazal, A.; Thankachan, D.; Harris, E.; Seipke, R.F. A chromatogram-simplified *Streptomyces albus* host for heterologous production of natural products. *Antonie Van Leeuwenhoek* **2020**, *113*, 511–520. [CrossRef]

109. Gust, B.; Challis, G.L.; Fowler, K.; Kieser, T.; Chater, K.F. PCR-targeted *Streptomyces* gene replacement identifies a protein domain needed for biosynthesis of the sesquiterpene soil odor geosmin. *Proc. Natl. Acad. Sci. USA* **2003**, *100*, 1541–1546. [CrossRef]

110. Komatsu, M.; Uchiyama, T.; Omura, S.; Cane, D.E.; Ikeda, H. Genome-minimized *Streptomyces* host for the heterologous expression of secondary metabolism. *Proc. Natl. Acad. Sci. USA* **2010**, *107*, 2646–2651. [CrossRef]

111. Fernández-Martínez, L.T.; Bibb, M.J. Use of the meganuclease I-SceI of *Saccharomyces cerevisiae* to select for gene deletions in *Actinomycetes. Sci. Rep.* **2014**, *4*, 7100. [CrossRef]

112. Tong, Y.; Weber, T.; Lee, S.Y. CRISPR/Cas-based genome engineering in natural product discovery. *Nat. Prod. Rep.* **2019**, *36*, 1262–1280. [CrossRef]
113. Huang, H.; Zheng, G.; Jiang, W.; Hu, H.; Lu, Y. One-step high-efficiency CRISPR/Cas9-mediated genome editing in *Streptomyces*. *Acta Biochim. Biophys. Sin.* **2015**, *47*, 231–243. [CrossRef]
114. Cobb, R.E.; Wang, Y.; Zhao, H. High-Efficiency Multiplex Genome Editing of *Streptomyces* Species Using an Engineered CRISPR/Cas System. *ACS Synth. Biol.* **2015**, *4*, 723–728. [CrossRef]
115. Salem, S.M.; Weidenbach, S.; Rohr, J. Two Cooperative Glycosyltransferases Are Responsible for the Sugar Diversity of Saquayamycins Isolated from *Streptomyces* sp. KY 40-1. *ACS Chem. Biol.* **2017**, *12*, 2529–2534. [CrossRef]
116. Yeo, W.L.; Heng, E.; Tan, L.L.; Lim, Y.W.; Lim, Y.H.; Hoon, S.; Zhao, H.; Zhang, M.M.; Wong, F.T. Characterization of Cas proteins for CRISPR-Cas editing in *Streptomycetes*. *Biotechnol. Bioeng.* **2019**, *116*, 2330–2338. [CrossRef]
117. Zhang, J.; Zhang, D.; Zhu, J.; Liu, H.; Liang, S.; Luo, Y. Efficient Multiplex Genome Editing in *Streptomyces* via Engineered CRISPR-Cas12a Systems. *Front. Bioeng. Biotechnol.* **2020**, *8*, 726. [CrossRef]
118. Sarokin, L.; Carlson, M. Optimization of gene expression in *Streptomyces lividans* by a transcription terminator. *Nucleic Acids Res.* **1987**, *15*, 4227–4240.
119. Horbal, L.; Siegl, T.; Luzhetskyy, A. A set of synthetic versatile genetic control elements for the efficient expression of genes in *Actinobacteria*. *Sci. Rep.* **2018**, *8*, 491. [CrossRef]
120. Curran, K.A.; Karim, A.S.; Gupta, A.; Alper, H.S. Use of High Capacity Terminators in *Saccharomyces cerevisiae* to Increase mRNA half-life and Improve Gene Expression Control for Metabolic Engineering Applications. *Metab. Eng.* **2013**, *19*, 88–97. [CrossRef]
121. Pulido, D.; Jiménez, A.; Salas, M.; Mellado, R.P. A *Bacillus subtilis* phage φ29 transcription terminator is efficiently recognized in *Streptomyces lividans*. *Gene* **1987**, *56*, 277–282. [CrossRef]
122. Garst, A.D.; Edwards, A.L.; Batey, R.T. Riboswitches: Structures and mechanisms. *Cold Spring Harb. Perspect. Biol.* **2011**, *3*, a003533. [CrossRef]
123. Rudolph, M.M.; Vockenhuber, M.P.; Suess, B. Synthetic riboswitches for the conditional control of gene expression in *Streptomyces coelicolor*. *Microbiology* **2013**, *159*, 1416–1422. [CrossRef]
124. Eyles, T.H.; Vior, N.M.; Truman, A.W. Rapid and Robust Yeast-Mediated Pathway Refactoring Generates Multiple New Bottromycin-Related Metabolites. *ACS Synth. Biol.* **2018**, *7*, 1211–1218. [CrossRef]
125. Horbal, L.; Luzhetskyy, A. Dual control system—A novel scaffolding architecture of an inducible regulatory device for the precise regulation of gene expression. *Metab. Eng.* **2016**, *37*, 11–23. [CrossRef]
126. Gondry, M.; Sauguet, L.; Belin, P.; Thai, R.; Amouroux, R.; Tellier, C.; Tuphile, K.; Jacquet, M.; Braud, S.; Courçon, M.; et al. Cyclodipeptide synthases are a family of tRNA-dependent peptide bond-forming enzymes. *Nat. Chem. Biol.* **2009**, *5*, 414–420. [CrossRef]
127. Meng, S.; Han, W.; Zhao, J.; Jian, X.H.; Pan, H.X.; Tang, G.L. A Six-Oxidase Cascade for Tandem C−H Bond Activation Revealed by Reconstitution of Bicyclomycin Biosynthesis. *Angew. Chem. Int. Ed.* **2018**, *57*, 719–723. [CrossRef]
128. Liu, J.; Kaganjo, J.; Zhang, W.; Zeilstra-Ryalls, J. Investigating the bifunctionality of cyclizing and "classical" 5-aminolevulinate synthases. *Protein Sci.* **2018**, *27*, 402–410. [CrossRef]
129. Zhu, L.; Qian, X.; Chen, D.; Ge, M. Role of two 5-aminolevulinic acid biosynthetic pathways in heme and secondary metabolite biosynthesis in *Amycolatopsis orientalis*. *J. Basic Microbiol.* **2018**, *58*, 198–205. [CrossRef]
130. Leskiw, B.K.; Lawlor, E.J.; Fernandez-Abalos, J.M.; Chater, K.F. TTA codons in some genes prevent their expression in a class of developmental, antibiotic-negative, *Streptomyces* mutants. *Proc. Natl. Acad. Sci. USA* **1991**, *88*, 2461–2465. [CrossRef]
131. Tercero, J.A.; Espinosa, J.C.; Jiménez, A. Expression of the *Streptomyces alboniger pur* cluster in *Streptomyces lividans* is dependent on the bldA-encoded tRNA^{Leu}. *FEBS Lett.* **1998**, *421*, 221–223. [CrossRef]
132. Garg, R.P.; Gonzalez, J.M.; Parry, R.J. Biochemical characterization of VlmL, a seryl-tRNA synthetase encoded by the valanimycin biosynthetic gene cluster. *J. Biol. Chem.* **2006**, *281*, 26785–26791. [CrossRef]
133. Jiang, M.X.; Yin, M.; Wu, S.H.; Han, X.L.; Ji, K.Y.; Wen, M.L.; Lu, T. GdmRIII, a TetR Family Transcriptional Regulator, Controls Geldanamycin and Elaiophylin Biosynthesis in *Streptomyces autolyticus* CGMCC0516. *Sci. Rep.* **2017**, *7*, 4803. [CrossRef]
134. Lyu, M.; Cheng, Y.; Han, X.; Wen, Y.; Song, Y.; Li, J.; Chen, Z. AccR, a TetR family transcriptional repressor, coordinates short-chain acyl coenzyme A homeostasis in *Streptomyces avermitilis*. *Appl. Environ. Microbiol.* **2020**, *86*, e00508–e00520. [CrossRef]
135. Shen, J.; Kong, L.; Li, Y.; Zheng, X.; Wang, Q.; Yang, W.; Deng, Z.; You, D. A LuxR family transcriptional regulator AniF promotes the production of anisomycin and its derivatives in *Streptomyces hygrospinosus* var. *beijingensis*. *Synth. Syst. Biotechnol.* **2019**, *4*, 40–48. [CrossRef]
136. Guo, J.; Zhang, X.; Lu, X.; Liu, W.; Chen, Z.; Li, J.; Deng, L.; Wen, Y. SAV4189, a MarR-family regulator in *Streptomyces avermitilis*, activates avermectin biosynthesis. *Front. Microbiol.* **2018**, *9*, 1358. [CrossRef]
137. Peng, Q.; Gao, G.; Lü, J.; Long, Q.; Chen, X.; Zhang, F.; Xu, M.; Liu, K.; Wang, Y.; Deng, Z.; et al. Engineered *Streptomyces lividans* strains for optimal identification and expression of cryptic biosynthetic gene clusters. *Front. Microbiol.* **2018**, *9*, 3042. [CrossRef]
138. Ish-Horowicz, D.; Burke, J.F. Rapid and efficient cosmid cloning. *Nucleic Acids Res.* **1981**, *9*, 2989–2998. [CrossRef]
139. Saraswathy, N.; Ramalingam, P. High capacity vectors. *Concepts Tech. Genom. Proteom.* **2011**, 49–56. [CrossRef]
140. Zimmer, R.; Gibbins, A.M.V. Construction and Characterization of a Large-Fragment Chicken Bacterial Artificial Chromosome Library. *Genomics* **1997**, *226*, 217–226. [CrossRef]

141. Jones, A.C.; Gust, B.; Kulik, A.; Heide, L.; Buttner, M.J.; Bibb, M.J. Phage P1-Derived Artificial Chromosomes Facilitate Heterologous Expression of the FK506 Gene Cluster. *PLoS ONE* **2013**, *8*, e69319. [CrossRef]

142. Kouprina, N.; Larionov, V. TAR cloning: Insights into gene function, long-range haplotypes and genome structure and evolution. *Nat. Rev. Genet.* **2006**, *7*, 805–812. [CrossRef]

143. Yamanaka, K.; Reynolds, K.A.; Kersten, R.D.; Ryan, K.S.; Gonzalez, D.J.; Nizet, V.; Dorrestein, P.C.; Moore, B.S. Direct cloning and refactoring of a silent lipopeptide biosynthetic gene cluster yields the antibiotic taromycin A. *Proc. Natl. Acad. Sci. USA* **2014**, *111*, 1957–1962. [CrossRef]

144. Fu, J.; Bian, X.; Hu, S.; Wang, H.; Huang, F.; Seibert, P.M.; Plaza, A.; Xia, L.; Müller, R.; Stewart, A.F.; et al. Full-length RecE enhances linear-linear homologous recombination and facilitates direct cloning for bioprospecting. *Nat. Biotechnol.* **2012**, *30*, 440–446. [CrossRef]

145. Wang, H.; Li, Z.; Jia, R.; Yin, J.; Li, A.; Xia, L.; Yin, Y.; Müller, R.; Fu, J.; Stewart, A.F.; et al. ExoCET: Exonuclease in vitro assembly combined with RecET recombination for highly efficient direct DNA cloning from complex genomes. *Nucleic Acids Res.* **2018**, *46*, e28. [CrossRef]

146. Du, D.; Wang, L.; Tian, Y.; Liu, H.; Tan, H.; Niu, G. Genome engineering and direct cloning of antibiotic gene clusters via phage φBT1 integrase-mediated site-specific recombination in *Streptomyces*. *Sci. Rep.* **2015**, *5*, 8740. [CrossRef]

147. Vior, N.M.; Lacret, R.; Chandra, G.; Dorai-Raj, S.; Trick, M.; Truman, A.W. Discovery and biosynthesis of the antibiotic bicyclomycin in distantly related bacterial classes. *Appl. Environ. Microbiol.* **2018**, *84*, e02828-17. [CrossRef]

148. Mevaere, J.; Goulard, C.; Schneider, O.; Sekurova, O.N.; Ma, H.; Zirah, S.; Afonso, C.; Rebuffat, S.; Zotchev, S.B.; Li, Y. An orthogonal system for heterologous expression of actinobacterial lasso peptides in *Streptomyces* hosts. *Sci. Rep.* **2018**, *8*, 8232. [CrossRef]

149. Duell, E.R.; D'Agostino, P.M.; Shapiro, N.; Woyke, T.; Fuchs, T.M.; Gulder, T.A.M. Direct pathway cloning of the sodorifen biosynthetic gene cluster and recombinant generation of its product in *E. coli*. *Microb. Cell Factories* **2019**, *18*, 32. [CrossRef]

150. D'Agostino, P.M.; Gulder, T.A.M. Direct Pathway Cloning Combined with Sequence- and Ligation-Independent Cloning for Fast Biosynthetic Gene Cluster Refactoring and Heterologous Expression. *ACS Synth. Biol.* **2018**, *7*, 1702–1708. [CrossRef]

151. Colloms, S.D.; Merrick, C.A.; Olorunniji, F.J.; Stark, W.M.; Smith, M.C.M.; Osbourn, A.; Keasling, J.D.; Rosser, S.J. Rapid metabolic pathway assembly and modification using serine integrase site-specific recombination. *Nucleic Acids Res.* **2014**, *42*, e23. [CrossRef]

152. Gao, H.; Taylor, G.; Evans, S.K.; Fogg, P.C.M.; Smith, M.C.M. Application of serine integrases for secondary metabolite pathway assembly in *Streptomyces*. *Synth. Syst. Biotechnol.* **2020**, *5*, 111–119. [CrossRef]

153. Wang, X.; Tang, B.; Ye, Y.; Mao, Y.; Lei, X.; Zhao, G.; Ding, X. Bxb1 integrase serves as a highly efficient DNA recombinase in rapid metabolite pathway assembly. *Acta Biochim. Biophys. Sin.* **2017**, *49*, 44–50. [CrossRef]

154. Enghiad, B.; Zhao, H. Programmable DNA-Guided Artificial Restriction Enzymes. *ACS Synth. Biol.* **2017**, *6*, 752–757. [CrossRef]

155. Basitta, P.; Westrich, L.; Rösch, M.; Kulik, A.; Gust, B.; Apel, A.K. AGOS: A Plug-and-Play Method for the Assembly of Artificial Gene Operons into Functional Biosynthetic Gene Clusters. *ACS Synth. Biol.* **2017**, *6*, 817–825. [CrossRef]

156. Jiang, W.; Zhao, X.; Gabrieli, T.; Lou, C.; Ebenstein, Y.; Zhu, T.F. Cas9-Assisted Targeting of CHromosome segments CATCH enables one-step targeted cloning of large gene clusters. *Nat. Commun.* **2015**, *6*, 8101. [CrossRef]

157. Lee, Y.; Lee, N.; Jeong, Y.; Hwang, S.; Kim, W.; Cho, S.; Palsson, B.O.; Cho, B.K. The Transcription Unit Architecture of *Streptomyces lividans* TK24. *Front. Microbiol.* **2019**, *10*, 2074. [CrossRef]

158. Ji, C.H.; Kim, J.P.; Kang, H.S. Library of Synthetic *Streptomyces* Regulatory Sequences for Use in Promoter Engineering of Natural Product Biosynthetic Gene Clusters. *ACS Synth. Biol.* **2018**, *7*, 1946–1955. [CrossRef]

159. Šmídová, K.; Ziková, A.; Pospíšil, J.; Schwarz, M.; Bobek, J.; Vohradsky, J. DNA mapping and kinetic modeling of the HrdB regulon in *Streptomyces coelicolor*. *Nucleic Acids Res.* **2019**, *47*, 621–633. [CrossRef]

160. Hwang, S.; Lee, N.; Jeong, Y.; Lee, Y.; Kim, W.; Cho, S.; Palsson, B.O.; Cho, B.K. Primary transcriptome and translatome analysis determines transcriptional and translational regulatory elements encoded in the *Streptomyces clavuligerus* genome. *Nucleic Acids Res.* **2019**, *47*, 6114–6129. [CrossRef]

161. Zhou, Q.; Ning, S.; Luo, Y. Coordinated regulation for nature products discovery and overproduction in *Streptomyces*. *Synth. Syst. Biotechnol.* **2020**, *5*, 49–58. [CrossRef]

162. Takano, E.; White, J.; Thompson, C.J.; Bibb, M.J. Construction of thiostrepton-inducible, high-copy-number expression vectors for use in *Streptomyces* spp. *Gene* **1995**, *166*, 133–137. [CrossRef]

163. Herai, S.; Hashimoto, Y.; Higashibata, H.; Maseda, H.; Ikeda, H.; Omura, S.; Kobayashi, M. Hyper-inducible expression system for *Streptomycetes*. *Proc. Natl. Acad. Sci. USA* **2004**, *101*, 14031–14035. [CrossRef]

164. Bibb, M.J.; Janssen, G.R.; Ward, J.M. Cloning and analysis of the promoter region of the erythromycin-resistance gene (*ermE*) of *Streptomyces erythraeus*. *Gene* **1985**, *38*, 215–226. [CrossRef]

165. Labes, G.; Bibb, M.; Wohlleben, W. Isolation and characterization of a strong promoter element from the *Streptomyces ghanaensis* phage 119 using the gentamicin resistance gene (*aacC1*) of Tn1696 as reporter. *Microbiology* **1997**, *143*, 1503–1512. [CrossRef]

166. Wang, W.; Li, X.; Wang, J.; Xiang, S.; Feng, X.; Yang, K. An engineered strong promoter for *Streptomycetes*. *Appl. Environ. Microbiol.* **2013**, *79*, 4484–4492. [CrossRef]

167. Wang, K.; Liu, X.F.; Bu, Q.T.; Zheng, Y.; Chen, X.A.; Li, Y.Q.; Mao, X.M. Transcriptome-Based Identification of a Strong Promoter for Hyper-production of Natamycin in *Streptomyces*. *Curr. Microbiol.* **2019**, *76*, 95–99. [CrossRef]

168. Farasat, I.; Kushwaha, M.; Collens, J.; Easterbrook, M.; Guido, M.; Salis, H.M. Efficient search, mapping, and optimization of multi-protein genetic systems in diverse bacteria. *Mol. Syst. Biol.* **2014**, *10*, 731. [CrossRef]
169. Na, D.; Lee, D. RBSDesigner: Software for designing synthetic ribosome binding sites that yields a desired level of protein expression. *Bioinformatics* **2010**, *26*, 2633–2634. [CrossRef]

Review

Bioactive Compounds and Nanodelivery Perspectives for Treatment of Cardiovascular Diseases

Rakesh K. Sindhu [1,*], Annima Goyal [1], Evren Algın Yapar [2] and Simona Cavalu [3,*]

[1] Chitkara College of Pharmacy, Chitkara University, Punjab 140401, India; annimagoyal@gmail.com
[2] Analysis and Control Laboratories Department, Turkish Medicines and Medical Devices Agency, MoH, Ankara 06520, Turkey; evren.yapar@yahoo.com
[3] Faculty of Medicine and Pharmacy, University of Oradea, P-ta 1 Decembrie 10, 410087 Oradea, Romania
* Correspondence: rakesh.sindhu@chitkara.edu.in (R.K.S.); simona.cavalu@gmail.com (S.C.)

Abstract: Bioactive compounds are comprised of small quantities of extra nutritional constituents providing both health benefits and enhanced nutritional value, based on their ability to modulate one or more metabolic processes. Plant-based diets are being thoroughly researched for their cardiovascular properties and effectiveness against cancer. Flavonoids, phytoestrogens, phenolic compounds, and carotenoids are some of the bioactive compounds that aim to work in prevention and treating the cardiovascular disease in a systemic manner, including hypertension, atherosclerosis, and heart failure. Their antioxidant and anti-inflammatory properties are the most important characteristics that make them favorable candidates for CVDs treatment. However, their low water solubility and stability results in low bioavailability, limited accessibility, and poor absorption. The oral delivery of bioactive compounds is constrained due to physiological barriers such as the pH, mucus layer, gastrointestinal enzymes, epithelium, etc. The present review aims to revise the main bioactive compounds with a significant role in CVDs in terms of preventive, diagnostic, and treatment measures. The advantages of nanoformulations and novel multifunctional nanomaterials development are described in order to overcome multiple obstacles, including the physiological ones, by summarizing the most recent preclinical data and clinical trials reported in the literature. Nanotechnologies will open a new window in the area of CVDs with the opportunity to achieve effective treatment, better prognosis, and less adverse effects on non-target tissues.

Keywords: bioactive compounds; cardiovascular; nanodelivery; bioavailability

Citation: Sindhu, R.K.; Goyal, A.; Algın Yapar, E.; Cavalu, S. Bioactive Compounds and Nanodelivery Perspectives for Treatment of Cardiovascular Diseases. *Appl. Sci.* **2021**, *11*, 11031. https://doi.org/10.3390/app112211031

Academic Editor: Ana M. L. Seca

Received: 15 October 2021
Accepted: 17 November 2021
Published: 21 November 2021

Publisher's Note: MDPI stays neutral with regard to jurisdictional claims in published maps and institutional affiliations.

1. Introduction

Cardiovascular disease (CVD) is one of the major disorders leading to death. It is mostly seen in both technologically advanced countries as well as in the evolving world, and it is accountable for high economic healthcare expenses [1]. Smoking, hypertension, hyperlipidemia, malnutrition, inadequate physical exercise, and obesity can all increase the risk of CVD. Coronary artery disease (CAD) is a condition caused by atherosclerosis as a result of the narrowing or blockage of coronary arteries. Atherosclerosis is a result of cholesterol and fatty deposits (plaques) inside the arteries.

Congestive heart failure, heart rhythm complications, congenital heart disease (heart disease that develops at birth), and endocarditis are just a few CV disorders [2,3]. The worldwide CVD deaths escalated by 21% from 2007 to 2017 because of population growth and aging; most deaths occur in low- and middle-income countries [4]. The World Health Organization (WHO) expresses that by 2030, CVD would be the major reason for around 23.3 million deaths [5]. The progress of CVD is accelerated by certain threat elements (smoking, dyslipidemia, hypertension, diabetes, and overweight). Fortunately, by adding prevention strategies, these can be avoided [6,7]. A transition to a healthier diet is a major key factor in preventing CVD and thus a shift to an ailment-free time [8].

An unhealthy diet is indeed one of the primary causes of CVD death, with about 72% of all CVD deaths [9]. Recent epidemiological data show that plant-based intakes are proficient and effective against CVD and cancer [10]. The term "plant-based diet" refers to a broad range of eating habits that include fewer animal products (such as meat and dairy products) and much more plant-source foods [11]. According to the recent literature [12], there is no consensus to describe "bioactive compound". However, it is broadly acknowledged that these "complexes have the proficiency and aptitude with one or more parts of active tissue by predicting most of the effects" [13]. They provide a beneficial health-boosting impact and are being investigated for deferent pathologies prevention: cancer, heart disease, diabetes mellitus, etc. Lycopene, resveratrol, lignan, tannins, and indoles are only a few examples of bioactive compounds [14]. Bioactive compounds are grouped according to their biochemical configuration and functions. These compounds have been shown to have a protective nature against certain pathologies correlated with the immune system, oxidative stress, and inflammation, and they also can lower LDL cholesterol oxidation and control endothelial nitric oxide amalgamation; some of them have poor estrogenic properties [15]. These compounds might contribute significantly to minimizing the onset of age-related chronic illnesses and to control the glucose metabolic rate [16].

Abundant epidemiologic studies show that enhancing vegetable and fruit intake is linked to a reduction in the rate of CVD [17]. Considerable data indicate the protective mechanism of fruits and vegetables, nuts and seeds, whole grains, and seafood in the prevention and treatment of various cancer and heart diseases [18,19]. The ocean's ecosystem accounts for half including all diversity in the world, rendering aquatic microbes a potential long-term reservoir of unique bioactive metabolites [20,21]. These dietary oral nutrients, when combined along with a regular meal, will help people get more of the components that are thought to have therapeutic outcomes [22]. It is well known that tea and coffee, the most consumed drinks worldwide, have certain beneficial properties. The caffeine present in the coffee seed is a purine alkaloid, i.e., 1,3,7-trimethylxanthine, with its latent properties, which are of some debated topics [23]. Meanwhile, components obtained from Allium sativum, also identified as garlic, is an herb belonging to the Alliaceae family that is frequently used as a seasoning in Southeast Asia and Europe. It comprises a high concentration of organosulfur complexes and flavonoids along with some other combinations that work together to impart a range of health benefits [24]. In the field of dietary supplements and universal healthcare, the bioactive constituents derived using environmental extraction techniques are gaining preference. The extraction acquiesces and pharmacological activities during the extraction method are also pressing issues that must be addressed. Scholarly research is being conducted on an ever-growing inventory of bioactive compounds [25].

However, in everyday life, it is difficult to ingest all the necessary nutrients to assure the proper function of the body or to complementary assist a drug-based treatment in CVDs. This is the reason why encapsulation techniques of nutraceuticals emerged as an effective approach designed to protect the bioactive compounds during the fabrication and storage, avoiding deterioration under environmental factors such as temperature, light, and UV exposure. Moreover, the development of encapsulated nutraceuticals into different carriers, overcomes the main drawback regarding their low bioavailability (such is the case of polyphenols), which greatly depends on several parameters, including solubility, digestibility, absorption, and metabolism. In addition, most bioactive compounds are unstable in alkaline conditions of biological fluids. By encapsulation in an adequate matrix, the incorporated compounds can be released with a specific concentration and time profile at a desirable site of action. At the nano-scale, the advanced nano-delivery systems have been demonstrated to boost the bioavailability and efficacy of bioactive compounds for therapeutic purpose [26]. On the other hand, effective nano-delivery systems have optimal characteristics for medicative agent-controlled release, long storage life, and enhanced therapeutic efficacy with no or minimal side effects.

In this context, the aim of this narrative review is to revise the main bioactive compounds with a significant role in CVDs in terms of preventive, diagnostic, and treatment measures. A discussion of the current evidence showing the advantages of nanoformulations and novel multifunctional nanomaterials able to overcome multiple obstacles, including the physiological ones, is made by summarizing the most recent preclinical data and clinical trials reported in the literature, describing the newer methodology and nanoformulation technologies that influence adequate drug or nutraceutical delivery.

2. Plant Bioactive Compounds

The phytoactive biocompounds can be isolated from plants using different extraction techniques. Based on specific functional group positions, bioactive compounds are categorized into primary and secondary metabolites [27]. Several metabolites are derived from fungi and vegetation based on their functional phase, tissue arrangement, ecological circumstances, and some other emphasis. Metabolites play a vital role in regulating cell maturation by functioning as plant development compounds. The most commonly found bioactive compounds are ethylene, auxin, gibberellins, plant development-abscisic acid, phytohormones–cytokinins, polyhydroxy steroids, and polyamines [28]. Glycosylated or covalently linked forms of bioactive compounds are engaged in the protection process that enables the plant to produce and store them in a harmless state [29]. Secondary metabolites used in traditional medicine can be extracted using a variety of methods, including Soxhlet extraction, maceration, and hydrodistillation. To minimize the use of solvents and introduce smoother extraction yields, ultrasonography, radiation, electromagnetic waves, high atmospheric pressure, and the use of supercritical fluids have already been examined [30–32]. Concentrated remains are acquired as the water drops are eradicated, and hence, the acquired part is known as an essential oil. Bioactive compounds are organized into three major classes based on their metabolic derivation: (a) phenolics; (b) terpenes; and (c) compounds containing nitrogen.

Fruit and vegetables consist of a wide range of bioactive compounds such as anthocyanins, betalains, carotenoids, flavonoids, glucosinolates, plant sterols, and tannins [33–36]. The occurrence and therapeutic properties of phytochemicals (phenolics, flavonoids, and carotenoids) have been collected in Table 1.

Table 1. Bioactive compounds in fruits and vegetables along with their therapeutic properties.

No.	Source	Bioactive Compounds	Therapeutic Potential	Ref.
1.	Abelmoschus esculentus	Quercetin	Anti-inflammatory, antioxidant, hypolipidemic.	[36]
2.	Ajuga iva	Naringein, apigenin-7-*O*-neohesperidoside.	Antioxidant, anti-inflammatory, antihypercholesterolemia.	[37]
3.	Anchusa italica Retz.	Rutin, hesperidin, quercetin, kaempferol, naringenin.	Anti-inflammatory, antioxidant, anticoagulant.	[38]
4.	Apples	Lutein, carotenoids, antioxidant: phlorizin, quercetin, catechin, procyanidin, epicatechin.	Antioxidant, antifungal, anti-proliferative.	[39]
5.	Barley leaves	2β-*O*-Glucosylisovitexin.	Antioxidant, membrane stabilizer, antitumor.	[40]
6.	Broccoli	Lutein, zeaxanthin, β-carotene, flavonoids.	Antioxidant, anti-inflammatory, anticarcinogenic.	[41]
7.	Carrots	Lignin, carotene.	Treat leukemia.	[42]
8.	Cereal crops	Orizanol, isovitexin, cyanidine-3-*O*-β-D-glycopyranoside, pinoresinol.	Prevent cardiovascular diseases and cancer diseases.	[43]
9.	Cocoa	Phytochemicals: methylxanthine, proantho-cyanidin, theobromin.	Antioxidant, anti-inflammatory, hypoglycemic, antiplatelet, antihypertensive.	[44]
10.	Corchorus olitorius Leaf and Corchorus capsularis	Luteolin	Antioxidant, hypotensive, diuretic.	[45,46]

Table 1. *Cont.*

No.	Source	Bioactive Compounds	Therapeutic Potential	Ref.
11.	Cotton seed oil	Quercetin, rutin, kaempferol, gossypeti, heracetin, dihydroquercetin, quercetrin, isoquercetrin.	Total cholesterol, low-density lipoprotein (LDL) cholesterol, and triglycerides (TG) lowered.	[47]
12.	Cymbopogon citratus (DC) Sapf.	Tannins, luteolin, apigenin.	Vasorelaxation, antioxidant, anti-inflammatory.	[48]
13.	Danshen (Salvia miltiorrhiza)	Tanshinones, salvianolic acids.	Angina, ischemic stroke, hyperlipidemia, antithrombotic.	[49]
14.	Dracocephalum moldavica L.	Tallianine, luteolin, apigenin, diosmetin.	Anticancer, anti-inflammatory, antioxidant.	[50]
15.	Edible wild fruit	Polyphenols: procyanidin, quercetin, phenolic acid, anthocyanin, carotenoids.	Anti-inflammatory, anti-obesity, antidiabetic.	[51]
16.	Ephedra herb	Epiafzelechin (flavanol), quercetin, gallocatechin, apigenin, luteolin.	Diuretic, anti-inflammatory, hypotensive, antioxidant.	[52]
17.	Equisetum arvense L. (Horsetail)	Resveratrol, apigenin, quercetin.	Antioxidant, anti-inflammatory.	[53]
18.	Flaxseed oil	Phytoestrogens lignans, coumestran, enterolactone, enterodial, coumestrol.	Reduce the growth rate of mammary cancer, lowered plasma LDL cholesterol.	[54]
19.	Foxglove (Digitalis species)	Glycoside digoxi.	Treat heart failure, atrial fibrillation.	[55]
20.	Garlic (Allium sativum)	Allicin.	Treat hypertension, hyperlipidemia, antithrombotic.	[56]
21.	Ginkgo (Ginkgo biloba)	Flavonol glycosides–quercetin and catechin and terpenoids–ginkgolides and bilobalides.	Cerebral insufficiency, peripheral vascular disease, antithrombotic.	[57]
22.	Ginseng (Panax species)	Triterpene saponins–ginsenosides.	Angina, hypertension, antidiabetic.	[58]
23.	Grapes	Polyphenols: resveratrol, carotenoids, flavonoids.	Antioxidant, anti-inflammatory, antihypertensive, antidiabetic.	[59]
24.	Hawthorn (Crataegus species)	Phenolic acids, quercetin, pyrocatechin, phlorodizin, terpenoids, lignans, steroids, organic acids, and sugars.	Heart failure, angina, hyperlipidemia.	[60]
25.	Heliotropium taltalense (Phil.)	Naringenin, pinocembrin, quercetin.	Anti-inflammatory, antioxidant, vasorelaxation.	[61]
26.	Lentil (Lens culinaris Medik.)	Quercetin, kaempferol.	Anticoagulant, antiplatelet.	[62]
27.	Lingzhi (Ganoderma lucidum)	Polysaccharides, triterpenes, polyphenols, proteins, amino acids, and organic germanium.	Hyperlipidemia, hypertension, antidiabetic.	[63,64]
28.	Moringa oleifera Lam.	Catechin, epicatechin, kaempferol, quercetin.	Antioxidant, anti-inflammatory.	[65]
29.	Mustard seed	Glucosinolates, sinigrin, phenolic acids, sinapic acid, methyl ester.	Anti-inflammatory, Anticancer, antioxidant, antihyperglycemia.	[66]
30.	Oats	β-Glucan, pectin, psyllium, esters of caffeic and ferulic acids.	Reduce both total cholesterol and LDL cholesterol.	[67]
31.	Olives	Phenolic compounds, hydroxy-tyrosol, oleuropein, polyphenols, flavonoids, theanine, quercetin.	Antioxidant, anti-inflammatory, antihypertensive, anti-obesity.	[68]
32.	Onions	Quercetin, myricetin.	Treat obesity, coagulation, inflammation, atherosclerosis, hyperlipidemia.	[69]
33.	Peanuts	Taxifolin, resveratrol.	Triacylglycerol (TAG) reduced.	[70]
34.	Polygonum minus (Persicaria minor)	Myricetin, quercetin, methyl-flavonol.	Antioxidant, anti-inflammatory.	[71]
35.	Psyllium seed	β-glucan, pectin, psyllium, soluble dietary fibers.	Antioxidant, laxative.	[72]
36.	Red wine	Resveratrol.	Anticancer activity.	[73]

Table 1. *Cont.*

No.	Source	Bioactive Compounds	Therapeutic Potential	Ref.
37.	Rice bran oil	Plant sterols, sitostanol, stigmasterol, campesterol.	Hypocholesterolemic effect, total cholesterol and LDL cholesterol reduced.	[74]
38.	Soybeans	Genistein, daidzein, isoflavones.	Reduced LDL-cholesterol level.	[75]
39.	Thai Perilla frutescens	Cyanidins, luteolin, phenolic acids.	Anti-inflammatory, antioxidant.	[76]
40.	Tomato	Phenols: phenolic acid, flavonoids, carotenoids.	Antioxidant, anti-inflammatory, antiplatelet, antihypertensive.	[77]
41.	Trichosanthes kirilowii	Luteolin.	Hypolipidemic, antioxidant, antiatherosclerotic.	[78]
42.	Wild rice	Phytic acid, luteolin glycoside, *p*-hydroxy acetophenone glycoside, 3,4,5-trimethoxycinnamin acid.	Health-promoting.	[79]

3. Types of Bioactive Compounds

To summarize, we will present the recent advances on bioactive compounds that are widely recognized as promising strategies in the prevention, adjuvant therapy, or even cure of different chronic diseases of the 21st century, with a focus on CVs. Their multi-facet features in terms of dietary lifestyles shows positive effects on treating and preventing CVDs being emphasized in the next sections.

3.1. Flavonoids

Flavonoids are a copious and distinct cluster of bioactive compounds that occur as the core elements of polyphenols. They are divided into flavonols, flavones, flavanonols, flavanones, flavans (catechins, anthocyanins, and proanthocyanidins) and isoflavones and flavanonols [80]. Each flavonoid subclass and category have its own arrangement of herbal sources, roles, and therapeutic effects. As for their recognized antioxidant and anti-inflammatory effects, this arrangement of herbal bioactive compounds was revealed to have wellness benefits for individuals [81]. The main properties, source, and structure of flavonoids are briefly presented in Table 2, while a list containing the types of flavonoids, their functional unit, source, and therapeutic properties are presented in Table 3.

Table 2. The main properties, source, and health benefits of flavonoids.

Flavonoids with Main Features	Natural Sources	Bountiful Health Benefits
- Flavonoids are hydrophilic. It has a C6–C3–C6 backbone. It consists a 15-carbon skeleton (two benzene rings linked with three-carbon chain i.e., oxygenated ring). - It is produced in response to microbial infection by plants. - The antioxidant activity is dependent on arrangement of functional groups. - Flavonoid basic structure [82,83]. 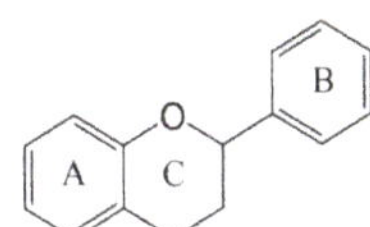 - Isoflavonoids basic structure. 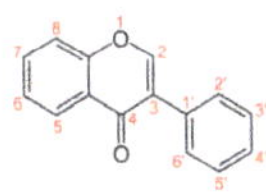	Berries, spices, leeks, ginger, herbs, grapefruit, carrot, grapes, apple, onion, broccoli, cabbage, kale, tomato, lemon, parsley, buckwheat, legumes, coffee, tea.	- Rich in phytonutrients, antioxidant. - Neutralize free radicals. - Limit cells damage. - Anti-inflammatory, anti-aging properties. - Improve the quality of blood vessels [84].

Table 3. Types of flavonoids, their functional unit, plant source, and therapeutic properties.

No.	Type of Flavonoid	Major Food Flavonoids	Functional Unit of B-Ring	Unsaturation in B-Ring	Common Plant Source	Therapeutic Effect	Ref.
1.	Anthocyanins	Cyanidin, delphinidin, malvidin, peonidin.	3-Hydroxy.	1-2, 3-4 Double bond.	Strawberries, red wine, and blueberries.	- Endothelial-dependent vasodilation. - Reduce risk of acute myocardial infarction.	[85]
2.	Flavan-3-ols [flavanols]	Cathechin, gallocatechin, epicatechin, epiallocatechin-3-gallate.	3-Hydroxy, 3-O-gallate.	None.	Red wine, blueberries, apples, chocolate, pears, cocoa, tea, and grape.	- Reduce systolic blood pressure. - Anti-inflammatory. - Antioxidant. - Antiatherogenic action.	[86]
3.	Flavanones	Eriodictyol, naringenin, hesperetin.	4-Oxo.	None.	Herbal tea, juice, fruit peels, and citrus fruit.	- Antioxidant. - Reduce blood pressure. - Modulate nitric oxide. - Antihypertensive effect.	[87]
4.	Flavones	Apigenin, luteolin, tangeretin, baicalein.	4-Oxo.	2-3 Double bond.	Herbal tea, garlic, celery, chamomile tea, and green peppers.	- Lower blood pressure. - Improve vasodilation. - Increase accumulation of camp. - Induces vascular relaxation by NO, regulated by Ca and K channels.	[88]
5.	Flavonols	Myricetin, quercetin, isorhamnetin, kaempferol.	3-Hydroxy, 4-Oxo.	2-3 Double bond.	Red wine, kale, broccoli, cherry, tomato, garlic, onions, tea, strawberries, beans, Spinach.	- Modulate the renin–angiotensin–aldosterone system. - Produces vasodilation. - Lowers blood pressure. - Decreases oxidative stress.	[89]
6.	Isoflavones	Daidzein, genistein, glycitein, biochaninA, glycitein.	4-Oxo.	2-3 Double bond.	Peanuts, legumes, and soy products.	- Act as estrogen receptor agonist. - Diminish oxidative stress. - Antihypertensive effects.	[90]

3.2. Anthocyanins

Anthocyanin is one of the subclasses of phenolic phytochemicals; they are produced in cell sap and are hydrophilic in nature. They occur in higher plants tissue, such as fruits, flowers, leaves, and roots. Anthocyanins are the major cause for their specific coloration. In a brief manner, Table 4 presents the structure, biological source, and main therapeutic effects of anthocyanins [91].

Table 4. The main source, benefits, and chemical structure of anthocyanins.

Chemical Structure and Main Features	Sources	Bountiful Health Benefits
- Phenolic entity of 15 C-atoms, which is made up of two benzene rings connected by a three-carbon chain. - Two benzyl rings A and B. it usually has a single glucoside unit, and many have two, three, or more sugar attached at multiple positions. - Sugar moiety at C-3 as 3-monoglycosides in the C-ring. - Sugar moiety at C-5, 7-position as diglycosides in the A-ring. - R3', R4', and R5' on B-ring–glycosylation - R3' and R5' positions of B-ring have different components attached and form different compounds. - 6 Anthocyanins ubiquitously distributed are cyanidin (Cy), delphinidin (Dp), petunidin (Pt), peonidin (Pn), pelargonidin (Pg), and malvidin (Mv).	Acai, blackcurrant, blueberry, bilberry cherry, red grape, and purple corn.	- Low stability and poor absorption—this is not beneficial! - Anticancer activity, reduce cancer cell proliferation, and inhibit tumor formation. - Anti-inflammatory properties. - Interact with other phytochemicals to potentiate biological effects. - Free-radical scavenging and antioxidant capacities.

3.3. Tannins

Tannins are water-soluble polyphenols that are astringent and form bonds with proteins, as well as other organic compounds and macromolecules. Secondary metabolites easily get sequestered in plant cell vacuoles and protect different cell constituents. Table 5 contains the brief information related to chemical structure, biological source, and therapeutic benefits of tannins [92,93].

3.4. Betalains

Betalains are nitrogen (N)-containing vacuole pigments, similar to anthocyanins and flavonoids in appearance, yet pigments are red and yellow; they are capable of dissolving in water, as they contain nitrogen, and they can be found almost exclusively in families of the Caryophyllales. They are commonly used as color additives in food, being toxicologically safe. In Table 6, a brief presentation of the chemical structure, biological source, and therapeutic effects of betalains are presented [94,95].

Table 5. The chemical structure, therapeutic benefits, and biological source of tannins.

Chemical Structure and Main Features	Natural Sources	Bountiful Health Benefits
- Gallic acid is the base unit in the complex mixtures of polymeric polyphenols. - It is classified into condensed (C-C linkage) and hydrolyzable (ester-like compounds) tannins. - Colorless to yellow or brown. - Causes astringency of food. - Gallic acid. - Phloroglucinol	Teas, coffee, pomegranates, persimmons, cranberries, strawberries, blueberries, grapes, red wine, cinnamon, vanilla, cloves, thyme, and oak gallnuts (tannin content of 50–70%).	- Acceleration of blood clotting. - Reduction of blood pressure. - Decrease serum lipid levels. - Modulation of immune response.

3.5. Carotenoids

Carotenoids are usually found in chloroplast in the form of red, yellow, and orange pigments. Carotenoids are in the category of lipid-soluble hydrocarbons. Xanthophylls are oxygenated derivatives of carotenoids. The red color of carrots denotes their name, but they are also present in green leaves, yellow fruits and red fruits, several fungi, and rhizomes. They are responsible for the color of egg yolk and some fish. Table 7 presents the main features related to carotenoids in a brief manner [96–99].

Table 6. Structure, biological source, and therapeutic benefits of betalains.

Chemical Structure and Main Features	Sources	Bountiful Health Benefits
- Indole derivatives, classified into red-violet betacyanins and yellow-orange betaxanthins. - The colors determine resonating double bonds in the betalain structures. - They are hydrophilic and therefore can be incorporated into aqueous food systems. 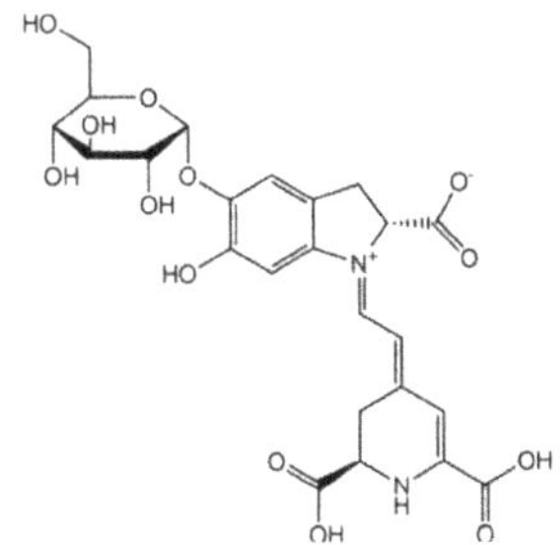	- Families of caryophyllales order and some higher-order fungi. - Beetroot (red and yellow), leafy amaranth, pear, swiss chard, red pitahaya, and cacti.	- Antioxidant. - Anticancer. - Antilipidemic. - Antimicrobial. - Nontoxic. - Promising agent for supplement therapies in oxidative stress, inflammation, and dyslipidemia-related. - Diseases such as stenosis of the arteries, atherosclerosis, hypertension, and some types of cancer.

Table 7. The structure, biological source, and therapeutic benefits of carotenoids.

Chemical Structure and Main Features	Sources	Bountiful Health Benefits
- Class of hydrocarbons and their oxygenated derivatives, composed of eight isoprenoid units linked in composition of isoprene entity, overturned in middle of the complexes. - Absorbs wavelength 400–550 nm and produces deeply yellow, orange, or red-colored complex. - It is broadly classified into two groups i.e., carotenes (orange color) and xanthophylls (yellow color). The main difference between the two group is that xanthophylls contain an oxygen group and carotenes are hydrocarbons without any oxygen group. - β-carotene (carotene) - β-Cryptoxanthin (Xanthophyll)	Apricots, carrots, cilantro, collard, cantaloupes, mango, plum, thyme, turnip, squash, kale.	- Precursor of vitamin A. - Anticancer, antioxidant. - Robust immune system. - Promote healthy skin. - Eye health and vision.

Carotenoids are further classified into following categories according to Table 8.

Table 8. Classification of carotenoids.

No	Types of Carotenoids	Composition	Source	Properties	Health Benefits	Ref.
1	α-Carotene	$C_{40}H_{56}$, carotene with β-ionone ring and α-ionone ring.	Carrots, pumpkin, broccoli, spinach, avocado, sweet potato, squash.	This is second most common carotenoid.	- Antioxidant properties. - Reduce risk of cancer. - Lower risk of CVD.	[100]
2	β-Carotene	$C_{40}H_{56}$, group consisting of isoprene units.	Carrots, apricots, mango, red pepper, greens (kale, spinach, broccoli).	This is the major and natural carotenoid present.	- Maintains eye health, helps in embryonic development and improves the immune system performance.	[101,102]
3	β-Cryptoxanthin	$C_{40}H_{56}O$, isoprene unit with a hydroxyl unit.	Orange, peaches, tangerines, papaya, egg yolk, butter, apples.	Natural carotenoid pigments.	- Improves respiratory function. - Pro-vitamin A activity.	[103]
4	Lutein and Zeaxanthin	$C_{40}H_{56}O_2$, they have similar isomers but differ in double bond location. Lutein shows the presence of three chiral canters, while zeaxanthin has two.	Kale, spinach, broccoli, sprouts, lettuce, egg yolk, yellow corns, and parsley.	These are dietary oxygenated carotenoids.	- Lowers risks of CHD and stroke. - Improves eye health. - Beneficial for skin health. - Prevents lipid peroxidation.	[104,105]
5	Lycopene	$C_{40}H_{56}$, in this a tetraterpene, is arranged with eight isoprene units of carbon and hydrogen.	Tomato, watermelon, grape, papaya, guava, rose.	These are hydrophilic, acyclic carotenoids with eleven conjugated double bonds.	- Effective antioxidant. - Reduce risk of CVD. - Prevents aging.	[106]

3.6. Plant Sterols

Plant sterols, also called phytosterols, are plant-derived fatty compounds. They are found in non-esterified and esterified forms of cinnamic/fatty acids. Plant sterol esters reduce coronary heart disease. Table 9 presents the data related to plant sterols in a brief manner [106,107].

3.7. Glucosinolates

Glucosinolates are natural, anionic plant secondary metabolites. These are sulfur rich and belong to the order Brassicaceae. Table 10 presents the structure, biological source, and therapeutic benefits of glucosinolates [108–111].

Table 9. The structure, biological source, and therapeutic benefits of plant sterols.

Chemical Structure and Main Features	Sources	Bountiful Health Benefits
- Cholesterol-like compounds influence cholesterol absorption and metabolism in both animals and humans. - It is a fused polycyclic structure. There is variation in the presence of the double bond and carbon side chain. - Nomenclature of plant sterol steroid skeleton 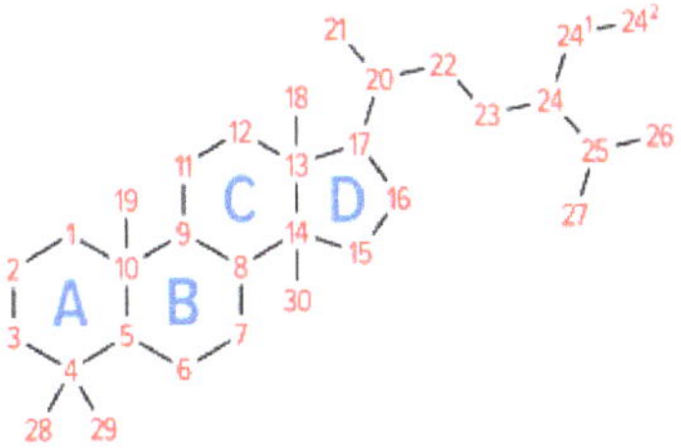 - Stigmasterol—by removing a hydrogen from C-22 and 23. 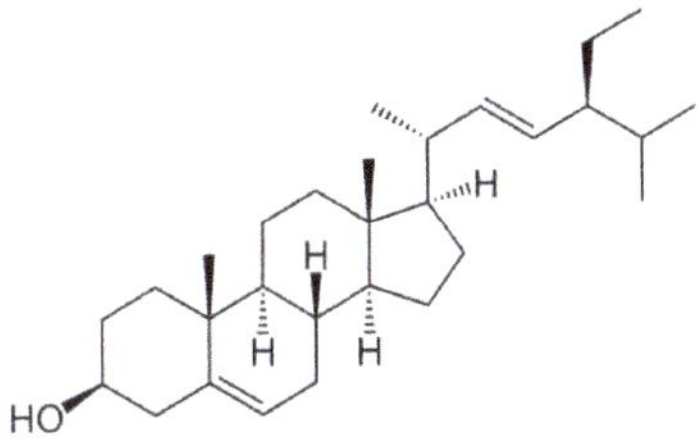	Vegetable oils, nuts, legumes, grains, cereals, wood pulp, margarine, milk, yogurt, soybean oil, and leaves.	Reduce the level of the LDL cholesterol in blood.

Table 10. Structure, biological source, and therapeutic benefits of glucosinolates.

Chemical Structure and Main Features	Sources	Bountiful Health Benefits
- Glucose and amino acid release substances such as sulfur and nitrogen that make up the glucosinolates. - They have a central carbon atom that links to the thioglucose group by a sulfur atom and to the sulfate group by a nitrogen atom. 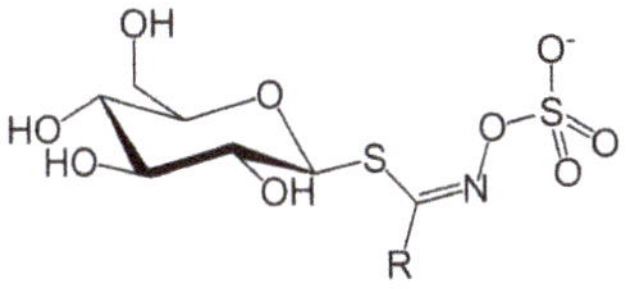	Cruciferous vegetables: wasabia japonica, cabbage, garden cress, kale, broccoli, and watercress.	- Protective role against cancer - Protective role in dementia risk.

4. Therapeutic Effect of Bioactive Compounds on Cardiovascular System

The advancement of disorders such as atherosclerosis and CVD is facilitated by oxidative stress. Diverse bioactive compounds' anti-oxidative, anti-inflammatory, and metabolic effects are linked to their defense against atherosclerosis and CVD [112].

4.1. Carotenoids

Carotenoids are found in high amount in fruits and vegetables. Their sub-categorization is as per the chemical structure they have; i.e., as carotenes and xanthophylls. Carotenoids have antioxidant properties that are beneficial for health. Human organs and tissues have carotenoids. In tissues, the level of carotenoids is high as they have high levels of low-density lipoprotein receptors. Carotenoids help in the prevention of chronic cardiovascular diseases such as stroke and coronary heart disease. They have a multifaceted metabolism and react to systemic forces. As an antioxidant, they improve superoxide dismutase, glutamate dehydrogenase, catalase, and glucan particles. Carotenoids also inhibit IGF-1 activity and have been proved to be effective as an anticancer agent [113–116].

4.2. Polyphenols (Anthocyanins)

Berries are high in polyphenols, especially anthocyanins, as well as micronutrients and fiber. The consumption of berries results in an improvement in heart health. Chokeberries, cranberries, blueberries, and strawberries contain distilled anthocyanin derivatives. They showed substantial progress in LDL oxidation, lipid peroxidation, total plasma antioxidant potential, dyslipidemia, and glucose metabolism. It causes the activation of endothelial nitric oxide synthase, the lowered activity of carbohydrate digestive enzymes, and reduced oxidative stress by increasing the endothelial functions and the plasma lipid profiles. This led to a reduction of abnormal platelet aggregation. The research recommends berries (anthocyanins) as an important fruit group in a heart-healthy diet [117].

Pomegranate juice contains anthocyanins, catechins, quercetin, rutin, and ellagitannins, and it results in reducing high blood pressure, which is a result of the ACE activity antioxidant activity due to the radical scavenging effect of anthocyanins and hydrolyzable tannins [118–121].

4.3. Lycopene

Lycopene is a hydrocarbon carotenoid that is oxygenated and has quite a similar structure to β-carotene. Lycopene has an antioxidant property due to the presence of conjugated double bonds. This fat-soluble pigment imparts red color to a good variety of food items such as tomato, guava, watermelon, and others. Smoking is a major CVD risk factor. Smoke introduces free radicals into the human body, causing LDL oxidation, foam formation, and leading to atherogenesis. The severity of atherosclerosis is linked with an increase in LDL oxidation inclination. Lycopene prevents the oxidation of LDL and protects humans from coronary heart disease (CHD) [122].

Lycopene and plasma levels in cardiovascular disease have been investigated by researchers. The findings were analyzed. The higher intake of lycopene was compared with reduced levels of estone. Here, a 17% of reduction of CVD was linked with lycopene. It works by several mechanism such as the reduction of oxidation of biomolecules, the antiangiogenic effects, the reduction of cholesterol levels, the stimulation of apoptosis, and the reduction of inflammation [123].

4.4. Flavonoids

Randomized trials and many cohort studies have shown that flavonoids reduce CVD risk. Flavonoids produce a suitable response to LDL cholesterol, sensitiveness to insulin, and endothelial function [123]. A comprehensive evaluation of many investigations has indicated that the nutritional consumption of different groups of flavonoids, specifically as flavonols, anthocyanidins, proanthocyanidins, flavones, flavanones, and flavan-3-ols, minimize any chances of CVD drastically [124]. The quality of flavonoid subcategories in foods will be more essential than total flavonoids. Furthermore, the inverse correlation between flavonoid consumption and CVD risk is more pronounced in females than males. Even so, owing to the complementary nature of randomized clinical trials, there is indeed a lot of variance in the evidence [125].

Flavonoid research has exploded in popularity since its inception [126]. Understanding the availability of flavonoids, both natural and synthetic, and finding ways to improve their bioavailability were two of the most difficult tasks [127]. The scarcity of flavonoids is well known, but recently, these problems have been addressed. According to new research, the gut and its microbiome play a significant function in the development of prebiotic and microbiota enhancers as phenolic metabolites [128].

Flavonoids have been shown to reduce gastric and intestinal inflammation, as the metabolites act as enhancers of gut immune function [129]. As a result, attempts to increase flavonoids' bioavailability aiming primarily on increasing their intestinal absorption. Borneol and methanol combination are traditional drug absorption enhancers [130]. Both borneol and methanol are toxic, and methanol causes blindness [131,132].

Effect on CV System

Regarding antiatherosclerotic effects, the pathogenesis of atherosclerosis begins with the oxidative alteration of low-density lipoproteins (LDL) by free radicals. Foam cells develop when oxidatively modified LDL is rapidly absorbed through a scavenger receptor. They act as antioxidant-chain breaking, in which flavonoids are radical species [133]. The capacity of quercetin and quercetin glycosides to shield LDL from oxidative activation has been shown to be successful [134]. A Japanese study found an inverted relationship between the flavonoid consumption and total plasma cholesterol levels [135].

Regarding antithrombogenic effects, platelet aggregation is critical in the composition of thrombotic disorders. Activated platelets bind to the vascular endothelium and develop lipid peroxides and oxygen free radicals, which prevent the formation of prostacyclin and nitrous oxide (NO) in the endothelium. Tea pigment shows the prevention of platelet adhesion or aggregation, and it reduces blood coagulation as well as increases fibrinolysis [136]. Flavonoids such as quercetin, kaempferol, and myricetin have been shown to inhibit platelet aggregation in animals [137]. Flavonols are particularly antithrombotic because they scavenge free radicals, keeping endothelial prostacyclin and nitric oxide concentrations constant [138].

Regarding cardioprotective effects, from both advanced and emerging economies, CVD is now the major cause of death. Atherosclerosis, coronary heart disease, arterial hypertension, and heart failure are all cardiovascular system (CVS) disorders. Oxidative stress is the primary cause of CVS disorders. Endogenous oxidants and reactive oxygen/nitrogen species (RONS) are in equilibrium in oxidative stress, with free radicals predominating. A prolonged dispensing of flavonoids has been shown to diminish or threaten to degrade the rate of CVD and its effects. They have an elevated inclination to transfer electrons, chelate ferrous ions, and sift reactive oxygen species [139]. Flavonoids are potential protectors against the persistent cardiotoxicity triggered by cytostatic medication such as doxorubicin [140,141].

As a vasorelaxant agent, flavonoids help to reduce endothelial dysfunction (ED) by improving the vasorelaxation mechanism, which lowers arterial blood pressure [142,143]. ED is a crucial occurrence in the progression of CVD as well as a significant consequence involving arteriosclerosis as well as the occurrence of arterial thrombosis [144]. Flavonoids help to avoid a variety of CVD, such as high blood pressure and atherosclerosis [145,146]. Current laboratory studies show that these polyphenols can lower arterial pressure and improve the vasodilating system. Flavonoids have long been known to cause an endothelium-reliant response. Moreover, researchers discovered that anthocyanin delphinidin has a major endothelium-dependent vaso-relaxing effect [147,148]. Some of the cardiovascular risk factors, with a description of the pathophysiological effects they cause, and some bioactive compounds that may reduce the severity of the risk factors and also the positive changes they induce are described in Figure 1.

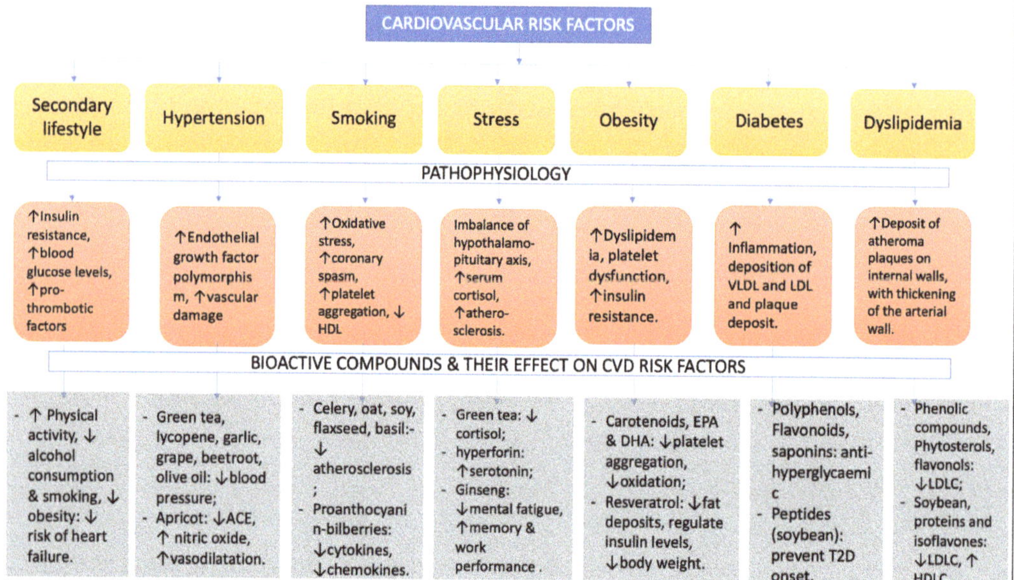

Figure 1. CVD risk factors: pathophysiology and beneficial bioactive compounds to reduce the risk factors.

5. Nano-Delivery Systems of Bioactive Compounds

The nano-delivery method allows for the regulation of food bio-active components' stability, solubility, and bioavailability, and it also maintains their targeted and controlled release. Food safety is a major issue because the usage of nano-delivery for food and drug delivery has grown in popularity [149–152]. Nano-delivery is carried out in two different ways: liquid and solid. Nano-emulsions, nano-liposomes, and nano-polymerases are the three different forms of liquid nano-delivery systems. Nano-liposomes are divided into SLVs (single lamellar vesicles) and MLVs (multilamellar large vesicles). Nanoparticles, polymeric nanoparticles, and nanocrystals are the three forms of solid nano-delivery systems. Solid lipid nanoparticles (SLNS) and nano-structured lipid carriers (NLCS) are two types of lipid particles. Polymeric nano-particles are of two different types, i.e., nano-spheres and nano-capsules. The types of nano-delivery system are presented in Figure 2.

Due to chemical and enzymatic barriers and the poor solubility of compounds in the GI tract, a significant decrease in the quality of orally delivered drugs was noticed. It is well known that the action of bioactive compounds depends upon the extent of the bioaccessibility and bioavailability to the organism. There are several conventional drug carrier systems currently used for the effective delivery of cardioprotective drugs, either as oral tablets, parenteral/intravenous administration, or transdermal patches. However, some CVD treatments such as medications against angina pectoris (nitrates, calcium channel blockers, β blockers) produce significant adverse effects such as rash, constipation, nausea, drowsiness, edema, low blood pressure, or headache [153].

In this context, nanotechnology plays a determinant role by manipulating bioactive compounds encapsulated into nano-carriers with dimensions of 1–100 nm, providing a longer half-life, longer circulation time, longer mean residence time, and better pharmacokinetic clearance from the body [154]. These systems particularly depend on physicochemical factors that will influence their absorption, distribution, metabolism, and excretion, which are critical for administering bio-active compounds with improved in vivo results [155]. As the high surface to volume ratio of nanoparticles allows the conjugation, absorption, or encapsulation of bioactive molecules for delivery to the target site, drug delivery vehicles are successfully employed owing to their ability to deliver poorly soluble or highly toxic drugs to the target areas. Nanoparticles bind to the gut mucosa and epithelial cells, enter the bloodstream, and then are distributed to tissues and organs such as the liver, kidneys, spleen, heart, lungs, and brain. Nanoparticles and their metabolites are mainly excreted into the liver, kidneys, and colon [156,157].

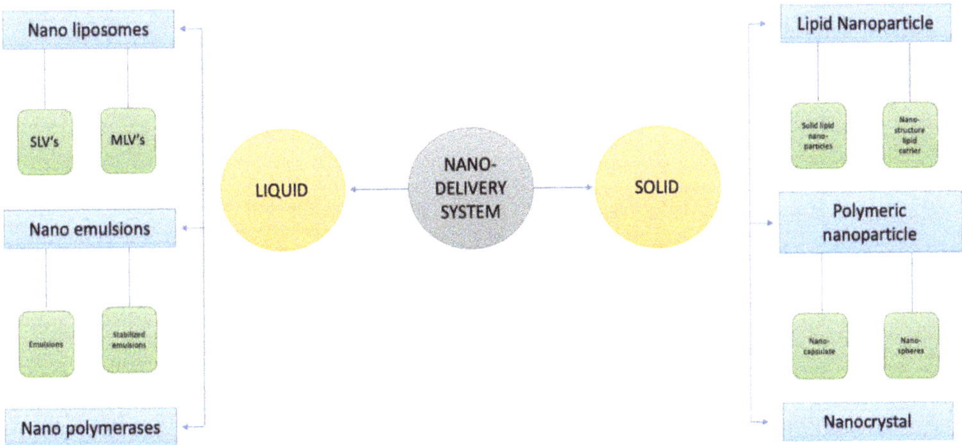

Figure 2. Types of nano-delivery systems.

Nowadays, conventional synthetic drugs are less present in practice owing to their costs and associated complications and side effects, while natural products have received much attention, as they are affordable for the majority of the population and possess multi-targeted effects with fewer side effects than synthetic drugs.

It is well known that polyphenols exhibit high antioxidant and anti-inflammatory properties, which are important for certain pathological conditions, including cardiovascular disease. Polyphenol consumption has been shown to improve endothelial function, blood pressure, and platelet function as well as the regulation of cellular processes such as inflammation and NO synthesis in vitro and in vivo [158]. The antioxidant mechanism is attributed to their ability to scavenge free oxygen and nitrogen species and to stimulate the expression of antioxidant enzymes (catalase, superoxide dismutase). When assessing the beneficial effects of polyphenols, it must be considered that only a small proportion of the ingested polyphenols are absorbed in the intestine, and therefore, a high quantity of polyphenols is required to achieve the expected improvement in terms of blood pressure lowering. The poor bioavailability is due not only to the low water solubility of polyphenols but also to the instability in alkaline conditions of biological fluids [159]. In order to overcome these drawbacks, nanoencapsulation technologies have emerged as a novel trend in drug carrier development, being nontoxic in nature, with the ability to escape from the host immune system, along with other advantages such as biodegradability, biocompatibility, and drug-targeting properties.

The encapsulation of active ingredients is nowadays a routine fabrication process and can be realized through different techniques: spray or freeze drying, coacervation, ionic gelation, extrusion, emulsion, electrospinning, electro-spraying, and liposomes formulation [160]. Encapsulation is also designed to protect the bioactive compounds during processing, storage, and transport from different undesirable factors such as temperature, light, and environmental oxidation. In this respect, biological materials such as polysaccharides, proteins, lipids, or low molecular surfactants are used [161], being highly biocompatible and nontoxic for human consumption. The release mechanisms might be related to diffusion, swelling, erosion, fragmentation, dissolution, or stimuli-controlled release. β-carotene, curcumin, quercetin, resveratrol, and epigallocatechin-3-gallate are only a few examples of nano-encapsulated bioactive compounds with improved bioavailability and metabolism when compared to non-encapsulated ones [26]. Based on literature research, Table 11 summarizes the most recent nanocarrier formulations developed for bioactive compounds with therapeutical indications in CVDs, along with experimental models (in vitro or in vivo) and the main outputs.

Table 11. Nanocarrier formulations of different bioactive compounds.

Bioactive Compound	Nanoformulation	In Vitro/In Vivo Experimental Model	Main Results	Ref.
Curcumin	Chitosan NPs	In vitro/Ehrlich ascites carcinoma and in vivo animal model (albino mice)	Protection against myocardial injury and cardiac function, ameliorates EAC-induced cardiotoxicity.	[162]
	PEG-PDLLA (polyethylene glycol methyl ether-block-poly (D, L lactide) NPs	In vitro/cardiomyocytes exposed to palmitate	Inhibition of cell apoptosis and NADPH-mediated oxidative stress; protective effect possibly mediated by endoplasmic reticulum stress-related signaling pathway.	[163]
	Nano emulsion/glyceryl monooleate oil phase	In vitro/HMG-CoA reductase assay	Increased not only the HMGR inhibition (showing antihypercholesterolemic effect) but also angiotensin-converting enzyme (antihypertensive effect).	[164]
Curcumin + resveratrol	Pluronic® F127 micelles	In vitro/embryonic rat cardiomyocytes	Cardioprotection, reduction in apoptosis and ROS of cardiomyocytes treated with doxorubicin.	[165]
Resveratrol NPs	Solid–lipid NPs/glycerol monostearate oil phase	In vivo pharmacodynamic study/male mice	Protecting the myocardium, improving the cardiomyocyte calcium cycling, inhibiting of doxorubicin cardiotoxicity, inhibiting the production of reactive oxygen species.	[166]
	Albumin NPs	In vivo/male Sprague–Dawley rats	Improved neurological score and decreased infarct volume at 24 h after administration in a dose-dependent manner; significantly attenuated oxidative stress due to prolonged circulation in blood and sustained release pattern.	[167]
Quercetin	PLGA NPs	In vitro/H9c2 cells, a surrogate model of cardiac cells	Improves cardioprotection during hypoxia–reoxygenation injury through the preservation of mitochondrial function; superior protection capability of PLGA–quercetin NP with respect to free quercetin.	[168]
	Mesoporous silica nanoparticles	In vivo rat model of myocardial ischemia reperfusion	Improves the apoptosis degree and oxidative stress level of myocardial cells by regulating the JAK2/STAT3 signaling pathway, promoting the recovery of cardiac blood flow.	[169]
Coenzyme Q10	Nano emulsion/soybean oil and egg lecithin surfactant	In vitro/cardiomyocytes and Fibroblasts model	Providing multiple molecular mechanisms of cardioprotection during doxorubicin and trastuzumab treatments; anti-inflammatory activities modulating the heart microenvironment.	[170]

5.1. Clinical Trials

Curcumin is probably the most studied polyphenol in treatment of CVD, being an inhibitor of p300 histone acetyltransferase activity, which is associated with heart failure. Highly bioavailable curcumin has been developed as a nanoformulation commercially known as Theracurmin® [171]. This revolutionary formulation is a submicron crystal solid dispersion of curcumin, consisting of 10 $w/w\%$ curcumin, 2% other curcuminoids (such as dimethoxy-curcumin and bisdemethoxycurcumin), 4% of gum ghatti, and 84% of water. In a clinical study performed with healthy participants, low (150 mg) and high (210 mg) doses of Theracurmin® were administrated in order to evaluate plasma curcumin levels in a dose-dependent manner [172]. The study evidenced that nanoformulation increases plasma curcumin levels in a dose-dependent manner without saturating the absorption system. Another clinical study revealed that the treatment of hypertensive patients using 60 mg/day of Theracurmin for 24 weeks significantly improved the parameters of diastolic function assessed by doppler echocardiography, which suggests that the nanoformulation improves left ventricular diastolic function without interfering with blood pressure in hypertensive patients [171].

However, it is difficult to interpret whether the observed effect is due to the nanoparticle formulation or to curcumin itself, as the authors did not include a non-encapsulated formulation of curcumin as a control group. A comparison between curcumin nanoformulation and powder curcumin was performed by Sasaki et al. [173] upon administration in a healthy participant, revealing that the bioavailability of Theracurmin® orally administrated was 27-fold higher than that of curcumin powder, even at a low dosage (30 mg). It was concluded that Theracurmin® shows higher bioavailability than currently available preparations (curcumin powder).

Despite the large number of reported preclinical studies related to polyphenol nanoformulations, their transition to the clinical sector been proven to have several limitations, as it is well known that the concentrations of polyphenols proved to be effective in vitro or in small animal models are much higher than the required levels in human subjects. Moreover, the effectiveness of nano-nutraceutical products strongly depends on preserving the bioavailability of the bioactive compounds.

5.2. Nanoparticles for Theranostic

The concept of a theranostic is derived from combined therapy and diagnostic tools assembled into a single platform [174], while the nano-theranostic embodies the most advanced technological approach with multifunctional attributes such as multimodal imaging, controlled, and localized drug targeting, allowing the development of personalized medicine. In this respect, engineered nanomaterials consisting of magnetic nanoparticles, liposomes, carbon-based nanomaterials, metal and/or polymeric based nanoparticles are good candidates for dual applications in terms of both diagnostics and therapeutic approaches [175–177]. For example, magnetic nanoparticles coated with natural compounds have proved their efficiency in the imaging of CVDs. Suzuki et al. [178] reported ultrasmall superparamagnetic iron oxide nanoparticles coated with a specific polysaccharide (fucoidan) which have been successfully employed in MRI as contrast agents for arterial thrombus and elastase-induced vascular injury in a rat model. Another example is the case of gold nanoparticles used as contrasting agents, for the efficient detection and diagnosis of myocardial infarction. The method is based on engineered gold NPs conjugated with collagen, which demonstrated high-resolution detection of myocardial and ischemic injuries, along with adequate therapeutic tools [179]. Liposomal platforms have also proved to be a successful tool for diagnostic and therapeutic methods for platelet targeting in CVDs. The surface of liposomes with natural peptides was demonstrated to facilitate the drug delivery of active compounds and at the same time to target the cardiovascular tissues or damaged areas [180]. Many evidences for nano-based theranostics in terms of prognosis of atherosclerosis, myocardial infarction, aneurysms, angiogenesis, and other CVDs [175] have been reported, highlighting the benefit of sensitive detection of pathophysiological conditions combined with concomitant therapeutic measures. Although there has been reported success of preclinical studies related to theranostic nanoplatforms, the translational approach to clinical sectors remains unexplored, and moreover, the innate toxicity and stability of the nanoformulations are less studied. Most of the in vivo models were considered with small animals, and hence, huge differences might be detected by comparison with human anatomical features. On the other hand, the costs/benefits ratio must be considered, as there is a lack of studies reporting these aspects.

Developing innovative multifunctional nanomaterials with qualities that allow them to deliver specialized therapies through various physiological obstacles and to target specific cell types, tissues, and organs in the body is a major challenge for this project. Effective nano-delivery systems have ideal characteristics for controlled release of medicament, long storage life, and enhanced therapeutic efficacy with no or minimal side effects [181,182].

6. Conclusions

Several bioactive compounds are being diagnosed and tested to see if they have the potential to improve human health. They have antioxidant, anti-inflammatory, and anticarcinogenic factors, in addition to physiological and cellular benefits that protect against infectious diseases and metabolic illnesses such as diabetes, cardiovascular disorder, and cancer. They are derived from plants, and their consumption in diets has been related to tremendous fitness outcomes, making them ideal assets for the manufacturing of new nutritional dietary supplements with extensive shielding and preservative abilities.

Considering the high prevalence of CVDs worldwide, with a high level of morbidity and mortality, and excessive side effects of current therapies, the optimized alternative approaches are necessary for the prevention and/or treatment of these diseases. A large group of plant-derived bioactive compounds are used as alternative therapies for CVDs, which are summarized in this paper. They can be tailor-made to match the character and cultural preferences, to meet those lofty objectives, global and country-wide tasks to sell healthier, primarily plant-based diets. The role and significance of bioactive compounds for CVD treatment is still being explored, and its consequences must be established. Some polyphenols and flavonols used as bioactive compounds have been shown to reduce CVD risk factors. Their antioxidant and anti-inflammatory properties are the most important characteristics that make them favorable candidates for CVDs treatment. Along with the detailed properties of bioactive compounds used in CVDs, the current review provides an overview of the therapeutic advantages of nanoformulations along with recent advancements in this field.

The advantages of nano-delivery systems of bioactive compounds, recent preclinical studies, clinical trials, and nano-theranostic approaches are also discussed in this review, in order to offer a clearer understanding of the connection between bioactive compounds and CVD prevention, diagnostics, and treatment. Although the reported success of preclinical studies is related to the effectivity of nanoformulations and theranostic nanoplatforms, the translational approach to clinical sectors remains poorly explored, as the therapeutic applications of nano-phytomedicines in CVDs are still in their initial clinical phases. Nanotechnologies will provide a new window in the area of CVDs with the opportunity to achieve effective treatment, better prognosis, and fewer adverse effects on non-target tissues. However, further research is required for the development of low-cost nanoformulations and their effective usage, along with research in the toxicological approach.

Author Contributions: Conceptualization, R.K.S. and S.C.; data curation, R.K.S., A.G., E.A.Y.; writing—original draft preparation, R.K.S., A.G., E.A.Y.; writing—review and editing, R.K.S. and S.C.; funding acquisition, S.C. All authors have read and agreed to the published version of the manuscript.

Funding: This research received no external funding.

Institutional Review Board Statement: Not applicable.

Informed Consent Statement: Not applicable.

Acknowledgments: The authors R.K.S. and A.G. are grateful to Madhu Chitkara, Chancellor, Chitkara University, Rajpura, Patiala, India and Ashok Chitkara, Chancellor, Chitkara University, Rajpura, Patiala, India, for support and institutional facilities.

Conflicts of Interest: The authors declare no conflict of interest, financial or otherwise.

References

1. Martínez-Augustin, O.; Aguilera, C.M.; Gil-Campos, M.; Sánchez de Medina, F.; Gil, A. Bioactive anti-obesity food components. *Int. J. Vitam. Nutr. Res.* **2012**, *82*, 148–156. [CrossRef] [PubMed]
2. NIH National Cancer Institute. What Is Cancer? Available online: https://www.cancer.gov/about-cancer/understanding/what-is-cancer (accessed on 5 May 2021).
3. Felman, A.; Kohli, P. What to know about cardiovascular disease? *Med. News Today* **2019**. Available online: https://www.medicalnewstoday.com/articles/257484 (accessed on 26 July 2019).
4. Hemler, E.C.; Hu, F.B. Plant-Based Diets for Cardiovascular Disease Prevention: All Plant Foods Are Not Created Equal. *Curr. Atheroscler. Rep.* **2019**, *21*, 18. [CrossRef] [PubMed]
5. Alwan, A.; World Health Organization. *Global Status Report on Noncommunicable Diseases 2010*; World Health Organization: Geneva, Switzerland, 2011; Volume 9, p. 162. Available online: https://www.who.int/nmh/publications/ncd_report_full_en.pdf (accessed on 5 May 2021).
6. Gensini, G.F.; Comeglio, M.; Colella, A. Classical risk factors and emerging elements in the risk profile for coronary artery disease. *Eur. Heart J.* **1998**, *19* (Suppl. A), A53–A61.
7. Mahmood, S.S.; Levy, D.; Vasan, R.S.; Wang, T.J. The Framingham Heart Study and the epidemiology of cardiovascular disease: A historical perspective. *Lancet* **2014**, *383*, 999–1008. [CrossRef]
8. Kris-Etherton, P.M.; Lefevre, M.; Beecher, G.R.; Gross, M.D.; Keen, C.L.; Etherton, T.D. Bioactive compounds in nutrition and health-research methodologies for establishing biological function: The antioxidant and anti-inflammatory effects of flavonoids on atherosclerosis. *Annu. Rev. Nutr.* **2004**, *24*, 511–538. [CrossRef]
9. GBD 2013 Mortality and Causes of Death Collaborators. Global, regional, and national age–sex specific all-cause and cause-specific mortality for 240 causes of death, 1990–2013: A systematic analysis for the Global Burden of Disease Study 2013. *Lancet* **2015**, *385*, 117–171. [CrossRef]
10. Bowen, K.J.; Sullivan, V.K.; Kris-Etherton, P.M.; Petersen, K.S. Nutrition and Cardiovascular Disease-an Update. *Curr. Atheroscler. Rep.* **2018**, *20*, 8. [CrossRef] [PubMed]
11. Satija, A.; Hu, F.B. Plant-based diets and cardiovascular health. *Trends Cardiovasc. Med.* **2018**, *28*, 437–441. [CrossRef] [PubMed]
12. Guaadaoui, A.; Benaicha, S.; Elmajdoub, N.; Bellaoui, M.; Hamal, A. What is a Bioactive Compound? A Combined Definition for a Preliminary Consensus. *Int. J. Nutr. Food Sci.* **2014**, *3*, 174–179. [CrossRef]
13. González, S. Dietary Bioactive Compounds and Human Health and Disease. *Nutrients* **2020**, *12*, 348. [CrossRef] [PubMed]
14. NCI Dictionary of Cancer Terms. Bioactive Compounds. Available online: https://www.cancer.gov/publications/dictionaries/cancer-terms/def/bioactive-compound (accessed on 24 February 2021).
15. Middleton, E., Jr.; Kandaswami, C.; Theoharides, T.C. The effects of plant flavonoids on mammalian cells: Implications for inflammation, heart disease, and cancer. *Pharmacol. Rev.* **2000**, *52*, 673–751. [PubMed]
16. Conti, M.V.; Guzzetti, L.; Panzeri, D.; De Giuseppe, R.; Coccetti, P.; Labra, M.; Cena, H. Bioactive compounds in legumes: Implications for sustainable nutrition and health in the elderly population. *Trends Food Sci. Technol.* **2021**. [CrossRef]
17. Bazzano, L.A.; He, J.; Ogden, L.G.; Loria, C.M.; Vupputuri, S.; Myers, L.; Whelton, P.K. Fruit and vegetable intake and risk of cardiovascular disease in US adults: The first National Health and Nutrition Examination Survey Epidemiologic Follow-up Study. *Am. J. Clin. Nutr.* **2002**, *76*, 93–99. [CrossRef] [PubMed]
18. Hu, F.B. Dietary pattern analysis: A new direction in nutritional epidemiology. *Curr. Opin. Lipidol.* **2002**, *13*, 3–9. [CrossRef] [PubMed]
19. Trumbo, P.; Schlicker, S.; Yates, A.A.; Poos, M. Food and Nutrition Board of the Institute of Medicine, The National Academies. Dietary reference intakes for energy, carbohydrate, fiber, fat, fatty acids, cholesterol, protein and amino acids. *J. Am. Diet. Assoc.* **2002**, *102*, 1621–1630. [CrossRef]
20. Jimeno, J.; Faircloth, G.; Sousa-Faro, J.M.F.; Scheuer, P.; Rinehart, K. New Marine Derived Anticancer Therapeutics—A Journey from the Sea to Clinical Trials. *Mar. Drugs.* **2004**, *2*, 14–29. [CrossRef]
21. Vignesh, S.; Raja, A.; James, A.R. Marine drugs: Implication and Future Studies. *Int. J. Pharmacol.* **2011**, *7*, 22–30. [CrossRef]
22. Food and Drug Administration. *FDA Basics—Dietary Supplements*; Office of Dietary Supplements Programs, HFS-810 (FDA). 2019, USA. Available online: https://www.fda.gov/food/dietary-supplements (accessed on 5 May 2021).
23. Scottish Intercollegiate Guidelines Network (SIGN). *Risk Estimation and the Prevention of Cardiovascular Disease: A National Clinical Guideline*. Edinburgh (Scotland): Scottish Intercollegiate Guidelines Network; 2017. *(SIGN publication no. 149)*; 1–118. Available online: http://www.sign.ac.uk (accessed on 5 May 2021).
24. Rahman, K.; Lowe, G.M. Garlic and cardiovascular disease: A critical review. *J. Nutr.* **2006**, *136* (Suppl. 3), 736S–740S. [CrossRef]
25. Zhang, J.; Wen, C.; Zhang, H.; Duan, Y.; Ma, H. Recent advances in the extraction of bioactive compounds with subcritical water: A review. *Trends Food Sci. Technol.* **2020**, *95*, 183–195. [CrossRef]
26. Gonçalves, R.F.S.; Martins, J.T.; Duarte, C.M.M.; Vicente, A.A.; Pinheiro, A.C. Advances in nutraceutical delivery systems: From formulation design for bioavailability enhancement to efficacy and safety evaluation. *Trends Food Sci. Technol.* **2018**, *78*, 270–291. [CrossRef]
27. Sharma, M.; Koul, A.; Sharma, D.; Kaul, S.; Swamy, M.; Dhar, M.K. Metabolic engineering strategies for enhancing the production of bio-active compounds from medicinal plants. In *Natural Bio-Active Compounds*; Akhtar, M., Swamy, M., Eds.; Springer: Singapore, 2019. [CrossRef]

28. Depuydt, S.; Van Praet, S.; Nelissen, H.; Vanholme, B.; Vereecke, D. How plant hormones and their interactions affect cell growth. In *Molecular Cell Biology of the Growth and Differentiation of Plant Cells*; Rose, R.J., Ed.; CRC Press: Boca Raton, FL, USA, 2016; pp. 174–195.

29. Chaves Lobón, N.; Ferrer de la Cruz, I.; Alías Gallego, J.C. Autotoxicity of Diterpenes Present in Leaves of *Cistus ladanifer* L. *Plants* **2019**, *8*, 27. [CrossRef] [PubMed]

30. Azmir, J.; Zaidul, I.S.M.; Rahman, M.M.; Sharif, K.M.; Mohamed, A.; Sahena, F.; Jahurul, M.H.A.; Ghafoor, K.; Norulaini, N.A.N.; Omar, A.K.M. Techniques for extraction of bioactive compounds from plant materials: A review. *J. Food Eng.* **2013**, *117*, 426–436. [CrossRef]

31. Ahmed, B.; Mohamed, S.; Adel, A.-R. Antioxidant Activities and Potential Impacts to Reduce Aflatoxins Utilizing Jojoba and Jatropha Oils and Extracts. *Int. J. Pharmacol.* **2017**, *13*, 1103–1114.

32. Loi, M.; Paciolla, C.; Logrieco, A.F.; Mulè, G. Plant Bioactive Compounds in Pre- and Postharvest Management for Aflatoxins Reduction. *Front. Microbiol.* **2020**, *11*, 243. [CrossRef] [PubMed]

33. Walden, R.; Tomlinson, B. Cardiovascular Disease. In *Herbal Medicine: Biomolecular and Clinical Aspects*, 2nd ed.; Benzie, I.F.F., Wachtel-Galor, S., Eds.; CRC Press/Taylor & Francis: Boca Raton, FL, USA, 2011; Chapter 16. Available online: www.ncbi.nlm.nih.gov/books/NBK92767/ (accessed on 5 May 2021).

34. Ciumărnean, L.; Milaciu, M.V.; Runcan, O.; Vesa, Ș.C.; Răchișan, A.L.; Negrean, V.; Perné, M.G.; Donca, V.I.; Alexescu, T.G.; Para, I.; et al. The Effects of Flavonoids in Cardiovascular Diseases. *Molecules* **2020**, *25*, 4320. [CrossRef]

35. McCullough, M.L.; Peterson, J.J.; Patel, R.; Jacques, P.F.; Shah, R.; Dwyer, J.T. Flavonoid intake and cardiovascular disease mortality in a prospective cohort of US adults. *Am. J. Clin. Nutr.* **2012**, *95*, 454–464. [CrossRef]

36. Durazzo, A.; Lucarini, M.; Novellino, E.; Souto, E.B.; Daliu, P.; Santini, A. Abelmoschus esculentus L.: Bioactive Components' Beneficial Properties-Focused on Antidiabetic Role-For Sustainable Health Applications. *Molecules* **2018**, *24*, 38. [CrossRef] [PubMed]

37. Saida, M.; Lamjed, B.; Amina, B.; Rym, E.; Soumaya, H.; Salem, E. Biological activities, and phytocompounds of northwest Algeria Ajuga iva (L) extracts: Partial identification of the antibacterial fraction. *Microb. Pathog.* **2018**, *121*, 173–178.

38. Wang, S.; Zhao, Y.; Song, J.; Wang, R.; Gao, L.; Zhang, L.; Fang, L.; Lu, Y.; Du, G. Total flavonoids from Anchusa italica Retz. Improve cardiac function and attenuate cardiac remodeling post myocardial infarction in mice. *J. Ethnopharmacol.* **2020**, *257*, 112887. [CrossRef]

39. Bondonno, N.P.; Bondonno, C.P.; Blekkenhorst, L.C.; Considine, M.J.; Maghzal, G.; Stocker, R.; Woodman, R.J.; Ward, N.C.; Hodgson, J.M.; Croft, K.D. Flavonoid-Rich Apple Improves Endothelial Function in Individuals at Risk for Cardiovascular Disease: A Randomized Controlled Clinical Trial. *Mol. Nutr. Food Res.* **2018**, *62*, 1700674. [CrossRef]

40. Panthi, M.; Subba, R.K.; Raut, B.; Khanal, D.P.; Koirala, N. Bioactivity evaluations of leaf extract fractions from young barley grass and correlation with their phytochemical profiles. *BMC Complement. Med. Ther.* **2020**, *20*, 64. [CrossRef]

41. Pereyra, K.V.; Andrade, D.C.; Toledo, C.; Schwarz, K.G.; Uribe-Ojeda, A.; RÃ-os-Gallardo, A.P.; Quintanilla, R.A.; Contreras, S.A.; Del Rio, R. Dietary supplementation of a sulforaphane-enriched broccoli extract protects the heart from acute cardiac stress. *J. Funct. Foods* **2020**, *75*, 104267. [CrossRef]

42. Zaini, R.G.; Brandt, K.; Clench, M.R.; Le Maitre, C.L. Effects of bioactive compounds from carrots (*Daucus carota* L.), polyacetylenes, beta-carotene and lutein on human lymphoid leukaemia cells. *Anticancer Agents Med. Chem.* **2012**, *12*, 640–652. [CrossRef] [PubMed]

43. Bartłomiej, S.; Justyna, R.K.; Ewa, N. Bioactive compounds in cereal grains—Occurrence, structure, technological significance and nutritional benefits—A review. *Food Sci. Technol. Int.* **2012**, *18*, 559–568. [CrossRef] [PubMed]

44. Martin, M.Á.; Ramos, S. Impact of cocoa flavanols on human health. *Food Chem. Toxicol.* **2021**, *151*, 112121. [CrossRef]

45. Neetu, K.; Choudhary, S.B.; Sharma, H.K.; Singh, B.K.; Kumar, A.A. Health-promoting properties of Corchorus leaves: A review. *J. Herb. Med.* **2019**, *15*, 100240.

46. Haidar, B.; Ferdous, M.; Fatema, B.; Ferdous, A.S.; Islam, M.R.; Khan, H. Population diversity of bacterial endophytes from jute (Corchorus olitorius) and evaluation of their potential role as bioinoculants. *Microbiol. Res.* **2018**, *208*, 43–53. [CrossRef]

47. Polley, K.R.; Oswell, N.J.; Pegg, R.B.; Paton, C.M.; Cooper, J.A. A 5-day high-fat diet rich in cottonseed oil improves cholesterol profiles and triglycerides compared to olive oil in healthy men. *Nutr. Res.* **2018**, *60*, 43–53. [CrossRef]

48. Simha, P.; Mathew, M.; Ganesapillai, M. Empirical modeling of drying kinetics and microwave assisted extraction of bioactive compounds from Adathoda vasica and Cymbopogon citratus. *Alex. Eng. J.* **2016**, *55*, 141–150. [CrossRef]

49. Orgah, J.O.; He, S.; Wang, Y.; Jiang, M.; Wang, Y.; Orgah, E.A.; Duan, Y.; Zhao, B.; Zhang, B.; Han, J.; et al. Pharmacological potential of the combination of Salvia miltiorrhiza (Danshen) and Carthamus tinctorius (Honghua) for diabetes mellitus and its cardiovascular complications. *Pharmacol. Res.* **2020**, *153*, 104654. [CrossRef]

50. Zhang, H.; Wang, S.; Liu, Q.; Zhang, H.; Wang, S.; Liu, Q.; Zheng, H.; Liu, X.; Wang, X.; Shen, T.; et al. Dracomolphin AE, new lignans from Dracocephalum moldavica. *Fitoterapia* **2021**, *150*, 104841. [CrossRef]

51. Bayang, J.P.; Laya, A.; Kolla, M.C.; Koubala, B.B. Variation of physical properties, nutritional value and bioactive nutrients in dry and fresh wild edible fruits of twenty-three species from Far North region of Cameroon. *J. Agric. Food Res.* **2021**, *4*, 100146. [CrossRef]

52. Miao, S.M.; Zhang, Q.; Bi, X.B.; Cui, J.L.; Wang, M.L. A review of the phytochemistry and pharmacological activities of Ephedra herb. *Chin. J. Nat. Med.* **2020**, *18*, 321–344. [CrossRef]

53. Graefe, E.U.; Veit, M. Urinary metabolites of flavonoids and hydroxycinnamic acids in humans after application of a crude extract from Equisetum arvense. *Phytomedicine* **1999**, *6*, 239–246. [CrossRef]

54. Shim, Y.Y.; Gui, B.; Arnison, P.G.; Shim, Y.Y.; Gui, B.; Arnison, P.G.; Wang, Y.; Reaney, M.J.T.; Jiang, M.; Wang, Y.; et al. Flaxseed (*Linum usitatissimum* L.) bioactive compounds and peptide nomenclature: A review. *Trends Food Sci. Technol.* **2014**, *38*, 5–20. [CrossRef]

55. Al-Snafi, A. Phytochemical constituents and medicinal properties of digitalis lanata and digitalis purpurea—A Review. *IAJPS* **2017**, *4*, 225–234.

56. Martins, N.; Petropoulos, S.; Ferreira, I.C. Chemical composition and bioactive compounds of garlic (*Allium sativum* L.) as affected by pre- and post-harvest conditions: A review. *Food Chem.* **2016**, *211*, 41–50. [CrossRef] [PubMed]

57. Hassan, I.; Ibrahim, W.; Yusuf, F.M.; Ahmad, S.A.; Ahmad, S. Biochemical Constituent of Ginkgo biloba (Seed) 80% Methanol Extract Inhibits Cholinesterase Enzymes in Javanese Medaka (Oryzias javanicus) Model. *J. Toxicol.* **2020**, *2020*, 8815313. [CrossRef]

58. Yu, H.; Zhao, J.; You, J.; Li, J.; Ma, H.; Chen, X. Factors influencing cultivated ginseng (Panax ginseng CA Meyer) bioactive compounds. *PLoS ONE* **2019**, *14*, e0223763. [CrossRef]

59. Rojas, R.; Castro-lópez, C.; Sánchez-Alejo, E.J.; Niño-Medina, G.; Martínez-Ávila, C.G. Phenolic compound recovery from grape fruit and by- products: An overview of extraction methods. *Grape Wine Biotechnol.* **2016**. [CrossRef]

60. Rababa'h, A.M.; Al Yacoub, O.N.; El-Elimat, T.; Rabab'ah, M.; Altarabsheh, S.; Deo, S.; Al-Azayzih, A.; Zayed, A.; Alazzam, S.; Alzoubi, K.H. The effect of hawthorn flower and leaf extract (*Crataegus* Spp.) on cardiac hemostasis and oxidative parameters in Sprague Dawley rats. *Heliyon* **2020**, *6*, e04617. [CrossRef] [PubMed]

61. Souza, J.S.N.; Machado, L.L.; Pessoa, O.D.; Braz-Filho, R.; Overk, C.R.; Yao, P.; Lemos, T.L. Pyrrolizidine alkaloids from heliotropium indicum. *J. Braz. Chem. Soc.* **2005**, *16*, 1410–1414. [CrossRef]

62. De Angelis, D.; Pasqualone, A.; Costantini, M.; Ricciardi, L.; Lotti, C.; Pavan, S.; Summo, C. Data on the proximate composition, bioactive compounds, physicochemical and functional properties of a collection of faba beans (*Vicia faba* L.) and lentils (*Lens culinaris* Medik.). *Data Brief* **2020**, *34*, 106660. [CrossRef]

63. Lu, J.; He, R.; Sun, P.; Zhang, F.; Linhardt, R.J.; Zhang, A. Molecular mechanisms of bioactive polysaccharides from Ganoderma lucidum (Lingzhi), a review. *Int. J. Biol. Macromol.* **2020**, *150*, 765–774. [CrossRef]

64. Sudheer, S.; Alzorqi, I.; Manickam, S. Bioactive Compounds of the Wonder Medicinal Mushroom "*Ganoderma lucidum*". In *Bioactive Molecules in Food. Reference Series in Phytochemistry*; Mérillon, J.M., Ramawat, K., Eds.; Springer: Cham, Switzerland, 2019.

65. Leone, A.; Spada, A.; Battezzati, A.; Schiraldi, A.; Aristil, J.; Bertoli, S. Moringa oleifera Seeds and Oil: Characteristics and Uses for Human Health. *Int. J. Mol. Sci.* **2016**, *17*, 2141. [CrossRef] [PubMed]

66. Tian, Y.; Deng, F. Phytochemistry and biological activity of mustard (*Brassica juncea*): A review. *CyTA-J. Food* **2020**, *18*, 704–718. [CrossRef]

67. Wu, J.R.; Leu, H.B.; Yin, W.H. The benefit of secondary prevention with oat fiber in reducing future cardiovascular event among CAD patients after coronary intervention. *Sci. Rep.* **2019**, *9*, 3091. [CrossRef] [PubMed]

68. Boskou, D. Table Olives as Sources of Bioactive Compounds. In Olive and Olive Oil Bioactive Constituents. *J. Agric. Food Chem.* **2015**, 217–259. [CrossRef]

69. Marrelli, M.; Amodeo, V.; Statti, G.; Conforti, F. Biological Properties and Bioactive Components of *Allium cepa* L.: Focus on Potential Benefits in the Treatment of Obesity and Related Comorbidities. *Molecules* **2018**, *24*, 119. [CrossRef] [PubMed]

70. Alper, C.M.; Mattes, R.D. Peanut consumption improves indices of cardiovascular disease risk in healthy adults. *J. Am. Coll. Nutr.* **2003**, *22*, 133–141. [CrossRef] [PubMed]

71. Mahmoudi, M.; Boughalleb, F.; Bouhemda, T.; Abdellaoui, R.; Nasri, N. Unexploited polygonum equisetiforme seeds: Potential source of useful natural bioactive products. *Ind. Crop. Prod.* **2018**, *122*, 349–357. [CrossRef]

72. Shah, A.R.; Sharma, P.; Gour, V.S.; Kothari, S.L.; Dar, K.B.; Ganie, S.A.; Shah, Y.R. Antioxidant, Nutritional, Structural, Thermal and Physico-Chemical Properties of Psyllium (Plantago Ovata) Seeds. *Curr. Res. Nutr. Food Sci.* **2020**, *8*. [CrossRef]

73. Ferraz da Costa, D.C.; Pereira Rangel, L.; Quarti, J.; Santos, R.A.; Silva, J.L.; Fialho, E. Bioactive compounds and metabolites from grapes and red wine in breast cancer chemoprevention and therapy. *Molecules* **2020**, *25*, 3531. [CrossRef] [PubMed]

74. Salehi, E.; Sardarodiyan, M. Bioactive phytochemicals in rice bran: Processing and functional properties. *Biochem. Ind. J.* **2016**, *10*, 101.

75. Zampelas, A. The Effects of Soy and its Components on Risk Factors and End Points of Cardiovascular Diseases. *Nutrients* **2019**, *11*, 2621. [CrossRef] [PubMed]

76. Ahmed, H.M.; Al-Zubaidy, A.M.A. Exploring natural essential oil components and antibacterial activity of solvent extracts from twelve *Perilla frutescens* L. Genotypes. *Arab. J. Chem.* **2020**, *13*, 7390–7402. [CrossRef]

77. Cheng, H.M.; Koutsidis, G.; Lodge, J.K.; Ashor, A.; Siervo, M.; Lara, J. Tomato and lycopene supplementation and cardiovascular risk factors: A systematic review and meta-analysis. *Atherosclerosis* **2017**, *257*, 100–108. [CrossRef] [PubMed]

78. Wu, S.; Xu, T.; Akoh, C. Effect of roasting on the volatile constituents of trichosanthes kirilowii seeds. *J. Food Drug Anal.* **2014**, *22*, 310–317. [CrossRef]

79. Melini, V.; Acquistucci, R. Health-promoting compounds in pigmented thai and wild rice. *Foods* **2017**, *6*, 9. [CrossRef]

80. Dai, W.; Zhang, Z.; Zhao, S. Baseline levels of serum high sensitivity C reactive protein and lipids in predicting the residual risk of cardiovascular events in Chinese population with stable coronary artery disease: A prospective cohort study. *Lipids Health Dis.* **2018**, *17*, 273. [CrossRef]

81. Ververidis, F.; Trantas, E.; Douglas, C.; Vollmer, G.; Kretzschmar, G.; Panopoulos, N. Biotechnology of flavonoids and other phenylpropanoid-derived natural products. Part I: Chemical diversity, impacts on plant biology and human health. *Biotechnol. J.* **2007**, *2*, 1214–1234. [CrossRef] [PubMed]

82. Lotito, S.B.; Frei, B. Consumption of flavonoid-rich foods and increased plasma antioxidant capacity in humans: Cause, consequence, or epiphenomenon? *Free Radic. Biol. Med.* **2006**, *41*, 1727–1746. [CrossRef] [PubMed]

83. Izzi, V.; Masuelli, L.; Tresoldi, I.; Sacchetti, P.; Modesti, A.; Galvano, F.; Bei, R. The effects of dietary flavonoids on the regulation of redox inflammatory networks. *Front. Biosci.* **2012**, *17*, 2396–2418. [CrossRef] [PubMed]

84. Chang, C.F.; Cho, S.; Wang, J. Epicatechin protects hemorrhagic brain via synergistic Nrf2 pathways. *Ann. Clin. Transl. Neurol.* **2014**, *1*, 258–271. [CrossRef] [PubMed]

85. Nishiumi, S.; Miyamoto, S.; Kawabata, K.; Ohnishi, K.; Mukai, R.; Murakami, A.; Ashida, H.; Terao, J. Dietary flavonoids as cancer-preventive and therapeutic biofactors. *Front. Biosci.* **2011**, *3*, 1332–1362. [CrossRef]

86. McKay, D.L.; Chen, C.Y.; Saltzman, E.; Blumberg, J.B. Hibiscus sabdariffa L. tea (tisane) lowers blood pressure in prehypertensive and mildly hypertensive adults. *J. Nutr.* **2010**, *140*, 298–303. [CrossRef] [PubMed]

87. Clark, J.L.; Zahradka, P.; Taylor, C.G. Efficacy of flavonoids in the management of high blood pressure. *Nutr. Rev.* **2015**, *73*, 799–822. [CrossRef] [PubMed]

88. Mahmoud, A.M.; Hernández Bautista, R.J.; Sandhu, M.A.; Hussein, O.E. Beneficial Effects of Citrus Flavonoids on Cardiovascular and Metabolic Health. *Oxid. Med. Cell. Longev.* **2019**, *2019*, 5484138. [CrossRef] [PubMed]

89. Su, J.; Xu, H.-T.; Yu, J.-J.; Gao, J.; Lei, J.; Yin, Q.S.; Li, B.; Pang, M.X.; Su, M.X.; Mi, W.J.; et al. Luteolin ameliorates hypertensive vascular remodeling through inhibiting the proliferation and migration of vascular smooth muscle cells. *Evid. Based Complement. Altern. Med.* **2015**, *2015*, 364876. [CrossRef]

90. Thamcharoen, N.; Susantitaphong, P.; Wongrakpanich, S.; Chongsathidkiet, P.; Tantrachoti, P.; Pitukweerakul, S.; Avihingsanon, Y.; Praditpornsilpa, K.; Jaber, B.L.; Eiam-Ong, S. Effect of N- and T-type calcium channel blocker on proteinuria, blood pressure and kidney function in hypertensive patients: A meta-analysis. *Hypertens. Res.* **2015**, *38*, 902. [CrossRef]

91. Jiang, R.W.; Lau, K.M.; Lam, H.M.; Yam, W.S.; Leung, L.K.; Choi, K.L.; Waye, M.M.; Mak, T.C.; Woo, K.S.; Fung, K.P. A comparative study on aqueous root extracts of Pueraria thomsonii and Pueraria lobata by antioxidant assay and HPLC fingerprint analysis. *J. Ethnopharmacol.* **2005**, *96*, 133–138. [CrossRef]

92. Tsuda, T. Dietary anthocyanin-rich plants: Biochemical basis and recent progress in health benefits studies. *Mol. Nutr. Food Res.* **2012**, *56*, 159–170. [CrossRef] [PubMed]

93. Srilakshami, B. *Food Science*, 7th ed.; New Age International Publishers: New Delhi, India, 2018.

94. Celestino, S.B.; Augustin, S. Proanthocyanidins and tannin-like compounds—Nature, occurrence, dietary intake and effects on nutrition and health. *J. Sci. Food Agric.* **2000**, *80*, 1094–1117.

95. Salisbury, F.; Ross, C. *Plant Physiology*, 4th ed.; Wadsworth: Belmont, CA, USA, 1991; pp. 325–326.

96. Rahimi, P.; Abedimanesh, S.; Mesbah-Namin, S.A.; Ostadrahimi, A. Betalains, the nature-inspired pigments, in health and diseases. *Crit. Rev. Food Sci. Nutr.* **2019**, *59*, 2949–2978. [CrossRef]

97. Antonescu, A.-I.; Miere, F.; Fritea, L.; Ganea, M.; Zdrinca, M.; Dobjanschi, L.; Antonescu, A.; Vicas, S.I.; Bodog, F.; Sindhu, R.K.; et al. Perspectives on the Combined Effects of Ocimum basilicum and Trifolium pratense Extracts in Terms of Phytochemical Profile and Pharmacological Effects. *Plants* **2021**, *10*, 1390. [CrossRef] [PubMed]

98. Leoncini, E.; Nedovic, D.; Panic, N.; Pastorino, R.; Edefonti, V.; Boccia, S. Carotenoid Intake from Natural Sources and Head and Neck Cancer: A Systematic Review and Meta-analysis of Epidemiological Studies. *Cancer Epidemiol. Biomark. Prev.* **2015**, *24*, 1003–1011. [CrossRef]

99. Voutilainen, S.; Nurmi, T.; Mursu, J.; Rissanen, T.H. Carotenoids and cardiovascular health. *Am. J. Clin. Nutr.* **2006**, *83*, 1265–1271. [CrossRef] [PubMed]

100. Granado-Lorencio, F.; Lagarda, M.J.; Garcia-López, F.J.; Sánchez-Siles, L.M.; Blanco-Navarro, I.; Alegría, A.; Pérez-Sacristán, B.; Garcia-Llatas, G.; Donoso-Navarro, E.; Silvestre-Mardomingo, R.A.; et al. Effect of β-cryptoxanthin plus phytosterols on cardiovascular risk and bone turnover markers in post-menopausal women: A randomized crossover trial. *Nutr. Metab. Cardiovasc. Dis.* **2014**, *24*, 1090–1096. [CrossRef] [PubMed]

101. Osganian, S.K.; Stampfer, M.J.; Rimm, E.; Spiegelman, D.; Manson, J.E.; Willett, W.C. Dietary carotenoids and risk of coronary artery disease in women. *Am. J. Clin. Nutr.* **2003**, *77*, 1390–1399. [CrossRef] [PubMed]

102. Ross, A.C.; Zolfaghari, R.; Weisz, J. Vitamin A: Recent advances in the biotransformation, transport, and metabolism of retinoids. *Curr. Opin. Gastroenterol.* **2001**, *17*, 184–192. [CrossRef] [PubMed]

103. Holden, J.M.; Eldridge, A.L.; Beecher, G.R.; Buzzard, I.M.; Bhagwat, S.; Davis, C.S.; Douglass, L.W.; Gebhardt, S.; Haytowiz, D.; Schakel, S. Carotenoid content of US foods: An update of the database. *J. Food Comp. Anal.* **1999**, *12*, 169–196. [CrossRef]

104. Stahl, W.; Sies, H. Lycopene: A biologically important carotenoid for humans? *Arch. Biochem. Biophys.* **1996**, *336*, 1–9. [CrossRef] [PubMed]

105. Sommerburg, O.; Keunen, J.E.; Bird, A.; Van Kuijk, F. J G M. Fruits and vegetables that are sources for lutein and zeaxanthin: The macular pigment in human eyes. *Br. J. Ophthalmol.* **1998**, *82*, 907–910. [CrossRef] [PubMed]

106. Jialal, I.; Norkus, E.P.; Cristol, L.; Grundy, S.M. beta-Carotene inhibits the oxidative modification of low-density lipoprotein. *Biochim. Biophys. Acta* **1991**, *1086*, 134–138. [CrossRef]

107. Ciccone, M.M.; Cortese, F.; Gesualdo, M.; Carbonara, S.; Zito, A.; Ricci, G.; De Pascalis, F.; Scicchitano, P.; Riccioni, G. Dietary intake of carotenoids and their antioxidant and anti-inflammatory effects in cardiovascular care. *Mediat. Inflamm.* **2013**, *2013*, 782137. [CrossRef] [PubMed]

108. Trautwein, E.A.; Vermeer, M.A.; Hiemstra, H.; Ras, R.T. LDL-Cholesterol Lowering of Plant Sterols and Stanols-Which Factors Influence Their Efficacy? *Nutrients* **2018**, *10*, 1262. [CrossRef]

109. Radojčić Redovniković, I.; Glivetić, T.; Delonga, K.; Vorkapic-Furac, J. Glucosinolates and their potential role in plant. *Period. Biol.* **2008**, *110*, 297–309.

110. Tse, G.; Eslick, G.D. Cruciferous vegetables and risk of colorectal neoplasms: A systematic review and meta-analysis. *Nutr. Cancer* **2014**, *66*, 128–139. [CrossRef] [PubMed]
111. Soundararajan, P.; Kim, J.S. Anti-Carcinogenic Glucosinolates in Cruciferous Vegetables and Their Antagonistic Effects on Prevention of Cancers. *Molecules* **2018**, *23*, 2983. [CrossRef] [PubMed]
112. Loef, M.; Walach, H. Fruit, vegetables and prevention of cognitive decline or dementia: A systematic review of cohort studies. *J. Nutr. Health Aging* **2012**, *16*, 626–630. [CrossRef]
113. Walia, A.; Gupta, A.K.; Sharma, V. Role of Bioactive Compounds in Human Health. *Acta Sci. Med. Sci.* **2019**, *3*, 25–33.
114. Greenberg, E.R.; Baron, J.A.; Karagas, M.R.; Greenberg, E.R.; Baron, J.A.; Karagas, M.R.; Stukel, T.A.; Nierenberg, D.W.; Stevens, M.M.; Mandel, J.S.; et al. Mortality associated with low plasma concentration of beta carotene and the effect of oral supplementation. *JAMA* **1996**, *275*, 699–703. [CrossRef] [PubMed]
115. Heart Protection Study Collaborative Group. MRC/BHF Heart Protection Study of antioxidant vitamin supplementation in 20,536 high-risk individuals: A randomised placebo-controlled trial. *Lancet* **2002**, *360*, 23–33. [CrossRef]
116. Liu, S.; Lee, I.M.; Ajani, U.; Cole, S.R.; Buring, J.E.; Manson, J.E. Physicians' Health Study. Intake of vegetables rich in carotenoids and risk of coronary heart disease in men: The Physicians' Health Study. *Int. J. Epidemiol.* **2001**, *30*, 130–135. [CrossRef]
117. Hennekens, C.H.; Buring, J.E.; Manson, J.E.; Stampfer, M.; Rosner, B.; Cook, N.R.; Belanger, C.; LaMotte, F.; Gaziano, J.M.; Ridker, P.M.; et al. Lack of effect of long-term supplementation with beta carotene on the incidence of malignant neoplasms and cardiovascular disease. *N. Engl. J. Med.* **1996**, *334*, 1145–1149. [CrossRef] [PubMed]
118. Basu, A.; Rhone, M.; Lyons, T.J. Berries: Emerging impact on cardiovascular health. *Nutr. Rev.* **2010**, *68*, 168–177. [CrossRef]
119. Basu, A.; Du, M.; Leyva, M.J.; Sanchez, K.; Betts, N.M.; Wu, M.; Aston, C.E.; Lyons, T.J. Blueberries decrease cardiovascular risk factors in obese men and women with metabolic syndrome. *J. Nutr.* **2010**, *140*, 1582–1587. [CrossRef] [PubMed]
120. Goszcz, K.; Duthie, G.G.; Stewart, D.; Leslie, S.J.; Megson, I.L. Bioactive polyphenols and cardiovascular disease: Chemical antagonists, pharmacological agents or xenobiotics that drive an adaptive response? *Br. J. Pharmacol.* **2017**, *174*, 1209–1225. [CrossRef] [PubMed]
121. Asgary, S.; Sahebkar, A.; Afshani, M.R.; Keshvari, M.; Haghjooyjavanmard, S.; Rafieian-Kopaei, M. Clinical evaluation of blood pressure lowering, endothelial function improving, hypolipidemic and anti-inflammatory effects of pomegranate juice in hypertensive subjects. *Phytother. Res.* **2014**, *28*, 193–199. [CrossRef] [PubMed]
122. Dohadwala, M.M.; Holbrook, M.; Hamburg, N.M.; Shenouda, S.M.; Chung, W.B.; Titas, M.; Kluge, M.A.; Wang, N.; Palmisano, J.; Milbury, P.E.; et al. Effects of cranberry juice consumption on vascular function in patients with coronary artery disease. *Am. J. Clin. Nutr.* **2011**, *93*, 934–940. [CrossRef]
123. Arab, L.; Steck, S. Lycopene and cardiovascular disease. *Am. J. Clin. Nutr.* **2000**, *71* (Suppl. 6), 1691S–1697S. [CrossRef]
124. Rees, A.; Dodd, G.F.; Spencer, J.P.E. The Effects of Flavonoids on Cardiovascular Health: A Review of Human Intervention Trials and Implications for Cerebrovascular Function. *Nutrients* **2018**, *10*, 1852. [CrossRef] [PubMed]
125. Wang, X.; Ouyang, Y.Y.; Liu, J.; Zhao, G. Flavonoid intake and risk of CVD: A systematic review and meta-analysis of prospective cohort studies. *Br. J. Nutr.* **2014**, *111*, 1–11. [CrossRef] [PubMed]
126. Kumar, P.; Mahato, D.K.; Kamle, M.; Mohanta, T.K.; Kang, S.G. Aflatoxins: A global concern for food safety, human health and their management. *Front. Microbial.* **2017**, *7*, 2170. [CrossRef] [PubMed]
127. Williamson, G.; Kay, C.D.; Crozier, A. The bioavailability, transport, and bioactivity of dietary flavonoids: A review from a historical perspective. *Compr. Rev. Food Sci. Food Saf.* **2018**, *17*, 1054–1112. [CrossRef] [PubMed]
128. Thilakarathna, S.H.; Rupasinghe, H.P. Flavonoid bioavailability and attempts for bioavailability enhancement. *Nutrients* **2013**, *5*, 3367–3387. [CrossRef]
129. Kawabata, K.; Yoshioka, Y.; Terao, J. Role of Intestinal Microbiota in the Bioavailability and Physiological Functions of Dietary Polyphenols. *Molecules* **2019**, *24*, 370. [CrossRef]
130. Pei, R.; Liu, X.; Bolling, B. Flavonoids and gut health. *Curr. Opin. Biotechnol.* **2020**, *61*, 153–159. [CrossRef]
131. Cinta-Pinzaru, S.; Cavalu, S.; Leopold, N.; Petry, R.; Kiefer, W. Raman and surface-enhanced Raman spectroscopy of tempyo spin labelled ovalbumin. *J. Mol. Struct.* **2001**, *565*, 225–229. [CrossRef]
132. Arriagada, F.; Günther, G.; Morales, J. Nanoantioxidant-Based Silica Particles as Flavonoid Carrier for Drug Delivery Applications. *Pharmaceutics* **2020**, *12*, 302. [CrossRef]
133. Zuo, L.; Ao, X.; Guo, Y. Study on the synthesis of dual-chain ionic liquids and their application in the extraction of flavonoids. *J. Chromatogr. A* **2020**, *1628*, 461446. [CrossRef]
134. De Whalley, C.V.; Rankin, S.M.; Hoult, J.R.; Jessup, W.; Leake, D.S. Flavonoids inhibit the oxidative modification of low-density lipoproteins by macrophages. *Biochem. Pharmacol.* **1990**, *39*, 1743–1750. [CrossRef]
135. Fuhrman, B.; Lavy, A.; Aviram, M. Consumption of red wine with meals reduces the susceptibility of human plasma and low-density lipoprotein to lipid peroxidation. *Am. J. Clin. Nutr.* **1995**, *61*, 549–554. [CrossRef]
136. Arai, Y.; Watanabe, S.; Kimira, M.; Shimoi, K.; Mochizuki, R.; Kinae, N. Dietary intakes of flavonols, flavones and isoflavones by Japanese women and the inverse correlation between quercetin intake and plasma LDL cholesterol concentration. *J. Nutr.* **2000**, *130*, 2243–2250. [CrossRef] [PubMed]
137. Lou, F.Q.; Zhang, M.F.; Zhang, X.G.; Liu, J.M.; Yuan, W.L. A study on tea-pigment in prevention of atherosclerosis. *Chin. Med. J.* **1989**, *102*, 579–583. [PubMed]

138. Osman, H.E.; Maalej, N.; Shanmuganayagam, D.; Folts, J.D. Grape juice but not orange or grapefruit juice inhibits platelet activity in dogs and monkeys. *J. Nutr.* **1998**, *128*, 2307–2312. [CrossRef]

139. Gryglewski, R.J.; Korbut, R.; Robak, J.; Swies, J. On the mechanism of antithrombotic action of flavonoids. *Biochem. Pharmacol.* **1987**, *36*, 317–322. [CrossRef]

140. Kandaswami, C.; Middleton, E., Jr. Free radical scavenging and antioxidant activity of plant flavonoids. *Adv. Exp. Med. Biol.* **1994**, *366*, 351–376. [PubMed]

141. Rodríguez-García, C.; Sánchez-Quesada, C.; Gaforio, J. Dietary Flavonoids as Cancer Chemopreventive Agents: An Updated Review of Human Studies. *Antioxidants* **2019**, *8*, 137. [CrossRef]

142. Bast, A.; Kaiserová, H.; den Hartog, G.J.; Haenen, G.R.; Van der Vijgh, W.J. Protectors against doxorubicin-induced cardiotoxicity: Flavonoids. *Cell Biol. Toxicol.* **2007**, *23*, 39–47. [CrossRef] [PubMed]

143. Fuhrman, B.; Aviram, M. Flavonoids protect LDL from oxidation and attenuate atherosclerosis. *Curr. Opin. Lipidol.* **2001**, *12*, 41–48. [CrossRef] [PubMed]

144. Bernátová, I.; Pecháňová, O.; Babál, P.; Kyselá, S.; Stvrtina, S.; Andriantsitohaina, R. Wine polyphenols improve cardiovascular remodeling and vascular function in NO-deficient hypertension. *Am. J. Physiol. Heart Circ. Physiol.* **2002**, *282*, H942–H948. [CrossRef] [PubMed]

145. Jayakody, R.L.; Senaratne, M.P.; Thomson, A.B.; Kappagoda, C.T. Cholesterol feeding impairs endothelium-dependent relaxation of rabbit aorta. *Can. J. Physiol. Pharmacol.* **1985**, *63*, 1206–1209. [CrossRef] [PubMed]

146. Hertog, M.G.; Feskens, E.J.; Hollman, P.C.; Katan, M.B.; Kromhout, D. Dietary antioxidant flavonoids and risk of coronary heart disease: The Zutphen Elderly Study. *Lancet* **1993**, *342*, 1007–1011. [CrossRef]

147. Khoo, H.E.; Azlan, A.; Tang, S.T.; Lim, S.M. Anthocyanidins and anthocyanins: Colored pigments as food, pharmaceutical ingredients, and the potential health benefits. *Food Nutr. Res.* **2013**, *61*, 1361779. [CrossRef]

148. Andriambeloson, E.; Kleschyov, A.L.; Muller, B.; Beretz, A.; Stoclet, J.C.; Andriantsitohaina, R. Nitric oxide production and endothelium-dependent vasorelaxation induced by wine polyphenols in rat aorta. *Br. J. Pharmacol.* **1997**, *120*, 1053–1058. [CrossRef]

149. Burns, J.; Gardner, P.T.; O'Neil, J.; Crawford, S.; Morecroft, I.; McPhail, D.B.; Lister, C.; Matthews, D.; MacLean, M.R.; Lean, M.E.; et al. Relationship among antioxidant activity, vasodilation capacity, and phenolic content of red wines. *J. Agric. Food Chem.* **2000**, *48*, 220–230. [CrossRef]

150. Trombino, S.; Cassano, R.; Muzzalupo, R.; Pingitore, A.; Cione, E.; Picci, N. Stearyl ferulate-based solid lipid nanoparticles for the encapsulation and stabilization of beta-carotene and alpha-tocopherol. *Colloids Surf. B Biointerfaces* **2009**, *72*, 181–187. [CrossRef]

151. Vicas, S.I.; Cavalu, S.; Laslo, V.; Tocai, M.; Costea, T.O.; Moldovan, L. Growth, Photosynthetic Pigments, Phenolic, Glucosinolates Content and Antioxidant Capacity of Broccoli Sprouts in Response to Nanoselenium Particles Supply. *Not. Bot. Horti Agrobot. Cluj Napoca* **2019**, *47*, 821–828. [CrossRef]

152. Vega-Villa, K.R.; Takemoto, J.K.; Yáñez, J.A.; Remsberg, C.M.; Forrest, M.L.; Davies, N.M. Clinical toxicities of nanocarrier systems. *Adv. Drug Deliv. Rev.* **2008**, *60*, 929–938. [CrossRef]

153. Pala, R.; Anju, V.T.; Dyavaiah, M.; Busi, S.; Nauli, S.M. Nanoparticle-Mediated Drug Delivery for the Treatment of Cardiovascular Diseases. *Int. J. Nanomed.* **2020**, *15*, 3741–3769. [CrossRef]

154. Borel, T.; Sabliov, C.M. Nanodelivery of bioactive components for food applications: Types of delivery systems, properties, and their effect on ADME profiles and toxicity of nanoparticles. *Annu. Rev. Food Sci. Technol.* **2014**, *5*, 197–213. [CrossRef] [PubMed]

155. Vaiserman, A.; Koliada, A.; Zayachkivska, A.; Lushchak, O. Nanodelivery of Natural Antioxidants: An Anti-aging Perspective. *Front. Bioeng. Biotechnol.* **2020**, *7*, 447. [CrossRef]

156. Naahidi, S.; Jafari, M.; Edalat, F.; Raymond, K.; Khademhosseini, A.; Chen, P. Biocompatibility of engineered nanoparticles for drug delivery. *J. Control. Release* **2013**, *166*, 182–194. [CrossRef]

157. Bertrand, N.; Leroux, J.C. The journey of a drug-carrier in the body: An anatomo-physiological perspective. *J. Control. Release* **2012**, *161*, 152–163. [CrossRef]

158. Tena, N.; Martín, J.; Asuero, A.G. State of the Art of Anthocyanins: Antioxidant Activity, Sources, Bioavailability, and Therapeutic Effect in Human Health. *Antioxidants* **2020**, *9*, 451. [CrossRef]

159. Dube, A.; Ng, K.; Nicolazzo, J.; Larson, I.C. Effective use of reducing agents and nanoparticle encapsulation in stabilizing catechins in alkaline solution. *Food Chem.* **2010**, *122*, 662–667. [CrossRef]

160. Oliveira, G.; Volino, M.; Conte-Junior, C.A.; Alvares, T.S. Food-derived polyphenol compounds and cardiovascular health: A nano-technological perspective. *Food Biosci.* **2021**, *41*, 101033. [CrossRef]

161. Cavalu, S.; Bisboaca, S.; Mates, I.M.; Pasca, P.; Vasile, L.; Costea, T.; Luminita, F.; Vicas, S.L. Novel Formulation Based on Chitosan-Arabic Gum Nanoparticles Entrapping Propolis Extract Production, physico-chemical and structural characterization. *Rev. Chim.* **2018**, *69*, 3756–3760. [CrossRef]

162. Alotaibi, B.; Tousson, E.; El-Masry, T.A.; Altwaijry, N.; Saleh, A. Ehrlich ascites carcinoma as model for studying the cardiac protective effects of curcumin nanoparticles against cardiac damage in female mice. *Environ. Toxicol.* **2021**, *36*, 105–113. [CrossRef] [PubMed]

163. Li, J.; Zhou, Y.; Zhang, W.; Bao, C.; Xie, Z. Relief of oxidative stress and cardiomyocyte apoptosis by using curcumin nanoparticles. *Colloids Surf. B Biointerfaces* **2017**, *153*, 174–182. [CrossRef]

164. Rachmawati, H.; Soraya, I.S.; Kurniati, N.F.; Rahma, A. In Vitro Study on Antihypertensive and Anti-hypercholesterolemic Effects of a Curcumin Nano emulsion. *Sci. Pharm.* **2016**, *84*, 131–140. [CrossRef]

165. Carlson, L.J.; Cote, B.; Alani, A.W.; Rao, D.A. Polymeric micellar co-delivery of resveratrol and curcumin to mitigate in vitro doxorubicin-induced cardiotoxicity. *J. Pharm. Sci.* **2014**, *103*, 2315–2322. [CrossRef]
166. Zhang, L.; Zhu, K.; Zeng, H.; Zhang, J.; Pu, Y.; Wang, Z.; Zhang, T.; Wang, B. Resveratrol solid lipid nanoparticles to trigger credible inhibition of doxorubicin cardiotoxicity. *Int. J. Nanomed.* **2019**, *14*, 6061–6071. [CrossRef]
167. Xu, H.; Hua, Y.; Zhong, J.; Li, X.; Xu, W.; Cai, Y.; Mao, Y.; Lu, X. Resveratrol Delivery by Albumin Nanoparticles Improved Neurological Function and Neuronal Damage in Transient Middle Cerebral Artery Occlusion Rats. *Front. Pharmacol.* **2018**, *9*, 1403. [CrossRef]
168. Lozano, O.; Lázaro-Alfaro, A.; Silva-Platas, C.; Oropeza-Almazán, Y.; Torres-Quintanilla, A.; Bernal-Ramírez, J.; Alves-Figueiredo, H.; García-Rivas, G. Nanoencapsulated Quercetin Improves Cardioprotection during Hypoxia-Reoxygenation Injury through Preservation of Mitochondrial Function. *Oxid. Med. Cell. Longev.* **2019**, *2019*, 7683051. [CrossRef] [PubMed]
169. Liu, C.J.; Yao, L.; Hu, Y.M.; Zhao, B.T. Effect of Quercetin-Loaded Mesoporous Silica Nanoparticles on Myocardial Ischemia-Reperfusion Injury in Rats and Its Mechanism. *Int. J. Nanomed.* **2021**, *16*, 741–752. [CrossRef] [PubMed]
170. Quagliariello, V.; Vecchione, R.; De Capua, A.; Lagreca, E.; Iaffaioli, R.V.; Botti, G.; Netti, P.A.; Maurea, N. Nano-Encapsulation of Coenzyme Q10 in Secondary and Tertiary Nano-Emulsions for Enhanced Cardioprotection and Hepatoprotection in Human Cardiomyocytes and Hepatocytes During Exposure to Anthracyclines and Trastuzumab. *Int. J. Nanomed.* **2020**, *15*, 4859–4876. [CrossRef] [PubMed]
171. Imaizumi, A. Highly bioavailable curcumin (Theracurmin): Its development and clinical application. *PharmaNutrition* **2015**, *3*, 123–130. [CrossRef]
172. Kanai, M.; Imaizumi, A.; Otsuka, Y.; Sasaki, H.; Hashiguchi, M.; Tsujiko, K.; Matsumoto, S.; Ishiguro, H.; Chiba, T. Dose-escalation and pharmacokinetic study of nanoparticle curcumin, a potential anticancer agent with improved bioavailability, in healthy human volunteers. *Cancer Chemother. Pharmacol.* **2012**, *69*, 65–70. [CrossRef]
173. Sasaki, H.; Sunagawa, Y.; Takahashi, K.; Imaizumi, A.; Fukuda, H.; Hashimoto, T.; Wada, H.; Katanasaka, Y.; Kakeya, H.; Fujita, M.; et al. Innovative preparation of curcumin for improved oral bioavailability. *Biol. Pharm. Bull.* **2011**, *34*, 660–665. [CrossRef] [PubMed]
174. Muthu, M.S.; Mei, L.; Feng, S.S. Nanotheranostics: Advanced nanomedicine for the integration of diagnosis and therapy. *Nanomedicine* **2014**, *9*, 1277–1280. [CrossRef] [PubMed]
175. Pala, R.; Pattnaik, S.; Busi, S.; Nauli, S.M. Nanomaterials as Novel Cardiovascular Theranostics. *Pharmaceutics* **2021**, *13*, 348. [CrossRef]
176. Cavalu, S.; Fritea, L.; Brocks, M.; Barbaro, K.; Murvai, G.; Costea, T.O.; Antoniac, I.; Verona, C.; Romani, M.; Latini, A.; et al. Novel Hybrid Composites Based on PVA/SeTiO$_2$ Nanoparticles and Natural Hydroxyapatite for Orthopedic Applications: Correlations between Structural, Morphological and Biocompatibility Properties. *Materials* **2020**, *13*, 2077. [CrossRef] [PubMed]
177. Fritea, L.; Banica, F.; Costea, T.O.; Moldovan, L.; Iovan, C.; Cavalu, S. A gold-nanoparticles—Graphene based electrochemical sensor for sensitive determination of nitrazepam. *J. Electroanal. Chem.* **2018**, *830*, 63–71. [CrossRef]
178. Suzuki, M.; Bachelet-Violette, L.; Rouzet, F.; Beilvert, A.; Autret, G.; Maire, M.; Menager, C.; Louedec, L.; Choqueux, C.; Saboural, P.; et al. Ultrasmall superparamagnetic iron oxide nanoparticles coated with fucoidan for molecular MRI of intraluminal thrombus. *Nanomedicine* **2015**, *10*, 73–87. [CrossRef] [PubMed]
179. Danila, D.; Johnson, E.; Kee, P. CT imaging of myocardial scars with collagen-targeting gold nanoparticles. *Nanomedicine* **2013**, *9*, 1067–1076. [CrossRef]
180. Lestini, B.J.; Sagnella, S.M.; Xu, Z.; Shive, M.S.; Richter, N.J.; Jayaseharan, J.; Case, A.J.; Kottke-Marchant, K.; Anderson, J.M.; Marchant, R.E. Surface modification of liposomes for selective cell targeting in cardiovascular drug delivery. *J. Control. Release* **2002**, *78*, 235–247. [CrossRef]
181. Bilia, A.R.; Piazzini, V.; Guccione, C.; Risaliti, L.; Asprea, M.; Capecchi, G.; Bergonzi, M.C. Improving on Nature: The Role of Nanomedicine in the Development of Clinical Natural Drugs. *Planta Med.* **2017**, *83*, 366–381. [CrossRef]
182. Piazzini, V.; Lemmi, B.; D'Ambrosio, M.; Luceri, C.; Cinci, L.; Ghelardini, C.; Bilia, A.R.; Di Cesare Mannelli, L.; Bergonzi, M.C. Nanostructured Lipid Carriers as Promising Delivery Systems for Plant Extracts: The Case of Silymarin. *Appl. Sci.* **2018**, *8*, 1163. [CrossRef]

applied sciences

MDPI

Review

Insights into the Bioactivities and Chemical Analysis of *Ailanthus altissima* (Mill.) Swingle

Débora Caramelo [1], Soraia I. Pedro [1,2], Hernâni Marques [3,4,†], Ana Y. Simão [3,4,†], Tiago Rosado [3,4,†], Celina Barroca [1,2,5], Jorge Gominho [1,*], Ofélia Anjos [1,2,5,*] and Eugenia Gallardo [3,4,*]

[1] Centro de Estudos Florestais, Instituto Superior de Agronomia, Universidade de Lisboa, Tapada da Ajuda, 1349-017 Lisboa, Portugal; dbrcaramelo@gmail.com (D.C.); soraia_p1@hotmail.com (S.I.P.); celinasbarroca@gmail.com (C.B.)

[2] Centro de Biotecnologia de Plantas da Beira Interior, 6001-909 Castelo Branco, Portugal

[3] Centro de Investigação em Ciências da Saúde (CICS-UBI), Universidade da Beira Interior, Av. Infante D. Henrique, 6200-506 Covilhã, Portugal; hefm976@gmail.com (H.M.); ana.simao@ubi.pt (A.Y.S.); tiagorosadofful@hotmail.com (T.R.)

[4] Laboratório de Fármaco-Toxicologia, UBIMedical, Universidade da Beira Interior, Estrada Municipal 506, 6200-284 Covilhã, Portugal

[5] Instituto Politécnico de Castelo Branco, 6001-909 Castelo Branco, Portugal

[*] Correspondence: jgominho@isa.ulisboa.pt (J.G.); ofelia@ipcb.pt (O.A.); egallardo@fcsaude.ubi.pt (E.G.); Tel.: +351-213-653-378 (J.G.); +351-272-339-600 (O.A.); +351-275-329-002/3 (E.G.)

[†] These authors contributed equally to this paper.

Abstract: Many species of the so-called exotic plants coexist with native species in a balanced way, but others thrive very quickly and escape human control, becoming harmful—these are called invasive alien species. In addition to overcoming geographic barriers, these species can defeat biotic and abiotic barriers, maintaining stable populations. *Ailanthus altissima* is no exception; it is disseminated worldwide and is considered high risk due to its easy propagation and resistance to external environmental factors. Currently, it has no particular use other than ornamental, even though it is used to treat epilepsy, diarrhea, asthma, ophthalmic diseases, and seborrhoea in Chinese medicine. Considering its rich composition in alkaloids, terpenoids, sterols, and flavonoids, doubtlessly, its use in medicine or other fields can be maximised. This review will focus on the knowledge of the chemical composition and the discovery of the biological properties of *A. altissima* to understand this plant better and maximise its possible use for purposes such as medicine, pharmacy, or the food industry. Methods for the extraction and detection to know the chemical composition will also be discussed in detail.

Keywords: *Ailanthus altissima*; biological properties; analytical techniques; potential applications

Citation: Caramelo, D.; Pedro, S.I.; Marques, H.; Simão, A.Y.; Rosado, T.; Barroca, C.; Gominho, J.; Anjos, O.; Gallardo, E. Insights into the Bioactivities and Chemical Analysis of *Ailanthus altissima* (Mill.) Swingle. *Appl. Sci.* **2021**, *11*, 11331. https://doi.org/10.3390/app112311331

Academic Editor: Susana Santos Braga

Received: 30 October 2021
Accepted: 26 November 2021
Published: 30 November 2021

Publisher's Note: MDPI stays neutral with regard to jurisdictional claims in published maps and institutional affiliations.

1. Introduction

Since ancient times humans have taken advantage of the existing flora, especially trees, for various food, wood and non-wood products. Over the years, trees have been planted to provide these services. Consequently, humans' ability to plant trees and transport goods worldwide has led to a greater expression of non-native trees [1]. The spread of many non-native tree species has shown in recent years an increase in the number of publications highlighting its potential constraints either regionally or globally (e.g., *Eucalyptus globulus* [2] and *Acacia* spp. [3]). Other studies have also noted the conflicts that some invasive alien species (IAS), including some non-native species, can cause in the ecosystem [4,5].

IAS are recognised for triggering various impacts, such as habitat modification, the alteration of community structure, and affecting ecosystem processes. Thus, they are considered one of the most significant global threats to biodiversity. However, these species can also be quite helpful to humans, providing a complex number of services, which makes it difficult to assess their positive and negative effects. For this reason, the importance of evaluating and investigating some IAS that have several adverse effects has been growing [6].

The IAS *Ailanthus altissima* (Mill.) Swingle, known as the "Tree of Heaven", was introduced into Europe and worldwide in the 18th century. This plant spreads and is distributed mainly in cities, agricultural fields, and transportation corridors. In cities, it was reported to cause problems from its roots such as damaging infrastructures plus allergic reactions and respiratory problems in humans. It is more frequent in the southern and submeridional zones whereas, in Europe, it has a closed distribution area in the Mediterranean [7]. Through this area, there can be modifications in the structure of the local vegetation and damage to the stability of the ecosystem [8]. In some warm regions of Europe, *A. altissima* can also cause damage to walls and other structures, as reported by Almeida et al. [9], where this species appeared and spread on a wall of a historic monument in Coimbra, Portugal.

Nowadays, this species has been the subject of several pharmacological investigations, showing its positive effects. Traditional Chinese medicine uses it to treat certain disorders such as epilepsy, diarrhea, asthma, ophthalmic diseases, and seborrhoea. Traditional medicine practitioners use different parts of the plant for different types of health problems. The bark, for example, is prescribed for dysentery, menorrhagia and spermatorrhea. For intestinal problems that last a few months, it is advisable to boil some bark with water and then drink the liquid together with gin. Other bark remedies can be made with other ingredients and plants, such as root onions and Chinese pepper (*Zanthoxylum simulans*) for rectal problems, especially after childbirth. In addition, it can also be used as an astringent, parasiticide, a narcotic substance, and a drug to relieve spasms [6,7]. On the other hand, *A. altissima* may have applications in agriculture as an environmentally friendly compound, since its extracts have strong herbicidal and insecticidal properties [10].

This work aims to review (1) the characteristics of the species *A. altissima*, (2) describe all its biological activities to date, and (3) detail studies performed on chemical extractions and consequently the analytical methodologies used. In addition, we intend to alert the possible uses of *A. altissima* biomass in the most diverse areas, showing that this IAS can benefit humans and the environment.

2. Taxonomy and Morphology

Ailanthus altissima is an accepted name with original publication details in J. Wash. Acad. Sci. 6: 495 (1916) and many synonyms [11]. It belongs to the family Simaroubaceae [6], which consists mainly of trees and shrubs, distributed in tropical and warm regions, comprising 30 genera and 150 species. The term for the genus *Ailanthus* is an Ambonese word derived from ailanto, meaning "tree of heaven" [12]. About 5–10 species of trees from this genus are known and characterised by being deciduous with rapid growth, large spreading branches, pinnate leaves with pointed leaflets, where the terminal leaflet is usually present, small yellow to greenish flowers, and the fruits samaras stretched into a long wing, where the seed lies in the middle (Figure 1C). In the root sprouts, the emerging leaves vary in the number of divisions (from unifoliate to compound pinnately) and are yellowish-green at first. In addition, the genus *Ailanthus* has a fine, satiny wood and is known for its medicinal values [12].

Regarding the morphology of the species in the study, *A. altissima* (Figure 1) is described as a tree growing up to 30 m in height with a trunk of greyish, smooth, or cracked bark, and in the oldest specimens and branches, a reddish-brown colour. Its leaves (30–100 cm) are alternate, compound, odd-pinnate (on the shoots) and paripinnate (arranged at the ends of the branches), and glabrous or with scattered hairs on the upper side of the margin. Each leaf can have 5–12 pairs of leaflets (or 4 to 35 leaflets [13]) of very variable size, petiolate, from narrowly lanceolate to ovate-lanceolate, with gradually narrowed apex, and an entire margin. Each of these leaflets has two to four glandular teeth near the base, with 1–8 openings as extrafloral nectaries—an important characteristic to identify *A. altissima* [6,13,14].

Figure 1. Some examples of *A. altissima* in the Castelo Branco region (Portugal); (**A**) an old plant; (**B**) a younger plant; (**C**) the samaras with seeds in the middle; (**D**) the trunk of the first plant and (**E**) part of a leaf with its leaflets.

A. altissima is a tree with both female and male terminally located inflorescences, where the latter are larger, pluriflor, and emit an unpleasant scent. In addition, unlike the female flowers, the male flowers have 10 well-functioning stamens, each with a fertile anther; the female flowers have abortive anthers. It is due to these facts that *A. altissima* is considered a dioecious tree [7,13,15]. Flowers have a tiny cupular and five-lobed calyx, a corolla with five sepals, lanceolate and green, five distinct petals that are hairy at the base, greenish to yellowish-green, and an annular and lobed gland disc (different types of nectaries); female flowers have five free carpels that generally develop into five separate winged fruits. The fruits (samaras) can vary from greenish-yellow to reddish-brown, and are arranged in clusters and dispersed by the wind [6,7,13,14]. In the samaras, the seeds are found and these, depending on the location of the trees, vary considerably in colour, size, weight, and thickness [6].

Concerning roots, the main root can develop several lateral roots, from pre-existing primordia or root sprouts. The root system has a highly variable and asymmetric spatial extent, in which lateral roots have been reported to reach a length of 27 m from the parent tree. The nectaries' primary function is to eliminate excess sugar from the plant. *A. altissima* has floral and extrafloral nectaries, which excrete different forms of sugar until October. These nectaries are located on the leaves, pseudostipules, and cataphylls. Trees of the genus *Ailanthus* have a rapid stem elongation, which leads to the fact that these trees can grow to heights of 25–45 m [7,12].

3. History and Distribution

The species *A. altissima* is a tree originating in northern and central China, which has spread to all continents except Antarctica. In China, it grows as a native tree, forming part of the broadleaf forests. At the same time, in other continents, namely North America and Europe, it has gained an expansion facilitated by seed transfer. This expansion was mediated by humans, which began in 1740 when a missionary, Pierre d'Incarville, introduced *Ailanthus* to France from Nanking to Paris. After that, it was introduced in London and then a little all over the European continent. In 1784, it was introduced to North America, more specifically to Philadelphia, from European seeds [6,7,13]. Subsequent to its introduction, *A. altissima* began to be planted in landscaped parks because of its tropical appearance, fast growth, and significant tolerance to urban life and pollution, which led to its use as an ornamental plant [6]. Although the male flowers release a nasty odour, they continue to promote shade as an ornamental plant. The species was also used for shelterbelts plantations in Austria in the 1950s to control erosion on hillsides, afforestation and reforestation in south-eastern Europe, and in some cases the reclamation of landfills and mine waste [7]. On the other hand, in China it is used in folk medicine, as firewood and wood, and as a food source for bees, where the honey is appetising but initially has a bad smell [6,7].

According to the review of Kowarik and Säumel [7] about the habitats, the tree of heaven can grow on three different sites: urban, transportation corridors, and forests. It grows on both anthropogenic and natural sites and on stony soils and rich alluvial bottoms. In urban areas, within the temperate to southern zone, i.e., Mediterranean cities, it has a substantial expansion, colonising walls, fence lines, and sidewalk cracks, among others [7]. Peculiarly in New York, *A. altissima* was less frequent in an open habitat than in soil sites with a limited surface [16]. *A. altissima* mainly colonises—outside the cities, road and railway edges, and medians of the highways. It can also invade the borders of agricultural fields and old fields from the roadside edge. In forests, *A. altissima* is invasive in several types, from riverine forests and some mesic and xeric forests in Europe to hemlock forest, oak, and maple forests in temperate North America. Furthermore, it has been reported to invade forests and riverbanks on the Danube, along streams and riverbeds in the most diverse areas of the world, such as southwestern France, southern Switzerland, Japan, and North America [7,17,18].

4. Biology and Ecology

According to Kowarik and Säumel's review [7], *A. altissima* is a short-lived tree that can reach more than 100 years. These authors also mentioned the existence of trees in Germany that were 130, 121, and 113 years old. In terms of germination and seed establishment, information varies widely in the literature. This species is characterised by not forming a long-term seed bank but can establish temporary seed banks in the soil. It has also been observed that a seed in contact with water influences its initiation, duration, and germination rate. In addition, the germination rate was reported to decrease with altitude. The same was observed on a germination experiment when comparing soil types (sand with gravel and peat substrates). The germination rate is much lower on sand and higher on peat substrates [7]. Considering seed establishment, the combination of litter, weed competition, and insect herbivory were studied. It was concluded that litter (without weed competition) delays germination but does not affect seedling biomass.

On the other hand, with weed competition, seedling growth was reduced but neutralised by the litter. However, the litter showed damage to the seedling performance by increasing herbivory [19,20]. In terms of stem elongation, *A. altissima* is one of the fastest-growing trees in North America and Great Britain. The height and diameter increase until the trees reach 5 to 10 years, and then from 10 to 20 years, it starts to decrease. In Hu's [13] review, he states that as young as 1 year old, seedlings can reach 1–2 m in height. On average, the height increment in trees between 20 and 25 years can be less than 8 cm per year [7].

A. altissima is a different tree from many other tree species. One of the requirements for the buds to open is high temperature, leading the seasonal development to start later and last longer (until late fall) [7]. As mentioned before, altitude influences the germination rate and in the case of the flowering stage this is also an essential factor to take into account. The flowers appear at different stages depending on altitude. In North America, they appear from mid-April to July; in the French Mediterranean, they start in mid-May; and in Central Europe, flowering occurs mainly in July. The ripening occurs in September-October, and soon after this, there is the abscission of the samaras, which can vary in time depending on the individual and the years. The fruits can be almost all released in some years before the end of February, and in others, a large part of them can remain on the tree until the beginning of May [6,7].

The reproduction of *A. altissima* occurs by seeds and by root suckers. In sexual reproduction, the seedlings give rise to flowers after 3 weeks of germination and large amounts of light. The flowers usually develop non-viable seeds, at this stage. However, *Ailanthus* is pollinated by bees, beetles, and other insects [7].

In vegetative reproduction, what occurs is the growth of new shoots from a mother plant. These shoots are known to have a great length and can reach up to 27 m or more. Alternatively, *A. altissima* can propagate itself through vegetative regeneration. It consists of originating shoots from pre-existing buds in the hypocotyl, axillary buds of the cataphylls, the roots, and adventitious buds on a cutting section [6,7]. The dispersal of *A. altissima* seeds, which give rise to new individuals, is achieved through the dispersal of samaras. This dispersion can occur from the following types of vectors: the wind, in which the samaras are moved individually or aggregated in groups by water (through river corridors) and animals (such as rodents and birds) [7].

The response to ecological factors is another critical issue to mention when it comes to this IAS. It was found that *A. altissima* is tolerant to several abiotic factors, contributing to its extensive spread. Among the several factors, the following stand out: temperature, frost, drought, soils, and air pollution [6]. First, this species is temperature tolerant, although it adapts better to high temperatures, equivalent to its distribution. Regarding low temperatures, it was observed in a study that old trees survived at $-33\,^{\circ}\mathrm{C}$; however, they suffered some damage. Kowarik and Säumel also mentioned that trees in the early stages of life might be more susceptible to the cold [7]. Drought is another factor to which *Ailanthus* is tolerant, presenting mechanisms that contest the lack of water, such as the closing of the stomata and the reallocation of food reserves to the lateral roots from the main root. Another fundamental factor is the soil. *A. altissima* grows in various soil types: from barren, rocky soils to saline and alkaline soils; it also tolerates dry, humid, and nutrient-rich or nutrient-poor soils. *A. altissima*'s tolerance to pollution is attributed to the high antioxidant capacity of the leaves. It is pretty tolerant of major components of air pollution such as SO_2. However, it is sensitive to ozone [7,21] and cold, which can inhibit some seedlings and limit their occurrence and distribution.

Furthermore, this species of *Ailanthus* has allelopathic effects on other plant species by producing a wide variety of active compounds and becoming resistant to herbivores and pathogens. All these ecological factors and the biological characteristics of *A. altissima*, make it a very competitive species and justify its wide distribution worldwide. Due to its tolerance to several factors, it is established in different habitats. It is widely dispersed and able to establish seedlings at long distances [6].

5. Chemical Characterization

Plants of the *Ailanthus* genus are phytochemically characterised by being rich in alkaloids, terpenoids, sterols, flavonoids, and other compounds. These compounds nowadays are widely studied since they are responsible for several pharmacological activities [12]. The bark of *A. altissima* has always been used in traditional medicine and, nowadays, it is known for being very rich in several compounds. It contains oleoresin, resin, ceryl alcohol, ailanthin, isoquercetin, tannins, ceryl palmitate, and saponins [7]. Two hundred and twenty-one compounds have been isolated and identified from the dry bark, such as alkaloids, quassinoids, phenylpropanoids, triterpenoids, and volatile oils. The main active compounds of dry bark are alkaloids (Figure 2), 32 of which have already been isolated. Quassinoids (Figure 3) are also one of the characteristic compounds of dry bark, with 40 isolated to date [22]. As reported by Pijush Kundu [12], the root bark essentially has alkaloids and terpenoids. Some ligands, coumarins, phenylpropanoids, and new terpenoids have also been identified and isolated [12,23–25]. The leaves of this species are characterised by having some percentage of tannins, quercetin, isoquercetin, alkaloids, and mainly flavonoids. These were widely used in traditional Indian medicine to treat seborrhoea and scabies [7,12].

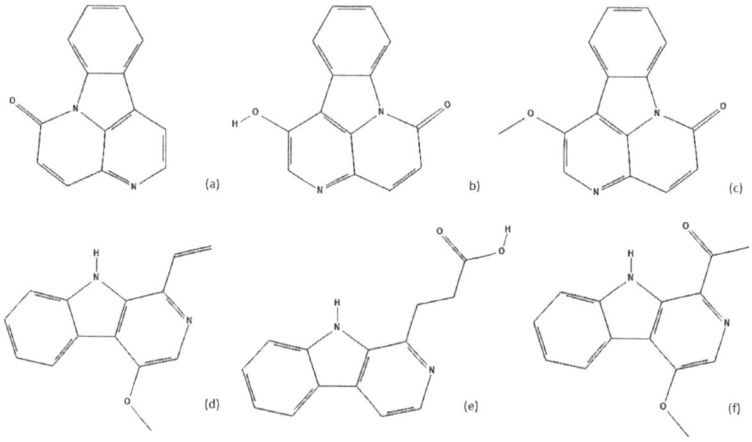

Figure 2. Some examples of alkaloids identified in *A. altissima*: (**a**) Canthin-6-one; (**b**) 1-Hydroxycanthin-6-one; (**c**) 1-Methoxycanthin-6-one; (**d**) 4-methoxy-1-vinyl-β-carboline; (**e**) β-carboline-1-propionic acid; (**f**) 1-Acetyl-4-methoxy-β-carboline.

Regarding the fruits of *A. altissima*, not much is known about its constituents, but they was widely used in China as a medicine for bleeding and as an antibacterial. On the other hand, in India it was used as an emmenagogue and to treat eye diseases [12,26]. Jian-Cheng Ni et al. [27] mentioned that previous studies have shown that only four quassinoid glycosides and several stigmasterols (Figure 4) [28] were identified from fruits. However, they elucidated the structure and isolated four new compounds: two phenylpropionamides, piperidine, and a phenolic derivative. In addition to these new compounds, 13 phenols, 10 flavonoids, and a phenylpropionamide were also isolated. According to Clair and Bory [29], the composition of *A. glandulosa* (syn. of *A. altissima*) extrafloral nectar diverge according to the nectary type. These authors also declared that the three essential nectars' sugars are sucrose, fructose, and glucose in the leaf nectaries. It is the high amount of fructose that characterizes this species. Furthermore, bound lipids (monogalactosyldiacylglycerol) and oleic, palmitic, and linoleic acids have been isolated in the secretion of glandular trichomes in cataphylls and in young stems [7,29].

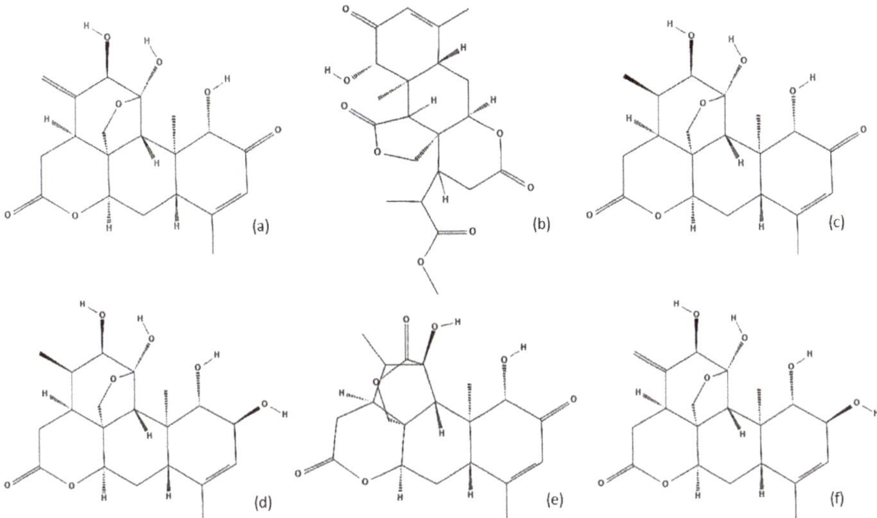

Figure 3. Some examples of the main *A. altissima* quassinoids identified: (**a**) Ailanthone; (**b**) Ailantinol A; (**c**) Chaparrinone; (**d**) Chaparrin; (**e**) Shinjudilactone; (**f**) Shinjulactone A.

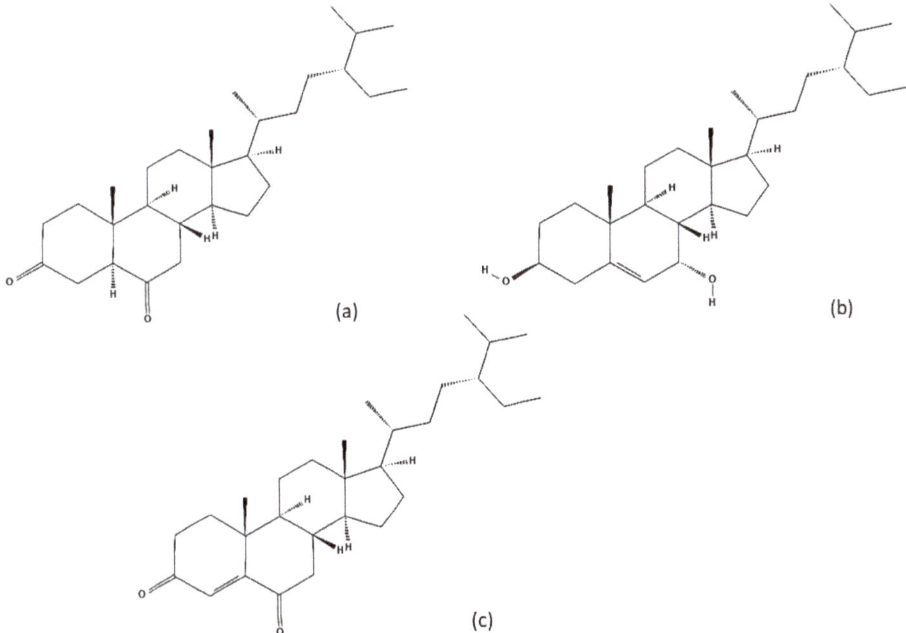

Figure 4. Some examples of steroids identified in *A. altissima*: (**a**) 5α-Stigmastane-3,6-dione; (**b**) Stigmast-5-ene-3β,7α-diol; (**c**) Stigmast-4-ene-3,6-dione.

The quassinoids class has been the main target of many studies due to their bioactivities and phytotoxic impacts. The most widely studied active compound in this group of terpenoids is ailanthone. This has been identified as an effective phytotoxic agent, capable of being used as a herbicide [7,30]. In addition to ailantthone, other derivatives have been isolated and studied (such as ailanthione, ailanthinol B, and chaparrin). There is an increasing demand for the isolation of new quassinoid derivatives to investigate their potential applications, mainly in pharmacology [31–33]. Table 1 summarises the main compounds by classes of compounds and which parts of the plants were used to identify the chemical compounds. These results are based on the review by Pijush Kundu and Subrata Laskar [12].

Table 1. The main compounds identified in *A. altissima* grouped by the major classes and which part of the plant was used for their isolation.

	Compounds	Parts Used (Source)
Alkaloids	1-Ethyl-4-methoxy-β-carboline	Root bark
	β-Carboline-1-propionic acid	Root bark
	Methyl-4-methoxy-β-caboline carboxylate	Root bark
	1-(1′-2′-Dihydroxyethyl)-4-methoxy-β-carboline	Root bark
	1-(2′-Hydroxyethyl)-4-methoxy-β-carboline/Crenatine	Root bark
	4-Methoxy-1-vinyl-β-carboline/Dehydrocrenatine	Dried leaves
	1-Methoxycarbonyl-4,8-dimethoxy-β-caboline	Leaves
	1-Methoxycarbonyl-β-carboline/1-caromethoxy-β-carboline	Leaves
	1-(1-Hydroxy-2-methoxy)-ethyl-4-methoxy-β-carboline	Root bark
	1-Carbamoyl-β-carboline	Root bark
	1-Acetyl-4-methoxy-β-carboline	Root bark
	Canthine-6-one	Root bark, leaves and wood
	1-Hydroxycanthine-6-one	Root bark
	1-Methoxy-canthine-6-one	Root bark, dried leaves and wood
	5-Hydroxymethylcanthine-6-one	Root bark
	Canthine-6-one-3-N-oxide	Wood
	1-Methoxycanthine-6-one-3-N-oxide	Root bark
Terpenoids	Ailanthone	Stem bark
	Chapparinone	Seed
	Chapparin	Aerial part
	Shinjulactone and Shinjulactone B	Stem bark
	Shinjulactone A	Seed
	Shinjulactone C, D, E, F, M, N, G, H, I, J, K and L	Root bark
	$\Delta^{13(18)}$-Dehydroglaucarubinone	Stem bark
	$\Delta^{13(18)}$-Dehydroglaucarubolone	Stem bark
	Ailantinol A e B	Stem bark
	Ailantinol C, D, E, F, G and H	Aerial part
	Shinjuglycoside A, B, C and D	Seed
	Shinjuglycoside E and F	Root bark
	Shinjudilactone	Root bark
	Cycloart-25-ene-3β-24R-diol and Cycloart-25-ene-3β-24S-diol	Fruits
	9,19-Cyclolanost-23(Z)ene-3β,25-diol	Fruits
	Ailantholide	Seed
	3-epi-ursolic acid	Fruits
	12β, 20 (S)-dihydroxy dammar-24-en-3-one	Fruits

Table 1. *Cont.*

	Compounds	Parts Used (Source)
Steroids	b-Sitosterol	Leaves and fruits
	b-Sitosterol glucoside/Daucosterol/Sitosterol-3-Ob-D-glucoside	Leaves and fruits
	Ailanthusterol A and B	Seed
	5α-Stigmastane-3,6-dione	Fruits
	5α-Stigmastane-3β, 6β-diol	Fruits
	6α-Hydroxy-stigmast-4-en-3-one	Fruits
	Stigmast-4-ene-3β, 6α-diol	Fruits
	Stigmast-4-ene-3,6-dione	Fruits
	6β-Hydroxy-stigmast-4-ene-3-one	Fruits
	Stigmast-4-ene-3β, 6β-diol	Fruits
	3β-Hydroxy-stigmast-5-en-7-one	Fruits
	Stigmast-5-ene-3β, 7α-diol	Fruits
	Stigmast-5-ene-3β, 7α, 20ζ-triol	Fruits
Flavonoids	Quercetin	Leaves
	Kaempferol	Leaves
	Isoquercetin/Quercetin-3-O-glucoside	Leaves
	Kaempferol-3-O-glucoside	Leaves
	Rutin	Leaves
	Luteolin-7-O-b-(600 galloylglucopyranoside)	Leaves
Miscellaneous compounds	1-O-b-D-Glucopyranosyl-(2S,3R,4E,9E)-2-(20-R-hydroxyhexadecenoy)-4,9-octadecadiene-1,3-diol	Fruits
	Ceryl alcohol	Leaves
	Ethyl gallate	Leaves
	Altissimacoumarin A and B	Bark
	Coumarin	Bark
	Isofraxidin	Bark
	Scopoletin	Bark

6. Biological Properties

For a long time, in the history of science, plants and parts of them were explored and used for medicinal purposes. *A. altissima* offers a promising natural alternative for food safety and bioconservation as well as for its antioxidant properties. Several studies report the pharmacological effects: antimicrobial (antibacterial and antiviral), antioxidant, cytotoxic, anti-inflammatory, antipyretic, analgesic, anti-progestogenic, and many others [34–37].

6.1. Antimicrobial Activity

Research on the bactericidal efficiency of phytochemicals as viable alternatives for chemical antibiotics has been conducted. An essential feature of extracts from plant origin is their antimicrobial activity, which contributes to alternative synthetic antibiotics [38].

Natural products derived from *A. altissima* may contribute to the development of new antimicrobial agents used as growth inhibitors of *Listeria monocytogenes*, *Staphylococcus aureus*, *Bacillus subtilis*, *Escherichia coli*, *Pseudomonas aeruginosa*, and some important foodborne pathogens and spoilage bacteria [39].

Methanolic extracts from leaves and hydrodistilled residues were strong and efficient against Gram-positive bacteria: *S. aureus*, *B. subtilis*, *Enterococcus faecium*, and *Streptococcus agalactiae* [40]. The same was reported by Rahman [39]; methanolic extracts of *A. altissima* leaves were most effective against Gram-positive bacteria, namely *L. monocytogenes* (ATCC 19,116, ATCC 19,118 and ATCC 19,166), *S. aureus* (ATCC 6538 and KCTC 1916), and *B. subtilis* (ATCC 6633), and two Gram-negative bacteria, *P. aeruginosa* (KCTC 2004) and *E. coli* (ATCC 8739). The zones of inhibition of methanol extract and its derived different polar subfractions against the tested bacteria were found in the 12.1–23.2 mm range, and the minimum inhibitory concentration (MIC) values were recorded between 62.5 and 500 mg/mL [39]. However, their extracts were ineffective against Gram-negative bacteria, such as *E. coli* (ATCC 43,888), *Salmonella enteritidis* (KCTC 12,021), and *Salmonella typhimurium* (ATCC 2525). The MIC values of the methanolic extract against the tested bacteria were found in the 125–500 mg/mL range. In addition, the MeOH extract and its ethyl acetate (EtOAc) subfraction were compared with the standard antibiotics, tetracycline and streptomycin; in some cases, these showed greater antibacterial activity compared to streptomycin, but in other cases tetracycline showed greater activity than the solvent fractions. Aissani et al. [41] observed in aqueous and methanolic extracts of the bark and wood strong antimicrobial activity against *P. aeruginosa* (ATCC 9027) strains and isolated strains with an inhibition zone of 12 ± 0.3 mm. Zhao et al. [28] showed that the extract from fruits was weakly active against *E. coli*, *S. aureus*, *P. aeruginosa*, and *S. typhiuriun*, with inhibition zones of 6.87–7.51 mm, using the concentration of ethanol extract of 1.2 mg/mL. The plant's chemical composition can be variable due to variations in origin, species, growth, harvesting, and processing conditions, thus altering the biological activities.

Several studies show that antibacterial activity can be attributed to the occurrence of some specific components such as total polyphenol contents [42], namely gallic acid, rutin [43], and epicatechin [44]. The resistance of Gram-positive bacteria towards plant extracts has been previously reported [44–46]. The results obtained encourage the use of species as a food preservative and for pharmaceutical purposes.

6.2. Antioxidant Activity

Scientific research is interested in quantifying and using antioxidants mainly due to their potent biological activity. [47]. Although antioxidant compounds may be synthesized, there is a growing interest in the natural compounds from different plants [48,49]. Taking this into account, Luis et al. [50] performed a research to determine the phenolic, flavonoid, and total alkaloid content on four different extracts (methanolic, ethanolic, hydroalcoholic, and acetone) of *A. altissima* (steams, stalks, and leaves) to establish a correlation to the antioxidant activity of these extracts. To assess their antioxidant activity, DPPH (2,2-diphenyl-1-picryl-hydrazyl-hydrate) assay and the β-carotene bleaching test were used.

It was possible to quantify the total amount of phenolic compounds, 268.15 mg/g of dry extract in all extracts. The most important finding of this research is the potential antioxidant properties of *A. altissima*, since it is abundant in phenolic, flavonoid, and alkaloid compounds. Lungu et al. [51] have identified the presence of flavonoids (ranging around 5.34 to 5.41 g of rutin equivalents per 100 g of dry weight) in leaf extracts of *A. altissima*. The extraction method by reflux or ultrasound did not have a significant effect on the final concentration.

Moreover, the antioxidant activity was performed by the DPPH assay and the 2,2-AzinoBis (3-ethylbenzthiazoline-6-Sulphonic acid) (ABTS) radical scavenging assay. Once again, the extraction method did not affect the obtained concentrations, meaning that values are approximately 0.16 mM Trolox equivalent of extract for DPPH and around 9 mM Trolox equivalent when ABTS was used. In summary, results showed that flavonoids were in higher amounts than other compounds and are responsible for the antioxidant properties of the ethanolic leaf extracts [51]. Another study with leaves from different localities in Tunisia found that the main phenolic compounds in methanolic extracts were gallic acid, chlorogenic acid, HHDP-galloylglucose, epicatechin, rutin, hyperoside, and quercetin-3-galloyl hexoside [40]. Once again, the typical antioxidant activity assays (DPPH and ABTS) were used, as well as the ferric reducing antioxidant power (FRAP) assay and the 2-deoxyribose method, which was used for the determination of the scavenging effects of the methanolic extracts on hydroxyl (OH) radicals. Results showed good antioxidant activity of the methanolic extracts.

Altogether, leaves from the Bousalem region showed better antioxidant activity than those from other regions, with EC_{50} values of 14.78 µg/mL, 8.64 µg/mL, and 4.42 µg/mL when using the DPPH assay, ABTS assay, and the OH scavenging test, respectively. Regarding the leaf hydrodistilled residues, results were similar in efficacy to the DPPH and ABTS test. However, when using the OH radical test, the hydrodistilled residues showed a better antioxidant activity for the species that came from Bab Saâdoun [40]. The DPPH test to assess the antioxidant activity of *A. altissima* leaves was also used by Rahman et al. [39]. Methanolic extracts presented an IC_{50} value of 35.46 µg/mL and its ethyl acetate fraction possessed an even lower IC_{50} value (16.45 µg/mL). Recently, research was conducted to evaluate the physicochemical parameters and pharmacological bioactivity of *A. altissima* seed oil. El Ayeb-Zakhama et al. [52] were able to show that the seed oil presents antioxidant activity. The DPPH assay determined that the IC_{50} was 24.57 µg/mL, although the concentration of polyphenols was considered low (1.067 mg gallic acid equivalent/100 g oil). In addition, the study also showed that this seed oil presents moderate antimicrobial activity against Gram-positive bacteria.

6.3. Anti-Inflammatory Effects

Inflammation can lead to detrimental effects of different pathologies such as neurodegenerative and cardiovascular diseases, cancer, diabetes, and others [21]. It is also essential to find compounds that can tackle this problem by inhibiting the pro-inflammatory molecules and associated pathways [53].

El Ayeb-Zakhama et al. [52] reported that *A. altissima* seed oil could induce anti-inflammatory effects on edema (in vivo experiment), by reducing it more than 60% after 3 h, via the administration of doses of 0.2 and 1 g/kg. The same study also demonstrates the analgesic capacity of the oil due to its complex composition. The acute toxicity study and the analgesic effect of the seed oil was compared with a widely marketed drug, acetylsalicylic acid (ASL), in which the average lethal dose of the oil was estimated to be more than 2 g/kg and the analgesic effect was almost as potent as ASL at a dose of 1 g/kg. Other research [54] has suggested that the ethanolic extract of *A. altissima* (leaves and branch) can inhibit inflammatory mediators, both in vitro and in vivo, mainly by the inhibition of cyclooxygenase-2 (COX-2). By extracting canthin-6-one, an alkaloid, from the stem barks of *A. altissima*, Cho et al. [55] has proven this compound's potential effect in exerting anti-inflammatory effects of macrophages dysregulating pro-inflammatory cytokines and pathways. This same family of compounds was the target of the research conducted by Kim et al. [56], amongst other compounds. The authors reached the same conclusion regarding the anti-inflammatory properties. Kang et al. [57] have shown the potent inhibitory effects of *A. altissima* decoction on the decrease of cytokine levels, such as tumour necrosis factor (TNF) and interleukin (IL)-6 and IL-8 as well as on the reduction of histamine levels, by using both in vitro and in vivo models.

Moreover, the nuclear factor kappaB (NF-κB) pathway can also be inhibited by *A. altissima* compounds. Kim et al. [58] used the ethanolic extract of *A. altissima* leaves to assess the anti-inflammatory properties in astrocytes, where inflammation was induced by lipopolysaccharide. Results showed that the anti-inflammatory mechanism functions through the inhibition of COX-2 and other cytokines and the inhibition of the NF-κB pathway. These findings may lead to the use of *A. altissima* as a new therapeutic approach to neuroinflammatory diseases.

Overall, studies about the anti-inflammatory effects of the *A. altissima* compounds, whether they come from the leaves, bark, or seed oil, all conclusions favour the beneficial effects against pro-inflammatory molecules.

6.4. Phytotoxic Effects/Phytochemical Activity

Recent studies have demonstrated the potential of plant phytotoxins, such as ailanthone from *A. altissima*, as bioherbicides.

A. altissima has produced phytotoxic compounds and the main toxin identified is the ailanthone quassinoid, first isolated by Heisey and Heisey [30]. Both tested the herbicidal effect of the methanol extracts from the stem and bark of *A. altissima* on 17 weed and crop species and found that their application provided a strong herbicidal effect on all tested species. Numerous studies confirmed that quassinoids have a wide range of biological activities, including antileukemic and anticancer, antiamoebal, antimalarial, insecticidal and antifeedant, antiviral, antifungal, antitubercular, and herbicidal activities [59]. It is considered that the bioactivity of quassinoids is based on the plasma membrane NADH oxidase inhibition [60]. The phytotoxic effect seemed to correlate better with the extracts of higher phenolic contents; such results have previously been reported in *Psychotria leiocarpa* [61]. Phytochemical studies have shown that the main allelochemicals are ailanthone, ailantinone, chaparrin, and ailantinol with the first component being the most potent allelochemical. Moreover, both the methanolic extract and the hydrodistilled residue presented a significant phytotoxic activity [40]. Lungu et al. [51] extracts showed an excellent inhibitory action on the germination of seeds and growth of lettuce seedlings.

Regarding inhibiting progestin, Ahmed et al. [62] analysed 13 Chinese plant species, including *A. altissima*, in which they used the extract of the stem part. With the treatment of the T47D human breast cancer cell line with 314.46 ng/mL of progesterone and with three different levels of concentration of 10 μg/mL, 20 μg/mL and 40 μg/mL of ethanolic extracts, it was possible to verify that these concentrations could significantly prevent the action of the hormone in a dose-dependent manner.

Despite all the beneficial effects, some reports depict the contrary. *A. altissima* has been reported to negatively affect human health, with effects such as allergies, dermatitis and myocarditis. In addition, this invasive tree can also produce toxic and noxious environments for other species in the neighbourhood from the soil since the main compound responsible for its allelopathy (ailanthone) is present in high concentrations in the root bark and when released into the soil [6,7,63].

6.5. Anti-Tumour and Anti-Viral Activity

Anti-tumour activity is one of the fields that has contributed to more research on the effects that the bioactive compounds of *A. altissima* exert. Several studies with the dried bark, show that its active compounds exert antitumor effects in several cell lines, for example in A549, MDA-MB-231, LAPC4, A375, B16, and SGC-7901 among others. These in vitro studies lead us to state that ailanthone specifically may be a good inhibitor for cancers such as melanoma, acute myeloid leukaemia, lung, liver, breast, bladder, osteosarcoma, and prostate cancer [22,64,65]. In the work of Ding et al. [65] the different mechanisms of the action of ailanthone that justify its inhibitory effects were described and explained. The main basis of these effects is the mechanism of apoptosis. Apoptosis can be triggered by regulation of Bcl-2 family proteins, transcription factors such as β-catenin, tumour suppressor genes, and by the PI3K/AKT/mTOR and JAK/STAT3 pathways. Arylanthone induces apoptosis by the downregulation of Bcl-2 and the downregulation of the Bax proteins. It also induces this cell death through the downregulation of p53 (tumour suppressor protein) and miR-195, which consequently leads to inhibition of the JAK/STAT3 signalling pathway. On the other hand, the signalling pathway which is responsible for cell proliferation, differentiation, and metabolism that can lead to anti-apoptosis is PI3K/AKT/mTOR. In this pathway ailanthone has a suppressive effect, inducing apoptosis through the phosphorylation of PI3K and AKT. The review work of these authors is very interesting in that they warn of the problem that there are few studies evaluating the cytotoxicity of ailanthone in normal cells. Importantly, a preclinical study by Tang et al. [66] assessed the safety of ailanthone in Kumming rats and classified the mean lethal dose of ailanthone (27.3 mg/kg) as level 2 (severe) from the Globally Harmonised System of Classification and Labelling of Chemicals.

Furthermore, they warn that research into this active compound from *A. altissima* is still very early, as there are no studies on the bioavailability and side effects of ailanthone. Researchers are advised to focus on comparing the efficacy of ailanthone with other existing chemotherapy drugs and also to conduct in vivo and clinical trials [65].

Regarding the antiviral activity, some studies demonstrated that the main compound responsible for this effect is the main compound of *A. altissima*: ailanthone. In the review by Li et al. [22] the authors summarize all the existing pharmacology considering the dried bark of this species. They state that the active compound shows moderate inhibitory effects against the tobacco mosaic virus (TMV). In a study conducted by Tan et al. [67] these effects reached IC_{50} values of 0.30 mmol/L. However, they mention that the mechanism of action leading to this activity is not yet well elucidated.

6.6. Potential and New Applications

Desami et al., 2019 [68], and Caser et al., 2020 [69], provide new insights into *A. altissima* extracts and their phytotoxicity as potential natural herbicide use as a sustainable solution for weed management in horticultural crops.

The methanolic extracts of *A. altissima* showed antibacterial activity against different food pathogens; simultaneously, antioxidant properties indicated a potential use of extracts from this plant as food preservatives [39]. However, toxicity studies must be developed concerning the uses in the food industry and given the presence of alkaloids. The paper-making potential of this species, and its use as a fiber alternative for pulp production for the paper, was also evaluated [70,71]. The results of the paper are close to those of the reference ones. Regarding the growth and longevity of the *A. altissima* tree, other wood applications could be studied for this species.

7. Analytical Techniques for the Determination of Active Compounds

Over the last years several extraction techniques associated with different analytical methods have been reported for the chemical characterisation of *A. altissima*. For this purpose, it is crucial to consider that this type of plant possesses a complex and high number of compounds among which there are several fatty acids, volatile compounds, antioxidants, and alkaloids, among others [50,52,72]. These compounds were studied to check if they had any biological activity; this could help understand mechanisms of action, develop pharmacological formulations, and apply them to different research fields. Therefore, it is essential to identify and quantify the main compounds to improve this species' knowledge and possible uses.

In a study carried out by El Ayeb-Zakhama et al. [52], compounds were extracted from the seeds in a powder format using a Soxhlet apparatus. The authors used hexane (400 mL) as a solvent and the extraction time was 8 h, after which the solvent was evaporated using a rotary evaporator set at 40 °C. The collected oil was analysed by (1) gas chromatography coupled with a flame ionisation detector (GC-FID) and (2) gas chromatography coupled with mass spectrometry with headspace solid-phase micro-extraction (HS-SPME-GC/MS). The GC-FID equipment was used to determine the sterols fraction and the profile of fatty acids after transformation into methyl esters by adding 1 mL of n-hexane for each 40 mL of previously extracted oil, to which was then added a solution of sodium methoxide (2M). After performing this step, the mixture was heated to 50 °C in a thermal bath, followed by the addition of HCl (2N).

On the other hand, the analysis of volatile compounds was carried out by HS-SPME-GC/MS. The fatty acids profile revealed a predominance of linoleic and oleic acids, observing a polyunsaturated/saturated ratio of 11.86. Regarding the level of sterols, a predominant presence of β-sitosterol was observed, which is associated with high antioxidant activity and the control of cholesterol levels. Lastly, concerning the volatile fraction, a significant percentage of non-terpenes were observed in seed oil which could not be observed in oils from other parts of the same plant species.

Luis et al. [50], developed an analytical method to determine phenolic compounds with antioxidant activity by reversed-phase HPLC coupled to an ultraviolet detector (UV) using a Phenomenex Kinetex Luna 2.6 μm PFP 100A reversed column, performing the monitoring of the compounds at a wavelength of 280 nm, with the extraction of the compounds of different parts of *A. altissima* being performed with Soxhlet equipment for methanol, acetone, and ethanol extractions, with hydroalcoholic extractions carried out through refluxing, followed by evaporation under vacuum to a final volume of 150 mL. The authors verified that a greater final concentration of phenolic compounds was obtained when methanol was used for extraction, with lower concentrations of phenols observed in stems and stalks compared with extracts from leaves. In addition, hydroxycinnamic acids were revealed to be predominant, ferulic acid being the one whose percentage was higher, finding that the antioxidant activity levels were closely related to the total phenolic content, with an R2 of 0.86.

He et al. [72] developed a method to evaluate the active components of *A. altissima* with the help of an HPLC-UV. The goal was to perform the fingerprinting and quantification from samples of different sources, proceeding to the extraction of the compounds after performing the drying and pulverization of the stem bark using ethanol (75%) for 1 kg of content, being later evaporated in a vacuum. In this study, concentrations ranged from 6.44 μg/mL to 825 μg/mL for ailanthone, one of the main quasinoids with anti-inflammatory, anti-microbial, and anti-allergic activity in *A. altissima*, used as quality control, with the UV detector set to a wavelength of 250 nm. The samples from different sources revealed 19 similar compounds among the 10 different samples studied with a degree of similarity never exceeding 1.5%, with the predominance of compounds differing in accordance with the source.

Mastelić et al. [73] described for the first time the application of GC-MS to identify volatile compounds present in leaves of different ages. For extraction, Clevenger-type equipment was used for the hydrodistillation of the plant for 3 h, adding 1 mL of pentane to the graduated part, after which the pentane layer was separated and dried with sodium sulfate before analysis. The authors report that the composition of volatile compounds was dependent on the plant's age and its drying process. A higher prevalence of oxygenated aliphatic compounds was observed in younger plants while older plants revealed a greater prevalence of sesquiterpene compounds.

To evaluate the herbicidal activity of ailanthone in *A. altissima* bark, Pedersini et al. [74] isolated this compound with solvents and purified it by Biogel P2. A higher concentration of ailanthone 92% *w/w* was obtained when dichloromethane was used. Three obtained fractions were subsequently fractionated by gel permeation chromatography using a Biorad Biogel P2 resin with a UV detector operating at 313 nm. Later, an HPLC coupled with a UV-VIS detector set at 256 nm was used to determine ailanthone in three different species.

Li et al. [75] isolated canthin-6-one to evaluate its antifungal activity. This study was performed by quantifying the different protein expressions in *Fusarium oxysporum* f. sp. *cucumerinum* (Foc) by liquid chromatography coupled to tandem mass spectrometry (LC-MS/MS). Extraction of the compound was carried out with three times ethanol (70%) at room temperature. The obtained extract was evaporated by vacuum and subsequently dissolved in water before partitioning with other solvents, namely n-hexane, ethyl acetate and butanol. Canthin-6-one present in 32 μg/mL revealed an elevated downregulating effect on the expression of glutamine synthetase, thus affecting the synthesis of glutamine and glutamate, apart from downregulating the transport of amino acids, affecting the growth of the fungus.

To characterise the volatile compounds from leaves and flowers, Yifan et al. [76] quantified them through a dynamic headspace sampling coupled with GC-MS. The compounds were extracted with 150 mL of distilled water for 4 h by pumping air through a SuperQ volatile collection trap equipment and then eluted with 100 mL methylene chloride. The authors observed the presence of fifty-three compounds belonging to groups such as terpenoids, fatty acids, benzenoids, and nitrogen-containing compounds. The percentage of fatty acids present in the total emissions from the leaves was higher than in flowers, which were found to have a higher percentage of terpenoids in the overall emissions for volatile compounds compared to leaves, namely monoterpenoids.

To evaluate the phytotoxic activity of leaves, Lungu et al. [51] studied the composition of this species with GC-MS after extracting water-ethanol (70%; V/V) by reflux and ultrasound for one hour. The extract was then centrifuged, and the supernatant was filtered. Twenty-seven different compounds were identified, of which some fatty acids and polyphenols were estimated to contribute to the phytotoxic activity.

Caboni et al. [77] identified the nematicidal activity of (E,E)-2,4-decadienal and (E)-2-decenal extracted from wood, leaf, bark, and root parts. These two compounds were extracted with methanol (1:10 w/w) with ultrasound for 15 min. After this process, the obtained solution was filtered and centrifuged. A GC-MS was used to determine the presence of these two compounds and the authors claimed their presence in wood parts only. The nematicidal activity revealed EC_{50} values of 11.70 mg/L for (E,E)-2,4-decadienal and 20.43 mg/L for (E)-2-decenal when applied against the nematode *Meloidogyne javanica*.

Additionally, Albouchi et al. [40] studied the composition of essential oils extracted from *A. altissima* leaves from three different regions of Tunisia. The dried material was extracted by hydrodistillation of Quickfit apparatus for 3 h, in which the distillate obtained was extracted with *n*-pentane (twice) and the organic layer was concentrated and consequently analysed by GC-FID and GC-MS. The three essential oils were characterised as containing a high percentage of non-terpenic compounds and some of the sesquiterpene hydrocarbons and oxygenated monoterpenes. The majority of 139 compounds identified were tetradecanol, heneicosane, tricosane, docosane, α-curcumene, α-gurjunene, methyl decanoate, α-terpinen-7-al, geranial, α-guaiene,α-humulene, and (E)-β-farnesene. In the same study, methanolic extracts of dried and ground leaves and the residual leaves (remained after hydrodistillation) were further analysed for their content of phenolic and flavonoid compounds. The total phenolic content was performed by the Folin–Ciocalteu method, and these were characterised by high-performance liquid chromatography with a photodiode array detector and tandem mass spectrometry (HPLC-PDA-MS/MS). Total flavonoid content was performed with the $AlCl_3$ colorimetric method. Comparing the two methods of total content determination, it was possible to verify a higher content of phenols in the dried leaves than the hydrodistilled residue and a higher content of flavonoids in the hydrodistilled residue compared with the dried leaves. Furthermore, from the methanolic extracts antioxidant, antimicrobial, and phytotoxic activities were analysed.

To quantify the phenolic compounds, Vidovic et al. [78] incubated the leaves in methanol with 0.1% chloride acid for 50 min. After this period, water and chloroform were added to the resulting supernatant and the sample was stirred for 45 min at 4 °C, followed by centrifugation and separation of the liquid phase. This process was repeated without the addition of chloroform. After extraction, the compounds were analysed by HPLC-DAD using wavelengths of 520 nm for anthocyanins, 340 nm for flavones and flavanones, 320 nm for resveratrol, coniferyl alcohol, hydroxycinnamic acid and its respective derivatives, 360 nm for flavonols and ellagic acid, and 280 nm for isoflavones, catechins, benzoic acids, and the respective derivatives.

The evaluation of the phenolic compounds' content, flavonoids, and non-flavonoids was also performed by Poljuha et al. [79], using a new HPLC procedure equipped with a C_{18} column using wavelengths of 360 nm and 310 nm to identify flavonoids and phenolic acids, respectively, which was faster (26 min) and used fewer solvents than other works. The extractions in deionised water were performed after maceration and respective extraction for 48 h, followed by filtration. The extractions in methanol were performed for a period of 72 h without maceration. The supernatants were evaporated to 140 mL, reconstituted with water to a final volume of 500 mL, and then stored at 4 °C until analysis. No flavonoid content was obtained from fresh leaves when water extraction was performed, while the content of both groups of compounds was higher when methanol was used for extraction.

8. Conclusions

Nowadays, *A. altissima* can be a valuable species to satisfy some human services, considering it is an IAS and presents some negative effects. There have been distinct investigations with different plant parts to explain its traditional use in medicine and unveil its potential for pharmacology and the food industry. It is a species that can present some variation in its morphology according to its location and is tolerant to diverse factors, which enable its propagation. In addition, biologically, it displays methods of reproduction that greatly favour its spread. Regarding antimicrobial activity, the studies performed so far have demonstrated that the methanolic extracts of the leaves show great effectiveness against Gram-negative bacteria. The methanolic and aqueous extracts of the bark and wood has also demonstrated strong antimicrobial activity. These results are very much in line with its traditional use in medicine, where it is used as an astringent and parasiticide. The antioxidant activity has been performed already with different parts of the plant extracted with different solvents. Most studies have shown that methanolic extracts of the leaves contain phenolic compounds and flavonoids. Furthermore, *A. altissima* showed anti-inflammatory effects due to the chemical composition of the leaves, bark, and seed oils. The results of some studies demonstrating the potent inhibitory effect of the extracts of this species, such as the decoction that reduces cytokine levels and interleukins and TNF, support the use of the plant boiled together with other ingredients to treat problems such as dysentery, intestinal bleeding, and others.

According to the literature, the main extracts for analysing the biological activities of *A. altissima* are essentially made from the leaves, seeds (e.g., essential oil), and bark. However, extracts from other plant parts have been rarely investigated, such as flowers and stems. Moreover, new compounds have been discovered, but few are known about their possible activities.

The novelty of this work is the detailed description of the major biological activities and analytical techniques that have been used to date using different plant extracts of *A. altissima*. Other articles describe only the characteristics of this species, or the chemical compounds, or only the bioactivities of one plant extract, while in this paper, we compiled all the information overall about *A. altissima* to better understand what possible investigations could be carried out.

This review also shows that knowledge about the phytochemical composition and biological activities of *A. altissima* has been growing, leading it to be a promising species for future research.

Author Contributions: Conceptualization, E.G., J.G. and O.A.; formal analysis, E.G., J.G. and O.A.; research and writing—original draft preparation, D.C., S.I.P., H.M., A.Y.S. and T.R.; writing—review and editing, D.C., S.I.P., H.M., A.Y.S., T.R., C.B., E.G., J.G. and O.A. All authors have read and agreed to the published version of the manuscript.

Funding: This work was partially supported by CICS-UBI, which is financed by National Funds from the Fundação para a Ciência e a Tecnologia (FCT) and by Fundo Europeu de Desenvolvimento Regional (FEDER) under the scope of PORTUGAL 2020 and Programa Operacional do Centro (CENTRO 2020), with project references UIDB/00709/2020 and UIDP/00709/2020. The Forest Research Centre was supported by the Fundação para a Ciência e a Tecnologia through UI/BD/151511//2021. D. Caramelo acknowledges the FCT in the form of fellowship (UI/BD/151511//2021).

Institutional Review Board Statement: Not applicable.

Informed Consent Statement: Not applicable.

Data Availability Statement: This manuscript not reports any data.

Conflicts of Interest: The authors declare no conflict of interest.

References

1. Castro-Díez, P.; Alonso, Á.; Saldaña-López, A.; Granda, E. Effects of widespread non-native trees on regulating ecosystem services. *Sci. Total. Environ.* **2021**, *778*, 146141. [CrossRef]
2. Krumm, F.; Vitkova, L. *Introduced Tree Species in European Forests: Opportunities and Challenges*; European Forest Institute: Joensuu, Finland, 2016; ISBN 9789525980318.
3. Castro-Díez, P.; Godoy, O.; Saldaña, A.; Richardson, D.M. Predicting invasiveness of Australian acacias on the basis of their native climatic affinities, life history traits and human use. *Divers. Distrib.* **2011**, *17*, 934–945. [CrossRef]
4. De Wit, M.P.; Crookes, D.J.; Van Wilgen, B.W. Conflicts of interest in environmental management: Estimating the costs and benefits of a tree invasion. *Biol. Invasions* **2001**, *3*, 167–178. [CrossRef]
5. Dickie, I.A.; Bennett, B.M.; Burrows, L.E.; Nuñez, M.A.; Peltzer, D.A.; Porté, A.; Richardson, D.M.; Rejmánek, M.; Rundel, P.W.; van Wilgen, B.W. Conflicting values: Ecosystem services and invasive tree management. *Biol. Invasions* **2014**, *16*, 705–719. [CrossRef]
6. Sladonja, B.; Sušek, M.; Guillermic, J. Review on Invasive Tree of Heaven (*Ailanthus altissima* (Mill.) Swingle) Conflicting Values: Assessment of Its Ecosystem Services and Potential Biological Threat. *Environ. Manag.* **2015**, *56*, 1009–1034. [CrossRef] [PubMed]
7. Kowarik, I.; Säumel, I. Biological flora of Central Europe: *Ailanthus altissima* (Mill.) Swingle. *Perspect. Plant Ecol. Evol. Syst.* **2007**, *8*, 207–237. [CrossRef]
8. Constán-Nava, S.; Bonet, A.; Pastor, E.; Lledó, M.J. Long-term control of the invasive tree *Ailanthus altissima*: Insights from Mediterranean protected forests. *For. Ecol. Manag.* **2010**, *260*, 1058–1064. [CrossRef]
9. Almeida, M.T.; Mouga, T.; Barracosa, P. The weathering ability of higher plants. The case of *Ailanthus altissima* (Miller) Swingle. *Int. Biodeterior. Biodegrad.* **1994**, *33*, 333–343. [CrossRef]
10. Kožuharova, E.; Lebanova, H.; Getov, I.; Benbassat, N.; Kochmarov, V. *Ailanthus altissima* (Mill.) Swingle—A terrible invasive pest in Bulgaria or potential useful medicinal plant? *Bothalia J.* **2014**, *44*, 213–230.
11. International Plant Nutrition Institute. International Plant Names Index. Available online: http://www.ipni.org (accessed on 1 October 2021).
12. Kundu, P.; Laskar, S. A brief resume on the genus *Ailanthus*: Chemical and pharmacological aspects. *Phytochem. Rev.* **2010**, *9*, 379–412. [CrossRef]
13. Hu, S.Y. Ailanthus. *Arnoldia* **1979**, *39*, 29–50.
14. Castroviejo, S. (coord. gen.). In *Flora Iberica*; Real Jardín Botánico, Spanish National Research Council: Madrid, Spain, 1986–2012.
15. Bashir, A.; Mohi-ud-din, R.; Farooq, S.; Bhat, Z.A. Pharmacognostic Standardization and Phytochemical Evaluation of *Ailanthus altissima* (Mill.) Swingle leaves. *J. Drug Deliv. Ther.* **2019**, *9*, 179–183.
16. Pan, E.; Bassuk, N. Establishment and Distribution of *Ailanthus altissima* in the Urban Environment. *J. Environ. Hortic.* **1986**, *4*, 1–4. [CrossRef]
17. Arnaboldi, F.; Conedera, M.; Maspoli, G. Distribuzione e potenziale invasivo di *Ailanthus altissima* (Mill.) Swingle nel Ticino centrale. *Boll. Soc. Ticin. Sci. Nat.* **2002**, *90*, 93–101.
18. Tabacchi, E.; Planty-Tabacchi, A.M. Recent changes in Riparian vegetation: Possible consequences on dead wood processing along rivers. *River Res. Appl.* **2003**, *19*, 251–263. [CrossRef]
19. Facelli, J.M.; Pickett, S.T.A.; Facelli, J.M. Indirect Effects of Litter on Woody Seedlings Subject to Herb Competition. *Oikos* **1991**, *62*, 129. [CrossRef]
20. Facelli, J.M. Multiple indirect effects of plant litter affect the establishment of woody seedlings in old fields. *Ecology* **1994**, *75*, 1727–1735. [CrossRef]
21. Ranft, H.; Dässler, H.-G. Rauchhärtetest an Gehölzen im SO$_2$-Kabinenversuch: Fume-Hardiness Test Carried out on Woods in a SO$_2$-Cabin-Trial. *Flora* **1970**, *159*, 573–588. [CrossRef]
22. Li, X.; Li, Y.; Ma, S.; Zhao, Q.; Wu, J.; Duan, L.; Xie, Y.; Wang, S. Traditional uses, phytochemistry, and pharmacology of *Ailanthus altissima* (Mill.) Swingle bark: A comprehensive review. *J. Ethnopharmacol.* **2021**, *275*, 114121. [CrossRef]
23. Du, Y.Q.; Yan, Z.Y.; Chen, J.J.; Wang, X.B.; Huang, X.X.; Song, S.J. The identification of phenylpropanoids isolated from the root bark of *Ailanthus altissima* (Mill.) Swingle. *Nat. Prod. Res.* **2021**, *35*, 1139–1146. [CrossRef]
24. Yang, X.L.; Yuan, Y.L.; Zhang, D.M.; Li, F.; Ye, W.C. Shinjulactone O, a new quassinoid from the root bark of *Ailanthus altissima*. *Nat. Prod. Res.* **2014**, *28*, 1432–1437. [CrossRef]
25. Zhang, D.D.; Bai, M.; Yan, Z.Y.; Huang, X.X.; Song, S.J. Chemical constituents from *Ailanthus altissima* (Mill.) Swingle and chemotaxonomic significance. *Biochem. Syst. Ecol.* **2020**, *93*, 104174. [CrossRef]
26. Ni, J.C.; Shi, J.T.; Tan, Q.W.; Chen, Q.J. Two new compounds from the fruit of *Ailanthus altissima*. *Nat. Prod. Res.* **2019**, *33*, 101–107. [CrossRef] [PubMed]
27. Ni, J.C.; Shi, J.T.; Tan, Q.W.; Chen, Q.J. Phenylpropionamides, piperidine, and phenolic derivatives from the fruit of *Ailanthus altissima*. *Molecules* **2017**, *22*, 2107. [CrossRef] [PubMed]
28. Zhao, C.C.; Shao, J.H.; Li, X.; Xu, J.; Zhang, P. Antimicrobial constituents from fruits of *Ailanthus altissima* SWINGLE. *Arch. Pharm. Res.* **2005**, *28*, 1147–1151. [CrossRef]
29. Bory, G.; Clair-Maczulajtys, D. Composition du nectar et rôle des nectaires extrafloraux chez l'*Ailanthus glandulosa*. *Can. J. Bot.* **1986**, *64*, 247–253. [CrossRef]

30. Heisey, R.M.; Heisey, T.K. Herbicidal effects under field conditions of *Ailanthus altissima* bark extract, which contains ailanthone. *Plant Soil* **2003**, *256*, 85–99. [CrossRef]

31. De Feo, V.; De Martino, L.; Quaranta, E.; Pizza, C. Isolation of phytotoxic compounds from tree-of-heaven (*Ailanthus altissima* swingle). *J. Agric. Food Chem.* **2003**, *51*, 1177–1180. [CrossRef]

32. Tan, Q.W.; Ni, J.C.; Shi, J.T.; Zhu, J.X.; Chen, Q.J. Two novel quassinoid glycosides with antiviral activity from the samara of *Ailanthus altissima*. *Molecules* **2020**, *25*, 5679. [CrossRef]

33. Wang, Y.; Wang, W.J.; Su, C.; Zhang, D.M.; Xu, L.P.; He, R.R.; Wang, L.; Zhang, J.; Zhang, X.Q.; Ye, W.C. Cytotoxic quassinoids from *Ailanthus altissima*. *Bioorg. Med. Chem. Lett.* **2013**, *23*, 654–657. [CrossRef]

34. Okunade, A.L.; Bikoff, R.E.; Casper, S.J.; Oksman, A.; Goldberg, D.E.; Lewis, W.H. Antiplasmodial activity of extracts and quassinoids isolated from seedlings of *Ailanthus altissima* (Simaroubaceae). *Phyther. Res.* **2003**, *17*, 675–677. [CrossRef] [PubMed]

35. Fukamiya, N.; Lee, K.H.; Muhammad, I.; Murakami, C.; Okano, M.; Harvey, I.; Pelletier, J. Structure-activity relationships of quassinoids for eukaryotic protein synthesis. *Cancer Lett.* **2005**, *220*, 37–48. [CrossRef] [PubMed]

36. Al-Snafi, A.E. The pharmacological importance of *Antirrhinum majus*—A review. *Asian J. Pharm. Sci. Technol.* **2015**, *5*, 313–320.

37. Swingle, M. Analgesic, antipyretic and antiulcer activities of *Ailanthus altissima*. *Phytopharmacology* **2012**, *3*, 341–350.

38. Ghannoum, M.A.; Rice, L.B. Antifungal agents: Mode of action, mechanisms of resistance, and correlation of these mechanisms with bacterial resistance. *Clin. Microbiol. Rev.* **1999**, *12*, 501–517. [CrossRef] [PubMed]

39. Rahman, A.; Kim, E.L.; Kang, S.C. Antibacterial and antioxidant properties of *Ailanthus altissima* swingle leave extract to reduce foodborne pathogens and spoiling bacteria. *J. Food Saf.* **2009**, *29*, 499–510. [CrossRef]

40. Albouchi, F.; Hassen, I.; Casabianca, H.; Hosni, K. Phytochemicals, antioxidant, antimicrobial and phytotoxic activities of *Ailanthus altissima* (Mill.) Swingle leaves. *S. Afr. J. Bot.* **2013**, *87*, 164–174. [CrossRef]

41. Aissani, N.; Jabri, M.; Mabrouk, M.; Sebai, H. Antioxidant Potential and Antimicrobial Activity of *Ailanthus altissima* (Mill.) Swingle Extracts against *Pseudomonas aeruginosa*. *J. Mol. Biol. Biotech.* **2018**, *3*, 3.

42. Končić, M.Z.; Kremer, D.; Gruz, J.; Strnad, M.; Biševac, G.; Kosalec, I.; Šamec, D.; Piljac-Žegarac, J.; Karlović, K. Antioxidant and antimicrobial properties of *Moltkia petraea* (Tratt.) Griseb. flower, leaf and stem infusions. *Food Chem. Toxicol.* **2010**, *48*, 1537–1542. [CrossRef]

43. Chanwitheesuk, A.; Teerawutgulrag, A.; Kilburn, J.D.; Rakariyatham, N. Antimicrobial gallic acid from *Caesalpinia mimosoides* Lamk. *Food Chem.* **2007**, *100*, 1044–1048. [CrossRef]

44. Betts, J.W.; Kelly, S.M.; Haswell, S.J. Antibacterial effects of theaflavin and synergy with epicatechin against clinical isolates of *Acinetobacter baumannii* and *Stenotrophomonas maltophilia*. *Int. J. Antimicrob. Agents* **2011**, *38*, 421–425. [CrossRef]

45. Tian, F.; Li, B.; Ji, B.; Yang, J.; Zhang, G.; Chen, Y.; Luo, Y. Antioxidant and antimicrobial activities of consecutive extracts from *Galla chinensis*:The polarity affects the bioactivities. *Food Chem.* **2009**, *113*, 173–179. [CrossRef]

46. Hosni, K.; Hassen, I.; Sebei, H.; Casabianca, H. Secondary metabolites from *Chrysanthemum coronarium* (Garland) flowerheads: Chemical composition and biological activities. *Ind. Crop. Prod.* **2013**, *44*, 263–271. [CrossRef]

47. Polumbryk, M.; Ivanov, S.; Polumbryk, O. Antioxidants in food systems. Mechanism of action. *Ukr. J. Food Sci.* **2013**, *1*, 15–40.

48. Olszowy, M. What is responsible for antioxidant properties of polyphenolic compounds from plants? *Plant Physiol. Biochem.* **2019**, *144*, 135–143. [CrossRef] [PubMed]

49. Pisoschi, A.M.; Pop, A.; Cimpeanu, C.; Predoi, G. Antioxidant Capacity Determination in Plants and Plant-Derived Products: A Review. *Oxid. Med. Cell. Longev.* **2016**, *2016*, 9130976. [CrossRef] [PubMed]

50. Luís, Â.; Gil, N.; Amaral, M.E.; Domingues, F.; Duarte, A.P. *Ailanthus altissima* (Miller) Swingle: A source of bioactive compounds with antioxidant activity. *BioResources* **2012**, *7*, 2105–2120. [CrossRef]

51. Lungu, L.; Savoiu, M.R.; Manolescu, B.N.; Farcasanu, I.C.; Popa, C.V. Phytotoxic and Antioxidant activities of leaf extracts of *Ailanthus altissima* swingle. *Rev. Chim.* **2016**, *67*, 1928–1931.

52. EL Ayeb-Zakhama, A.; Chahdoura, H.; Ziani, B.E.C.; Snoussi, M.; Khemiss, M.; Flamini, G.; Harzallah-Skhiri, F. *Ailanthus altissima* (Miller) Swingle seed oil: Chromatographic characterization by GC-FID and HS-SPME-GC-MS, physicochemical parameters, and pharmacological bioactivities. *Environ. Sci. Pollut. Res.* **2019**, *26*, 14137–14147. [CrossRef]

53. Yeung, Y.; Aziz, F.; Guerrero-Castilla, A.; Arguelles, S. Signaling Pathways in Inflammation and Anti-inflammatory Therapies. *Curr. Pharm. Des.* **2018**, *24*, 1449–1484. [CrossRef]

54. Jin, M.H.; Yook, J.; Lee, E.; Lin, C.X.; Quan, Z.; Son, K.H.; Bae, K.H.; Kim, H.P.; Kang, S.S.; Chang, H.W. Anti-inflammatory Activity of *Ailanthus altissima* in Ovalbumin-Induced Lung Inflammation. *Biol. Pharm. Bull.* **2006**, *29*, 884–888. [CrossRef]

55. Cho, S.-K.; Jeong, M.; Jang, D.; Choi, J.-H. Anti-inflammatory Effects of Canthin-6-one Alkaloids from *Ailanthus altissima*. *Planta Med.* **2018**, *84*, 527–535. [CrossRef] [PubMed]

56. Kim, H.M.; Lee, J.S.; Sezirahiga, J.; Kwon, J.; Jeong, M.; Lee, D.; Choi, J.-H.; Jang, D.S. A New Canthinone-Type Alkaloid Isolated from *Ailanthus altissima* Swingle. *Molecules* **2016**, *21*, 642. [CrossRef]

57. Kang, T.-H.; Choi, I.-Y.; Kim, S.-J.; Moon, P.-D.; Seo, J.-U.; Kim, J.-J.; An, N.-H.; Kim, S.-H.; Kim, M.-H.; Um, J.-Y.; et al. *Ailanthus altissima* swingle has anti-anaphylactic effect and inhibits inflammatory cytokine expression via suppression of nuclear factor-kappaB activation. *In Vitro Cell. Dev. Biol. Anim.* **2010**, *46*, 72–81. [CrossRef] [PubMed]

58. Kim, S.R.; Park, Y.; Li, M.; Kim, Y.K.; Lee, S.; Son, S.Y.; Lee, S.; Lee, J.S.; Lee, C.H.; Park, H.H.; et al. Anti-inflammatory effect of *Ailanthus altissima* (Mill.) Swingle leaves in lipopolysaccharide-stimulated astrocytes. *J. Ethnopharmacol.* **2021**, 114258. [CrossRef] [PubMed]

59. Fiaschetti, G.; Grotzer, M.A.; Shalaby, T.; Castelletti, D.; Arcaro, A. Quassinoids: From traditional drugs to new cancer therapeutics. *Curr. Med. Chem.* **2011**, *18*, 316–328. [CrossRef] [PubMed]

60. Morré, D.J.; Grieco, P.A.; Morré, D.M. Mode of action of the anticancer quassinoids—Inhibition of the plasma membrane NADH oxidase. *Life Sci.* **1998**, *63*, 595–604. [CrossRef]

61. Corrêa, L.R.; Soares, G.L.G.; Fett-Neto, A.G. Allelopathic potential of *Psychotria leiocarpa*, a dominant understorey species of subtropical forests. *S. Afr. J. Bot.* **2008**, *74*, 583–590. [CrossRef]

62. Ahmed, H.M.; Yeh, J.-Y.; Tang, Y.-C.; Cheng, W.T.-K.; Ou, B.-R. Molecular screening of Chinese medicinal plants for progestogenic and anti-progestogenic activity. *J. Biosci.* **2014**, *39*, 453–461. [CrossRef]

63. Lawrence, J.; Colwell, A.; Sexton, O. The Ecological Impact of Allelopathy in *Ailanthus altissima* (Simaroubaceae). *Am. J. Bot.* **1991**, *78*, 948–958. [CrossRef]

64. Chen, Y.; Zhu, L.; Yang, X.; Wei, C.; Chen, C.; He, Y.; Ji, Z. Ailanthone induces G2/M cell cycle arrest and apoptosis of SGC-7901 human gastric cancer cells. *Mol. Med. Rep.* **2017**, *16*, 6821–6827. [CrossRef]

65. Ding, H.; Yu, X.; Hang, C.; Gao, K.; Lao, X.; Jia, Y.; Yan, Z. Ailanthone: A novel potential drug for treating human cancer. *Oncol. Lett.* **2020**, *20*, 1489–1503. [CrossRef] [PubMed]

66. Tang, S.; Ma, X.; Lu, J.; Zhang, Y.; Liu, M.; Wang, X. Preclinical toxicology and toxicokinetic evaluation of ailanthone, a natural product against castration-resistant prostate cancer, in mice. *Fitoterapia* **2019**, *136*, 104161. [CrossRef] [PubMed]

67. Tan, Q.W.; Ouyang, M.A.; Wu, Z.J. A new seco-neolignan glycoside from the root bark of *Ailanthus altissima*. *Nat. Prod. Res.* **2012**, *26*, 1375–1380. [CrossRef]

68. Demasi, S.; Caser, M.; Vanara, F.; Fogliatto, S.; Vidotto, F.; Negre, M.; Trotta, F.; Scariot, V. Ailanthone from *Ailanthus altissima* (Mill.) Swingle as potential natural herbicide. *Sci. Hortic.* **2019**, *257*, 108702. [CrossRef]

69. Caser, M.; Demasi, S.; Caldera, F.; Dhakar, N.K.; Trotta, F.; Scariot, V. Activity of *Ailanthus altissima* (Mill.) Swingle Extract as a Potential Bioherbicide for Sustainable Weed Management in Horticulture. *Agronomy* **2020**, *10*, 965. [CrossRef]

70. Baptista, P.; Costa, A.P.; Simões, R.; Amaral, M.E. *Ailanthus altissima*: An alternative fiber source for papermaking. *Ind. Crop. Prod.* **2014**, *52*, 32–37. [CrossRef]

71. Ferreira, P.J.T.; Gamelas, J.A.F.; Carvalho, M.G.V.S.; Duarte, G.V.; Canhoto, J.M.P.L.; Passas, R. Evaluation of the papermaking potential of *Ailanthus altissima*. *Ind. Crop. Prod.* **2013**, *42*, 538–542. [CrossRef]

72. He, Q.; Xiao, H.; Li, J.; Liu, Y.; Jia, M.; Wang, F.; Zhang, Y.; Wang, W.; Wang, S. Fingerprint analysis and pharmacological evaluation of *Ailanthus altissima*. *Int. J. Mol. Med.* **2018**, *41*, 3024–3032. [CrossRef]

73. Mastelić, J.; Jerković, I. Volatile constituents from the leaves of young and old *Ailanthus altissima* (Mill.) swingle tree. *Croat. Chem. Acta* **2002**, *75*, 189–197.

74. Pedersini, C.; Bergamin, M.; Aroulmoji, V.; Baldini, S.; Picchio, R.; Pesce, P.G.; Ballarin, L.; Murano, E. Herbicide activity of extracts from *Ailanthus altissima* (Simaroubaceae). *Nat. Prod. Commun.* **2011**, *6*, 593–596. [CrossRef]

75. Li, Y.; Zhao, M.; Zhang, Z. Quantitative proteomics reveals the antifungal effect of canthin-6-one isolated from *Ailanthus altissima* against *Fusarium oxysporum* f. sp. *cucumerinum* In Vitro. *PLoS ONE* **2021**, *16*, e0250712. [CrossRef] [PubMed]

76. Jiang, Y.; Ye, J.; Jiang, B. Characterization of volatile constituents from the flowers and leaves of *Ailanthus altissima* by dynamic head-space collection/GC-MS. *Asian J. Chem.* **2014**, *26*, 6737–6739. [CrossRef]

77. Caboni, P.; Ntalli, N.G.; Aissani, N.; Cavoski, I.; Angioni, A. Nematicidal activity of (E,E)-2,4-decadienal and (E)-2-decenal from *Ailanthus altissima* against *Meloidogyne javanica*. *J. Agric. Food Chem.* **2012**, *60*, 1146–1151. [CrossRef] [PubMed]

78. Vidović, M.; Morina, F.; Milić, S.; Jovanović, S.V. An improved HPLC-DAD method for simultaneously measuring phenolics in the leaves of *Tilia platyphyllos* and *Ailanthus altissima*. *Bot. Serbica* **2015**, *39*, 177–186.

79. Poljuha, D.; Sladonja, B.; Šola, I.; Dudaš, S.; Bilić, J.; Rusak, G.; Motlhatlego, K.E.; Eloff, J.N. Phenolic composition of leaf extracts of *Ailanthus altissima* (simaroubaceae) with antibacterial and antifungal activity equivalent to standard antibiotics. *Nat. Prod. Commun.* **2017**, *12*, 1609–1612. [CrossRef]

Review

Comparison of the Biological Potential and Chemical Composition of Brazilian and Mexican Propolis

Norma Patricia Silva-Beltrán [1], Marcelo Andrés Umsza-Guez [2,*,†], Daniela Méria Ramos Rodrigues [2], Juan Carlos Gálvez-Ruiz [3], Thiago Luiz de Paula Castro [2] and Ana Paola Balderrama-Carmona [4,*]

1   Departamento de Ciencias de la Salud, Universidad de Sonora, Campus Cajeme, Ejido Providencia, Ciudad Obregón C.P. 85010, Sonora, Mexico; norma.silva@unison.mx
2   Curso de Biotecnologia, Universidade Federal da Bahia, Av. Reitor Miguel Calmon, s/n-Canela, Salvador 40110-902, BA, Brazil; danmeria@gmail.com (D.M.R.R.); thiago.luiz@ufba.br (T.L.d.P.C.)
3   Departamento de Ciencias Químico Biológicas, Universidad de Sonora, Blvd. Luis Encinas y Rosales S/N, Colonia Centro, Hermosillo C.P. 83000, Sonora, Mexico; juan.galvez@unison.mx
4   Departamento de Ciencias Químico-Biológicas y Agropecuarias, Universidad de Sonora, Unidad Regional Sur, Blvd. Lazaro Cárdenas 100, Colonia Francisco Villa, Navojoa C.P. 85880, Sonora, Mexico
*   Correspondence: marcelo.umsza@ufba.br (M.A.U.-G.); paola.balderrama@unison.mx (A.P.B.-C.); Tel.: +55-713-283-8999 (M.A.U.-G.); +52-642-425-9969 (A.P.B.-C.)
†   Shared first authorship.

Citation: Silva-Beltrán, N.P.; Umsza-Guez, M.A.; Ramos Rodrigues, D.M.; Gálvez-Ruiz, J.C.; de Paula Castro, T.L.; Balderrama-Carmona, A.P. Comparison of the Biological Potential and Chemical Composition of Brazilian and Mexican Propolis. *Appl. Sci.* **2021**, *11*, 11417. https://doi.org/10.3390/app112311417

Academic Editor: Ana M. L. Seca

Received: 28 October 2021
Accepted: 23 November 2021
Published: 2 December 2021

Publisher's Note: MDPI stays neutral with regard to jurisdictional claims in published maps and institutional affiliations.

Abstract: Propolis is a resinous substance collected by bees from plants and its natural product is available as a safe therapeutic option easily administered orally and readily available as a natural supplement and functional food. In this work, we review the most recent scientific evidence involving propolis from two countries (Brazil and Mexico) located in different hemispheres and with varied biomes. Brazil has a scientifically well documented classification of different types of propolis. Although propolis from Brazil and Mexico present varied compositions, they share compounds with recognized biological activities in different extraction processes. Gram-negative bacteria growth is inhibited with lower concentrations of different types of propolis extracts, regardless of origin. Prominent biological activities against cancer cells and fungi were verified in the different types of extracts evaluated. Antiprotozoal activity needs to be further evaluated for propolis of both origins. Regarding the contamination of propolis (e.g., pesticides, toxic metals), few studies have been carried out. However, there is evidence of chemical contamination in propolis by anthropological action. Studies demonstrate the versatility of using propolis in its different forms (extracts, products, etc.), but several potential applications that might improve the value of Brazilian and Mexican propolis should still be investigated.

Keywords: propolis contamination; propolis cytotoxicity; antimicrobial activity; antiprotozoal activity; artepelin C; formononetin; pinocembrin; quercetin; kaempferol; propolis extract

1. Introduction

The etymology of the word "propolis" comes from the Greek "pro" (in defense of) and "polis" (city/community), which defines this word as a "natural product in defense of the community" [1]. Examples and evidence of ancient use of propolis ("black wax" or "balsam") for therapeutic purposes are found in biblical records [2]. Use by ancient civilizations aimed different purposes such as embalming antiputrefactive (Egyptians) and medical (Greek and Roman) and antipyretic agent (Incas) [3]. In the 17th century, propolis became popular in Europe for its antimicrobial activity, so London pharmacopeias listed it on the official drug list [2].

In 1908, the first scientific work [4] appeared on the chemical properties and composition of propolis, indexed in the Chemical Abstracts. In 1968, the summary of the first patent, from Romania, appeared in Chemical Abstracts, using propolis for the production of bath lotions (RO 48101) [5]. During wars (South Africa, Russia), propolis was widely used as a wound healing agent and also for tuberculosis treatment. In the 1980s, with the knowledge of its pharmacological properties, propolis began to be industrially incorporated into food and pharmaceutical products (topical applications) as a disease preventative [6].

Mexican propolis is a product that has been used since ancient times and has maintained its appeal status through the years due to its medicinal properties. Since pre-Hispanic times, beekeeping was as important as the cultivation of corn, considering honey as an essential food product for the Maya culture. In Puebla, Mexico, the Nahua community continues to market products such as sweets, eye drops, soap, shampoo, and other cosmetic products whose main ingredients are honey, pollen, and propolis [7]. However, the first scientific studies on Mexican propolis did not begin until the early 2000s. Scientific research is motivated by the great potential and positioning of propolis. In 2017, to establish the quality standards of Mexican propolis, the Secretary of Agriculture, Livestock, Rural Development, Fishing and Feeding established the Official Mexican Standard for the Production and Specifications of Propolis (NOM-003-SAG/GAN-2017-Propolis, production and specifications for its processing) [8].

In Brazil, the first scientific study on propolis was published in 1981 and is related to antibacterial capabilities. Regarding patents, the first deposit in Brazil was made in 1992 (Japanese patent—JP 92033270), while the first Brazilian patent was deposited in 1994 (BR 94042861). As of 2006, the number of patent deposits has increased; however, there are still few deposits made [9–11].

Brazil (8,516,000 km^2) and Mexico (1,973,000 km^2) are large countries in surface area and present different types of biomes (humid jungles, coastal jungle, deserts, mangroves, savannah, and others). In this diversity of ecosystems, there are different species of bees and, consequently, different types of propolis [12]. Both Mexico and Brazil have carried out researches on biological properties of propolis from different locations (Figure 1); however, contamination in Brazilian propolis by herbicides and heavy metals is also reported.

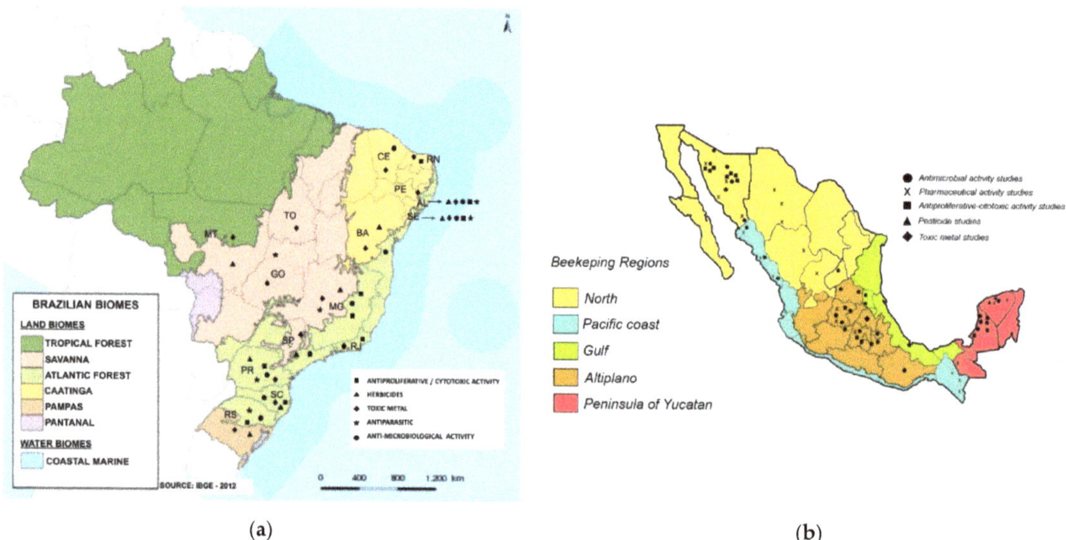

(a)   (b)

Figure 1. Maps with the biomes of (**a**) Brazil, beekeeping regions of (**b**) Mexico and the collection points of the different articles evaluated.

Due to the different beneficial properties attributed to propolis from different geographical origins, the scientific community has been researching on this matrix in different areas in the last decades. Between 2013 and 2017, over 2000 scientific articles related to the study of propolis were published, with investigations on Brazilian propolis corresponding to 16% among a total of 94 countries responsible for all the studies [13].

In this review, the scientific production on native propolis of two countries (Brazil and Mexico) is assessed considering different geographic locations and ecosystems and collecting relevant information on the antimicrobial and anticancer activities of propolis, as well as chemical composition and potential contaminants.

2. Methods

This work gave priority to relevant scientific articles (published from 2010) referring to research conducted using propolis from Brazil and Mexico. For the preparation of tables and figures, data (antimicrobial activity, cytotoxicity, contaminants) reported in the articles were extracted. Graphs were plotted to summarize propolis' inhibitory activities against microorganisms and cancer cell lines.

Table 1 was generated to classify the data obtained from the different articles validated in this study. For general antibacterial activity (Table 1), the species inhibited by different extracts of Brazilian and Mexican propolis were counted and displayed in 2 graphs, one for the Brazilian and another for Mexican propolis. Figures to show the distribution of Minimal Inhibitory Concentrations (MICs) of each extract were also made (Table 1). For that purpose, all MICs from 13 papers (for Brazilian and Mexican Propolis) were converted to μg/mL and added to a spreadsheet, and then divided in 5 intervals to facilitate visualization. Papers that did not determinate the MIC were excluded. Following, the frequency of a MIC in each interval was measured by counting every time a MIC was found in each interval. In Table 1, the frequency is shown, for example, as follows: the ethanolic extract of Brazilian green propolis had 4 MIC in the interval of 0–31.3, 9 in the 50–250, 5 in the 400–1600, 3 in the 3000–10,000, and none in the 29,000–300,000 μg/mL interval.

Table 1. Number of occurrences of inhibitory concentrations divided in intervals.

Distribution of Brazilian Propolis Minimal Inhibitory Concentrations					
Type of Propolis Extract	Concentration (μg/mL)				
	0–31.3	50–250	400–1600	3000–10,000	29,000–300,000
Ethanolic Extract of Green Propolis (EtGP)	4	9	5	3	0
Supercritical Extract of Green Propolis (ScGP)	0	2	8	0	0
Ethanolic Extract of Red Propolis (EtRP)	8	11	3	4	3
Supercritical Extract of Red Propolis (ScRP)	0	9	5	1	0
Ethanolic Extract of Brown Propolis (EtBP)	0	2	9	1	0
Supercritical Extract of Brown Propolis (ScBP)	0	0	9	0	0
Distribution of Mexican Propolis Minimal Inhibitory Concentrations					
	0–64	100–512	750–3750	7500–15,000	20,000–30,000
Aqueous Extract (Aq)	0	0	20	37	60
Ethanolic Extract (Et)	12	31	54	19	15
Methanolic Extract (Met)	10	15	0	0	0
Distribution of Brazilian and Mexican Propolis Inhibitory Concentration (IC$_{50}$)					
	1–15	15–60	60–100	100–550	
Brazilian Propolis	2	23	7	3	
Mexican Propolis	8	9	2	3	

The same methodology was used for the cytotoxic activity figures. Cell line and inhibitory concentration (IC$_{50}$) frequency graphs were made for both countries (Table 1). The different types of cell lines and the times they appeared in 16 papers of Brazilian propolis and in 5 papers of Mexican propolis were listed. Then, the cells were grouped by the human body system they belong to and 2 graphs were made for each country: one for cells grouped in systems and another for the cell lines. The IC$_{50}$ frequency graph followed the same methodology as the MIC frequency graph, except that it was not divided by the type of extract.

Classification of propolis (color, botanical sources, and chemical composition), extraction methods, antifungal, antiviral, antiprotozoal activity, veterinary application, and propolis contamination were discussed based on tables assembled or discussed directly in the text, based on information obtained from the articles analyzed.

3. Discussion

3.1. Propolis Composition

Propolis is an apicultural product obtained mainly from *Apis mellifera* bees that has compounds such as wax, resins, and minor constituents, including pollen and minerals [1]. One cannot fail to mention the contaminants commonly found in propolis, such as toxic metals and herbicides [14,15], among others. Salatino and Salatino [16] point out that the composition of propolis declared in many scientific articles for many years (resins and vegetable balsam 50%, bee wax 30%, pollen 5%, and essential and aromatic oils 10%) are generic (without details of the methodologies used to determine these parameters), considering that the composition varies mainly with the geographic location and ecosystem. Propolis samples from Italy [17], Brazil [18], Guinea-Bissau [19], Ethiopia [20], and Morocco [21] had different percentages in this composition. Propolis with wax content above 25% is rejected for commercial use [22].

3.2. Propolis Production in Mexico and Brazil

There is no official information on crude propolis trade. Mexico is among the primary honey producers worldwide and ranks the eighth in exports, being Germany its leading importer. However, Mexico is not among the exporting countries of propolis globally, and the leading exporters of this product are China, Brazil, Argentina, Cuba, Chile, Uruguay, and Canada [23]. Brazilian propolis is highly valued in the international market, and Brazil is one of the largest exporters of propolis globally. Brazilian propolis can reach a value up to three times higher (between 100 and 150 euros per kg) than Chinese propolis, mainly due to the differential in quality [24].

Minas Gerais (MG), Rio Grande do Sul (RS), Bahia (Ba), Santa Catarina (SC) and São Paulo (SP) are the Brazilian states that stand out for their production of propolis. The main market for Brazilian propolis is Asia, 92% of all fresh propolis consumed in Japan is from Brazil. European countries and the United States are also consumers of Brazilian propolis [9,14,25–27].

Unlike Brazil, Mexico does not have reliable statistics indicating propolis production, nor does it have updated patents on this beekeeping product. In all regions of Mexico, propolis production is potential, and the propolis producing regions are classified as follows: the Northern Region, Pacific Coast Region, Gulf Region, Altiplano Region, and Peninsula of Yucatán. Compared to the other regions, the Peninsula of Yucatan is more organized, producing 6 tons per year of propolis [28].

In order to encourage the beekeeping industry for the production and regulation of propolis, the Mexican Official Standard for the Production and Specifications of Propolis, NOM-003-SAG/GAN-2016 [8], was approved by the Federal Commission for Regulatory Improvement (COFEMER) and recently promoted, regulating the product's quality and its sale at competitive prices in the international market. In Brazil, regulation of propolis production is promoted by a specific Normative Instruction—SDA No. 03, of 19 January 2001 [22]—Technical Regulation of Identity and Quality of Apitoxin, Bee Wax, Royal Jelly, Lyophilized Royal Jelly, Bee Pollen, Propolis and Propolis Extract (https://www.gov.br/agricultura/pt-br (accessed on 23 October 2021). However, this regulation lacks many specifications regarding the production and quality of propolis.

3.3. Propolis: Color, Botanical Sources, and Chemical Composition

The coloration of propolis will depend on the plant's origin and its ecosystem since bees collect substances from plants. For example, the Brazilian green propolis is derived mainly from the species *Baccharis dracunculifolia*. On the other hand, in the Gulf Region of Mexico, the majority of propolis found is from the green variety, unlike the northern region, where the ecosystem is more desertic than the other regions, and mesquite *(Prosopis laevigata)* is a predominant vegetable source [29]. Each plant species has its secondary metabolites responsible for its biological activities, and propolis contains the metabolites of the plant that will define its biological activity. In Mexico, the most remarkable locations with diversity of tropical vegetation are in the Gulf and Pacific Coasts of the country; for this reason, a diversity of propolis colors is seen (chestnut-green, red, yellow-red, dark yellow, dun, or black) [28].

In Brazil, the classification of propolis by color is strongly associated with its botanical source [30]. Park et al. [31] classified propolis samples collected in the country (except the north region) into 12 groups, according to the appearance and color of the extracts. Subsequently, a new propolis variation was found in hives along the coast and mangroves in northeastern Brazil, which was classified as group 13, standing for red propolis [32].

In the southern region of Brazil, five types of propolis were described according to the classification of color (yellow, light brown, dark brown, greenish-brown, and reddish brown). In the northeast region, 6 were described (greenish-brown, dark brown, yellow, dark yellow, and red), and in the southeast region one type of propolis was described as green (or greenish-brown). It is noteworthy that these regions have diverse and varied ecosystems (Figure 1a), so although the name of the color may be similar, the composition of propolis is different mainly due to the different botanical sources. Propolis is a complex mixture, and more than 300 chemical moieties have been identified [33].

According to the verified articles (Tables S1 and S2 of supplementary material) [33], Brazilian and Mexican propolis composition analyses found many shared compounds, even considering different geographic regions, ecosystems, and types (color differences) [34–43]. Many of these chemical compounds stand out for their biological activity, for example, phenolic acids, phenolic acid esters, flavonoids, terpenoids, luteolin, galantine, trans ferulic acid, caffeic acid, chrysin, pinobanksin 5-methyl ether, quercetin, apigenin, kaempferol, naringenin, rutin, catechin, p-coumaric, pinobanksin, pinocembrin, and pinobanksin 3-acetate, and many others [21,25–32]. On the other hand, some compounds that are considered propolis-type markers, such as in propolis from Brazil, are artepillin C (3,5-diphenyl-4-hydroxycinnamic acid) for green propolis [44–47] and formononetin (3-hydroxy-8-9-dimethoxypterocarpan) for red propolis [32,48–50]. These chemical markers described for each propolis are phytologically associated with the main botanical source. The primary source of green propolis is wild rosemary (alecrim do campo—*Baccharis dracunculifolia*) and rabo-de-bugio or marmelo-do-mangue (*Dalbergia ecastaphyllum* (L.) Taub), with 35.68% of artepillin C [51], while the red propolis presents 67.59% formononetin in its chemical composition [49,50].

In the case of brown propolis, its composition varies according to several botanical sources (Eucalyptus, *Mimosa caesalpiniaefolia*, *Mimosa scabrella*, Cecropia, Anacardiaceae, Asteraceae, Citrus, Cocos and Poaceae, Populus), without a specific marker described [52,53].

In Mexico, among the botanical sources described for red propolis, in the Gulf, Pacific Coast, and Peninsula of Yucatan regions are *Bursera simaruba* (L.) Sarg, *Dalbergia glabra*, *Cordia alliodora*, *Cardiospermum halicacabum*, *Dombeya wallichii*, *Antigonon leptopus*, *Sapindus saponaria*, and *Dalbergia species* [54–56]. Valencia et al. [57] refer to brown-green-ocher (chestnut-green) propolis that differs in composition from the Brazilian green propolis, despite the similar color-based nomenclature. Distinctive chemical compounds are present due to botanical sources such as *Encelia farinosa*, *Ambrosia deltoidea*, *Ambrosia ambrosioides*, *Bursera laxiflora*, *Populus fremontii* S. *Prosopis laevigata*, and *Acacia greggii*. However, this propolis contains compounds also found in other types of propolis, such as gallic acid, cinnamic acid, p-coumaric acid, naringenin, quercetin, luteolin, kaempferol, apigenin, pinocembrin, pinobanksin 3-acetate, CAPE, chrysin, galangin, acacetin, and pinos-trobin [29].

3.4. Extraction Methods Commonly Used to Produce Propolis Extracts

Usually extractions are carried out with different solvents [58]. These extracts can be prepared by maceration of crude propolis with solvents. Solvents such as water [59], methanol [60], ethanol (50–80%) [26,34,61], chloroform, and ethyl acetate [60] are commonly used, with ethanol (70–80%) being the most used extraction method, whether in research or industrially [12]. In recent years, the process of extraction with supercritical fluid (SFE) has been used as an alternative to obtaining propolis extracts, usually applying as the solvent CO_2 (which has low cost, is easily available in high purity levels, non-toxic, non-flammable, and non-explosive) [62].

This process has high selectivity and reduced use of organic solvents, allowing the obtention of extracts with high biological value compared to conventional ethanolic extraction [34,35,39,63]. To improve the performance of SFE, many authors use co-solvents (methanol, ethanol) to intensify the extraction of compounds with a lower lipophilic nature in propolis, increasing the yield of the extractive process with CO_2 and obtention of relevant compounds [62,63]. Thus, the data reported on the evaluation of propolis extracts in various biological functions are generally associated with these extraction forms. Ultrasound-assisted extraction (UAE) is considered a green and economically viable alternative to conventional SC-CO_2 techniques for propolis extraction [64]. Promising results have been reported for successfully extracting bioactive constituents from propolis using microwave irradiation [65,66]. Commercially, most propolis extracts are ethanolic.

In most studies, SC-CO_2 extracts have been compared to ones obtained with ethanol using maceration and UAE. The lower polarity of SC-CO_2 expectedly resulted in the low content of bioactive compounds in extracts from different propolis types [34,35,67–69]. On the other hand, in experiments with propolis from different origins, some selectivity of SC-CO_2 to flavonoids [70,71], artepillin C, and p-coumaric acid from green Brazilian propolis [34,72] has been detected. SC-CO_2 extraction has been suggested as a pretreatment of propolis before a second classic extraction with ethanol, and this two-stage process resulted in an ethanol extract with high concentrations of active components [73,74].

While several scientific works applied mainly ethanol and supercritical extraction in Brazil, the works in Mexico are based on ethanol extractions, as seen in the evaluation of antimicrobial activity (Figures 2–5).

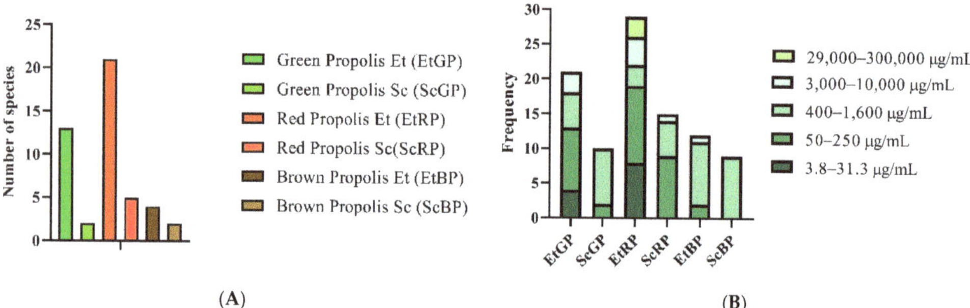

Figure 2. Antimicrobial activity of Brazilian propolis against different microorganisms. (**A**) Several microbial species inhibited by different propolis extracts: Ethanolic (Et) and Supercritical (Sc). (**B**) Distribution of the Minimal Inhibitory Concentrations (MICs) found in the 13 studies evaluated. Frequency: number of MICs found in that interval. EtGP, ScGP, EtRP, ScRP, EtBP and ScBP.

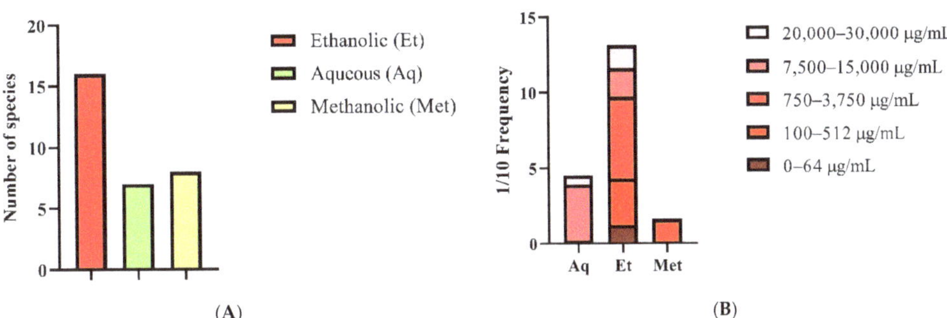

Figure 3. Antimicrobial activity of Mexican Propolis against different bacteria. (**A**) Some bacterial species are inhibited by different propolis extracts: Et, Aq, and Met. (**B**) Distribution of the minimal inhibitory concentrations (MICs) found in the 13 studies utilized. 1/10 Frequency: number of MICs found in that interval divided by 10.

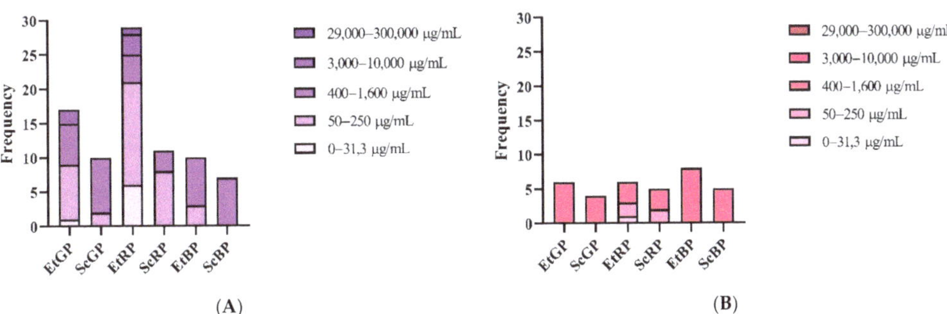

Figure 4. Antibacterial activity of Brazilian propolis against Gram-positive and negative bacteria. Distribution of the MICs found in the 13 studies utilized. (**A**) Gram-positive bacterial species (**B**) Gram-negative bacterial species. Frequency: number of times a MICs was found in that interval. EtGP, ScGP, EtRP, ScRP, EtBP, and ScBP. Gram-positive microorganism: *Staphylococcus* (spp. *aureus* ATCC 25923, 25913, 3359, *epidermidis* 25/4, 194/2), *Enterococcus* spp, *Enterococcus faecalis*, *Corynebacterium pseudotuberculosis*, *Eubacterium lentum*, *Peptostreptococcus anaerobius*, *Streptococcus* (*mutans*, *sobrinus*, *pyogenes* 93007, 75194, *sanguinis*, *salivarius*), *Actinomyces naeslundii*, *Lactobacillus casei*, *Paenibacillus larvae*. Gram-negative microorganism: *Escherichia coli*, *Klebsiella* spp., *Aggregatibacter actinomycetemcomitans*, *Fusobacterium* (*nucleatum*, *necrophorum*), *Porphyromonas gingivalis*, *Prevotella* (*intermedia*, *nigrescens*), *Pseudomonas aeruginosa* [26,34,35,37,59,60,75–80].

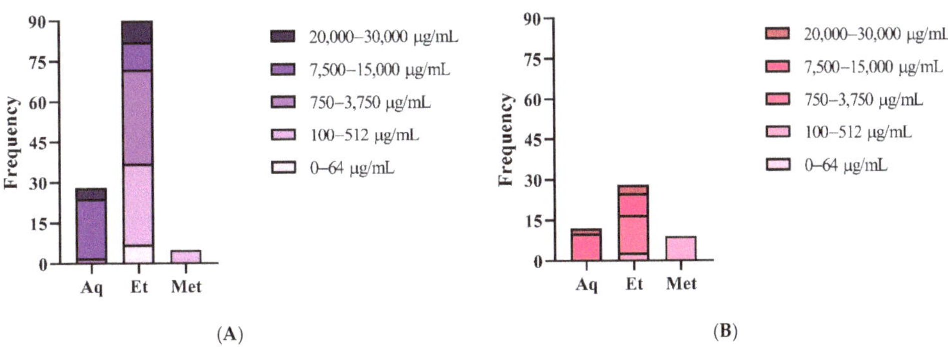

Figure 5. Antibacterial activity of Mexican Propolis against Gram-positive and negative bacteria. Distribution of the Minimal Inhibitory Concentrations (MICs) found in the 13 studies utilized. Frequency: number of MICs found in that interval.

(**A**) Gram-positive bacterial species (**B**) Gram-negative bacterial species. Et, Aq, and Met extract of propolis. Gram-positive microorganism: *Staphylococcus (aureus, epidermidis), Streptococcus (mutans, oralis, sanguinis, pyogenes, agalactiae), Listeria (monocytogenes, innocua), Mycobacterium tuberculosis, Salmonella typhi* [29,40,41,56,81–87]. Gram-negative microorganism: *Porphyromonas gingivalis, E. coli, Klebsiella pneumoniae, Vibrio (cholerae, alginolyticus, vulnificus), Pseudomonas aeruginosa* [81–86,88].

3.5. Antibacterial Activity

From the analysis of articles published in the last ten years, graphs were generated to show the number of bacterial species treated with different concentrations of extracts (Figures 2 and 3).

Figure 2A shows that among the 3 colors of Brazilian propolis represented, the red propolis had a broader antibacterial activity, affecting more species. This type of propolis was the most studied one in the last decade, followed by the green and brown propolis extracts, respectively. Red propolis was characterized in the first decade of the 2000s; thus, more related scientific works emerged. Brown propolis had a very low spectrum of species affected, and the ethanolic extract of green propolis had an intermediate range. All supercritical extracts had a low variability of species affected. Figure 2B shows the MICs grouped in 5 different intervals.

Regarding the distribution of the MICs, it was verified that the supercritical extract (Sc) of brown propolis inhibited microorganisms only in the 400–1600 μg/mL concentration range (Figure 2B). The ethanolic extract (Et) of green propolis had a good inhibitory effect in low concentrations, as more than half of occurrences was on the 3.8–250 μg/mL interval. The red propolis extracts, mainly the ethanol extracts, had a higher MIC spectrum (3.8 to 250 μg/mL), followed by the green and brown propolis extracts, although 2 thirds of occurrences were in the 3.8–250 μg/mL range. The Sc extract was not so effective in all the 3 colors of propolis.

Scientific work related to the Mexican propolis extracts is limited for alcoholic (methanolic and ethanolic) and aqueous extracts. The ethanol extract was the most used one to verify the antibacterial activity of propolis, reaching a total of 16 species (Figure 3A). The ethanol extract showed a higher MIC spectrum when compared to other types of extracts (Figure 3B). The methanolic extract had all its MIC on the 100–512 μg/mL range, and the aqueous extract was only effective in high concentrations (7500–30,000 μg/mL). In comparison with the Brazilian propolis, the ethanolic extract of Mexican propolis was also the most used one, but Mexican propolis was effective in higher concentrations.

Many variables interfere with the antibacterial capacity or potential of a propolis extract, mainly geographic region, flora, and the species of bee. Some studies indicated that the compounds found in propolis extract act on the cell wall of bacteria and the function of the ribosome-inhibiting protein synthesis and consequently bacterial growth (cell division), and disorganize the cytoplasmic membrane and cell wall, causing partial bacteriolysis (permeability alteration). The antibacterial mechanism is associated with chemical compounds of propolis already described in the literature as presenting biological activities, such as phenolic compounds/flavonoids (pinocembrin, galantine, caffeic acid, pinobanksin, artepillin C, ferulic acid, umbellic acid, p-coumaric acid, kaempferol, catechin, epicatechin, formononetin, isoformononetin, luteolin, naringenin, calycosin, quercetin, and others) [12,35,37,44,56,89,90].

In Figures 4 and 5, the antimicrobial effects of Brazilian and Mexican propolis in Gram-positive and -negative bacteria were compared according to the distribution of the minimal inhibitory concentrations (MICs). In scientific articles carried out with Brazilian propolis, it was observed that the Gram-negative bacteria were significantly affected by the action of different extracts applied at concentrations up to 1600 μg/mL. The Gram-positive bacteria show a greater variation in MICs, where supercritical extracts reach the concentration range of 400–1600 μg/mL. EtGP, EtRP, and ScRP extracts have a higher frequency of inhibition concentrations between the range of 50 to 250 μg/mL.

In the case of Mexican propolis extracts, it is verified that ethanol extracts have a greater concentration spectrum for inhibition of both classes of bacteria, and the concentration range from 0 to 3750 µg/mL presents a higher frequency of inhibition. Aqueous extracts are the least efficient in terms of MICs.

As expected, the extracts from both Brazilian and Mexican propolis samples showed higher activity against the Gram-positive strains than against the Gram-negative strains. These results following those from several authors, which can easily be explained by the structural differences of the bacterial cell wall [91–97]. Compared to the extraction method, EtOH extracts showed the best antimicrobial activities, and as previously shown, these extracts also had the best antioxidant activities and the highest content of total phenolic acids and flavonoids. Although the mechanism of action of these compounds in the antimicrobial function of propolis is poorly understood, some studies suggest that certain isolated constituent compounds have antimicrobial activity [98]. Among the more than 300 compounds already identified in different propolis, the following are examples. (1) Gallic acid and ferulic acid can cause irreversible changes in membrane properties and consequently the occurrence of local rupture or formation of pores in cell membranes, causing leakage of essential intracellular constituent substances. Furthermore, ferulic acid enhances the antibacterial activity of quinolone antibiotics against *A. baumannii* [99,100]. (2) Artepillin C has bacteriostatic activity with membrane blebbing [101]. (3) Cinnamic acid and its derivatives inhibit bacteria by division of the cell membrane, inhibiting ATPases, cell and biofilm formation [102]. (4) Catechins in vitro studies have demonstrated antimicrobial effects in bacteria (G+ and G−) and have been reported as effective antivirulence agents [103].

3.6. Antifungal Activity

The increase in mycosis and the appearance of resistant fungal strains in patients has been increasingly observed throughout the world. This has generated the search and development of new compounds that meet the requirements of an antifungal and in that sense, Mexico and Brazil have studied the antifungal properties of ethanolic extracts of propolis in different species of yeast [104] and filamentous fungi [105,106].

Table 2 shows these various studies and as can be seen, most studies have been restricted to the study of ethanolic extracts of propolis against *Candida albicans*. Likewise, when comparing the antifungal activity of propolis from both countries, it is observed that the extracts from Mexico require a lower concentration to achieve antifungal activity.

Table 2. Antifungal properties of propolis of different regions of Brazil and Mexico.

Microorganism	Color	Region of the Country	Extraction	MIC (µg/mL)	Reference
Fungi Mexico					
	ND	South	Ethanolic	32	[106]
Candida albicans	ND	Southeast	Ethanolic	1.6–2.30	[107]
	Brown	Southeast	Ethanolic	750	
		Center	Ethanolic	750–2400	[85]
Cryptococcus neoformans	ND	South	Ethanolic	32	[106]
Candida albicans (ATCC 14065) *Aspergillus flavus*	Green, Brow, Yellow	Southeast	Ethanolic	1.6–2.30	[108]
Aspergillus fumigatus	ND	Center	Ethanolic	32	[106]
Fungi Brazil					
	Red	Northeast	Ethanolic/Sc-CO$_2$	≥1000	[26]
Candida albicans	Red	Northeast	Ethanolic/Sc-CO$_2$	1.56	[76]
	Brown	Northeast	Ethanolic/Sc-CO$_2$	≥1000	[105]
Malassezia pachydermatis	Red	Northeast	Ethanolic/Sc-CO$_2$	4000–8000	[105]
	Green	Southeast	Ethanolic	4000–8000	
	Browns	South	Ethanolic	≥16,000	

ND. Not declared.

3.7. Antiviral Activity

Around the world, many types of research with antiviral activity (human immunodeficiency virus (HIV), adenovirus, HSV-1, HSV-2, Newcastle virus disease, bovine rotavirus, pseudorabies virus, canine adenovirus type 2, feline calicivirus, bovine viral diarrhea virus, influenza virus types A and B, parainfluenza virus, infectious bursal disease virus, and avian reovirus) have already been conducted [109–114].

Few scientific articles with the antiviral activity of propolis extracts have been verified. For example, González (102) studied the antiviral effect of Mexican propolis, when it has been evaluated different treatments with propolis, flavonoids individually, and a mixture of the three flavonoids that are normally present in propolis (quercetin, naringenin, and pinocembrin). They demonstrated that the best antiviral activity was the administration of propolis or the mixture of flavonoids than the individual compounds. On the other hand, Brazilian propolis has been verified in MS2 and Av-08 bacteriophages showing antiviral reduction ~3 and ~4.5 Log 10 PFU/mL respectively [115,116]. In addition, in the current COVID-19 pandemic, propolis from Brazil, specifically a standardized green propolis extract (EPP-AF®), has been evaluated in clinical treatment, using doses of 400–800 mg/day of the preparation, and a reduction in hospitalization time has been observed [117].

3.8. Antiprotozoan Activity

Some scientific articles have identified the effect of different concentrations of extracts (methanolic, ethanolic, and supercritical) of different propolis (brown, green, and red) against protozoa were identified. The extracts showed good inhibition against these organisms (*Trypanosome brucei*, *Trypanosoma cruzi* Y, *Leishmania* (V.) *braziliensis*, *Leishmania amazonenses*) particularly when using the green and red propolis extracts from Brazil [26,34,35,37]). Red propolis extract showed a better reduction of L. (V) *braziliensis* comparing with green propolis extract, corroborating the results that red propolis had more effective antioxidant activity [26,34,35,64,118].

For Mexican propolis, only one scientific report on the effect of propolis on protozoa was found, and a good inhibition of *Giardia lamblia* was verified.

3.9. Antiproliferative and Cytotoxic Activity

Propolis cytotoxic activity has been extensively studied, particularly in cancer research [119]. The chemical composition of propolis is well known and quite varied. Some compounds found in different types of propolis have been reported and associated with cytotoxic activity [120,121]. This cytotoxic and antiproliferative activity against tumor cells may not be due to isolated compounds nonetheless to the synergism between all the compounds in propolis [98]. Many of these compounds that are associated with these activities have previously been identified with antioxidant capabilities [122].

Figures showing the comparison of the cell lines used in researches carried out with Brazilian propolis are presented (Figure 6A,B). A greater variety of cell types was observed where the cytotoxicity of the extracts was tested, especially the cells of the reproductive system (≈40%). Among the articles verified, more than 80% of the cytotoxicity tests with the different cells were carried out with different types of extracts (ethanolic and supercritical) of red propolis, originating from the northeast region of Brazil (SE, AL, BA, RN). As mentioned before, in the last decade, studies with this type of propolis gained notoriety. The most used cell lines in the tests were HCT-116, OVCAR-8, SF-295, HL-60, PC3 [50,64,123–125].

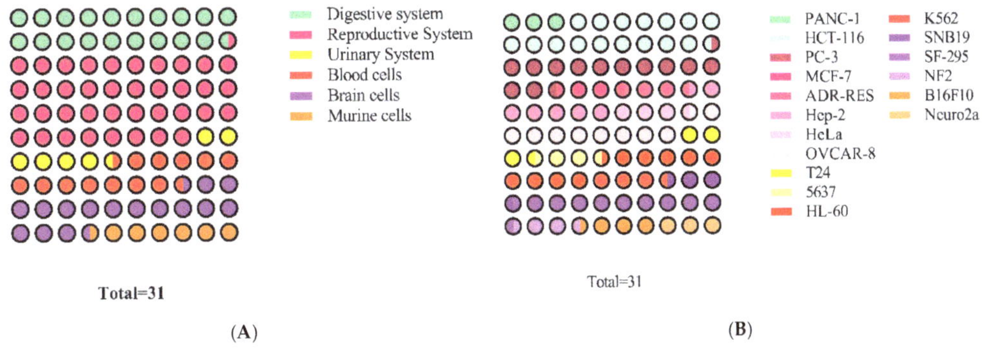

Figure 6. Variety of cell lines inhibited by Brazilian propolis. (**A**) The cell lines are divided into systems, and (**B**) into cell lines. All cell lines are shown in shades of the color of the system they belong. Sixteen papers were used. Brazilian propolis anticancer activity: 5637 (human bladder carcinoma), B16F10 (murine), PC3 (prostate adenocarcinoma), PANC-1, MCF-7, ADR-RES, Hep-2 (human laryngeal epidermoid carcinoma cells), HeLa (human cervical adenocarcinoma) cancer cell lines, HL-60 (leukemia), K562 (erythroleukemia), SNB19 (glioblastoma), SF-295 (glioblastoma-human), OVCAR-8 (breast), HCT-116 (colon), Neuro2a, NF2 (tumor in mice), skin carcinogenesis/papilloma, T24, proteinuria, inhibiting TNF-α, stimulating IL-10 production [26,50,64,76,123,125–136].

In studies related to the use of Mexican propolis extracts (Figure 7A,B) with cytotoxic activity, it appears that most of the research was carried out on murine cells (40%) and cells of the reproductive system (40%). The M12.C3.F6 murine cell line was the most used on tests.

Altiplano region propolis was its effectiveness proven compared with anti-cancer drugs and show an antiproliferative effect on glioma cells better than temozolomide, despite this proliferation and viability in cervical cancer cells (HeLa, SiHa, and CaSki) is lower than cisplatin [41]. The method used in the study was the MTT test, being this the most frequently used test to analyze the metabolic activity of a cell and evaluate its cytotoxic activity.

In the northern region of Mexico, murine B cell lymphoma (M12.C3.F6 cancer cell line) has been reported to evaluate the antiproliferative activity on cancer cells in Sonora desert propolis due to its high sensitivity to anti-cancer drugs [57,137,138] having excellent results. It has been proven that propolis and its components show proapoptotic activity inducing extrinsic and intrinsic pathways for cell apoptosis [119]. However, desert propolis samples showed a low antiproliferative activity on the murine normal cell line L-929, demonstrating that propolis shows a much lower antiproliferative effect in murine non-cancerous than in murine cancerous cell lines [137]. Another main result is that the antiproliferative effect is affected by quantitative fluctuations in its desert propolis polyphenolic profile due to its collection time [138,139]. Sonora desert propolis has been extensively studied by Li et al. [140,141]. Results showed more cytotoxicity against human lung adenocarcinoma (A549 cells) than with other five cancer cell lines, and PANC-1 human pancreatic cancer cells observed resolute preferential cytotoxic activity with similar results in propolis from Brazil [126]. The compounds with potent and preferential toxicity in PANC-1 is (6aR,11aR)-3,8-dihydroxy-9-methoxypterocarpan and (7″R)-8-[1-(4′-hydroxy-3′-methoxyphenyl)prop-2-en-1-yl]galangin in Brazilian and Mexican propolis, respectively.

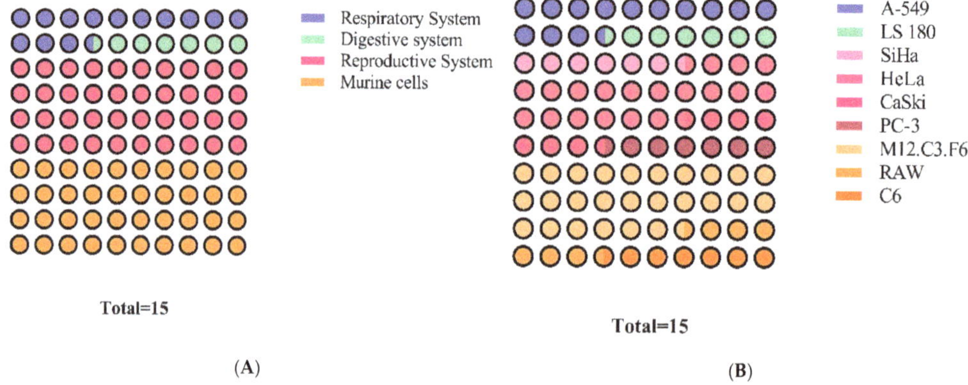

Figure 7. Variety of cell lines inhibited by Mexican propolis. (**A**) The cell lines are divided into systems, and (**B**) into cell lines. All cell lines are shown in shades of the color of the system they belong. Five papers were used. A-549 cell line, effect antiproliferative in LS 180 cell line Hela, M12.C3.F6, PC-3, RAW, C6, SiHa, CaSki, activity against cancer cell line M12.C3. F6 [41,57,137,138,142].

The antiproliferative effect of propolis extracts generally depends on their chemical composition, as described above. Chemical compounds present in various types of propolis, to which biological activities are related, could act as powerful cytocidal effects and induced levels of apoptosis in all the cellular lines [122,143–145].

The antitumor action of propolis (from the northern region, specifically from Sonora, Mexico) has been reported in several studies showing antiproliferative effects in humans (pancreatic cell lines, B-cell lymphoma, human lung carcinoma, human colon adenocarcinoma). Likewise, various studies have shown that propolis from southern Mexico has anti-inflammatory activities, inhibiting TNF alpha and stimulating interleukin 10 system [137,138,142].

In the case of studies carried out with different Brazilian propolis, 35 different inhibitory concentrations were found (in some cases, the same article tests several different cells and propolis) and 22 with Mexican propolis. Figure 8 shows that 63% of the IC50 studies were in the concentration range of 15–60 µg/mL, whereas for Mexican propolis this concentration range represented 41%. It is observed that Mexican propolis in the verified articles presented an inhibition range of 1 to 60 µg/mL for 77.3% of the tested samples, thus demonstrating a better IC50 for this type of studies.

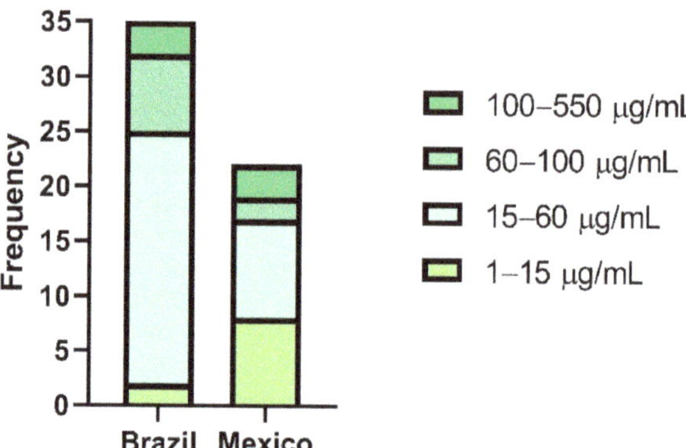

Figure 8. Anticancer activity of propolis from Brazil and Mexico. Distribution of the half-maximal Inhibitory Concentrations found in the 16 Brazilian propolis studies and the 5 Mexican propolis studies utilized. Frequency: number of times a IC50 was found in that interval.

3.10. Veterinary Application

With scientific proof of the positive effects of propolis associated with antimicrobial, anticancer activities, and others, its application is also successful in the veterinary field [146]. The application of propolis extracts is associated with producing a vehicle, usual formulations of topical ointments, and soap, which are mainly used with an antimicrobial function. Examples of the recent application of propolis in the veterinary field can be seen in Table 3.

Table 3. Veterinarian products with propolis extract.

Veterinary Diseases	Microorganisms	Product with Propolis Extract	Author
Effect canine dermatophytosis effect in dogs	*Microsporum gypseum* *Microsporum canis*	Soap	[147]
Treatment of Dermatomycosis in Horses	*Trichophyton mentagrophytes* *Candida albicans*	Shampoo	[148,149]
Post-surgical Treatment of Caseous Lymphadenitis in Sheep	*Corynebacterium pseudotuberculosis*	Ointment	[79]

4. Propolis Contamination

4.1. Xenobiotics in Propolis

Honeybees and their products are considered bioindicators of environmental pollution [150–152]. Xenobiotic substances, such as inorganic compounds, bactericides, fungicides, insecticides, herbicides are deposited on the soil when sprayed in the air, and can be transported to the hive through bee food or in other necessary compounds (water, nectar, pollen from trees and flowers, etc.) where they can remain for years [153]. Hence, propolis can be polluted directly or indirectly for pesticides and toxic metals of anthropogenic origin [154].

4.1.1. Pesticides in Brazilian and Mexican Propolis

Propolis can be polluted in direct or indirect form principally for pesticides and heavy metals of anthropogenic origin. Pesticides and heavy metals have been reported in many Hispanic and Portuguese countries.

Much of the bee production is found close to regions where conventional crops (soybeans, corn, wheat) that use herbicides regularly are found [155]. These crops can serve as a source of raw material for propolis production by bees, or they can be close to their range of action. According to Moreira et al. [156], a significant part of the agricultural production areas are close to the different biomes of each region. Due to the contamination of these productive areas and the volatilization of pesticides, such as contaminating precipitation.

Pesticides (mainly systemic herbicides) can be deposited on plants and trees. These plants can absorb toxic compounds through their leaves or roots and translocate them to other parts of the plant, such as buds [157–159], so bees will come into contact and can transfer these compounds to bee products, including propolis.

Studies of pollutants in Mexican and Brazilian propolis are very scarce. As shown in Table 4, there are two scientific works in each country regarding the verification of the presence of pesticides. Among the various pesticides investigated, only Organophosphate [160] was found in a propolis sample from southern Mexico. In a study carried out with propolis from Brazil [15], they verified the presence of AMPA and Atrazine in 32% of the samples evaluated.

Pesticides in Chilean raw propolis detected triadimefon on concentrations in mg/kg^{-1} until to 0.95 and Dicofol 0.54 and in Spanish propolis triadimefon 2.65, dicofol 1.2, dichlofluanid 0.38, procymidone 0.36, Folpet 3.74, and metazachlor 1.3 $mg\ kg^{-1}$ [161]. In processed propolis (capsules, tablets, tinctures, candies, and syrups) from Spain, Portugal, and Chile report pesticide in mg/kg of quintozene 1.06, procymidone 0.11, metazachlor 6.09, folpet 11.31, dichlofluanid 0.29, and chlorfenson 1.05 [162]. Other study found coumaphos in an

Argentinean propolis candy sample of 0.36 mgkg^{-1}, whereas chlorpyrifos was detected at 0.02 mg/kg in one Uruguayan sample [163].

Table 4. Pesticides research in Brazilian and Mexican propolis.

Mexican Propolis					
Number of Samples	**Geographical Origin**	**Elements Analyzed**	**Toxic Elements Found**	**Analytical Methods**	**References**
N/E	South	Dichlorvos, Diazinon, Methyl parathion, Malathion, and Coumaphos	No found	GC/MS-SIM	[164]
4	South	Dichlorvos, Diazinon, Methyl parathion, Malathion, and Coumaphos	Organophosphate pesticides ($\leq$025 µg/mL)	GC-MS	[160]
Brazilian Propolis					
19	Northeast (Ba, Se, Al); South (SC, Pr). Southeast (MG) and Centersouth (MT)	AMPA Atrazine Glyphosate Picloram	AMPA (10.2 $\pm$ 1.39–11.3 $\pm$ 2.65) Atrazine (9.7 $\pm$ 0.11–17.4 $\pm$ 2.60)	HPLC	[15]
50	South (SP)	Organochlorines, Organophosphates, Pyrethroids, Carbamates, Herbicides, Fungicides, and Acaricides any pesticide	ND	(GC/ECD), (GC/NPD)	[165]

ND. Not detected.

Generally, if there is a presence of toxic compounds in propolis, other bee products, mainly honey, can be contaminated. Valdominos–Flores et al. [166] present a complete pesticide analysis in honey and wax from Altiplano, Northern, and Peninsula of Yucatan in the highest concentrations were the phenyl phenol and the organochlorine 2,4′-DDT in the south; malathion, chlorpyrifos, phenyl phenol, and thiabendazole in the Northern Mexico; and chlorpyrifos and imidacloprid in Altiplano. These results demonstrate the pesticide exposure in the Mexican beehives. In Brazil in 1% and 44% of honey samples evaluated were found pesticides [167,168], respectively. Oliveira et al. [167] found pesticides in 7% of pollen samples analyzed.

Overall, environmental contamination by agrochemicals is being associated as a significant factor in the decline of insects, including bees and other pollinators [169–171]. Researchers relate to evidence of a decrease in the population of bees and even their mortality to exposure to pesticides [172–176].

4.1.2. Toxic Metals in Brazilian and Mexican Propolis

As described above, propolis is composed basically of a mixture of substances that can be contaminated by metals of different sources such as bees, air, water, plants, and soil [26–28]. Anthropogenic actions (industrial activities, mining, increased urbanization, fertilizers, and pesticides) or natural process (eruptions, leaching) also contribute to the contamination of the environment and consequently bees and their products [9,28,30].

In the survey of studies referring to toxic metals in samples of Brazilian propolis, six studies were found (Table 5). Lead and cadmium are the toxic metals found in most conducted researches. Other toxic compounds found were As, Cr, and Cd. Hodel et al. [14] found As, Cd, and Pb in 26.3%, 5.2%, and 73.9% of the 19 samples evaluated.

Table 5. Brazilian propolis toxic metals contamination.

Number of Samples	Toxic Elements Found	Analytical Methods	References
19	Pb, Cd, As, Cu	GFAAS and FAAS	[14]
106	Hg, Cd, Pb, Sn	FAAS	[177]
6	Pb, Cd, Cr	FAAS	[178]
42	Cd, Cr, Pb	FAAS	[179]

FAAS—flame atomic absorption spectrometry; GFAAS—graphite furnace atomic absorption spectrometry; AS–atomic absorption spectrophotometry.

Substances originating from plants as raw material for propolis production (resins) can be sources of contamination, as mentioned above. Researchers have shown that the primary botanical sources (*Baccharis dracunculifolia*) for the production of propolis can accumulate cadmium [180], turning this plant into a possible source of contamination. Other sources of this toxic metal can be agricultural inputs, such as fungicides, used in plantations in Santa Catarina (SC) and Rio Grande do Sul (RS) [31,34,181,182], where Cd was found in pollen samples.

Heavy metals in Argentinian raw propolis have been detected concentrations in ppm of Zn (15.2), Pb (7.73), and Cu (2.45) [183]. Although, in the research by González–Martín et al. [162], in Chilean and Spanish processed propolis have been reported heavy metals, such as, Cr (17.7), Ni (7.01), Cu (6.44), Zn (6.44), and Pb (7.21 ppm).

In Brazil, the maximum levels for metals in crude propolis are not defined in this regulation, but it states that inorganic contaminants, such as Cd, As, and Pb, should not be present in propolis in higher amounts as defined by RDC regulation number 42 by ANVISA for honey [184].

In Mexico, publications on heavy metals in propolis are limited to that of Montiel et al. [185], where the highest concentrations were reported in ppm in an agricultural area Pb (4.75), in urban area Cd (3.87), and in rural area Cr (4.74), all in the altiplano region of Mexico. Compared with the Brazilian propolis, these results are higher in Cd, but Cr and Pb present lower results [177,179]. The Mexican and Brazilian regulations do not specify permissible limits for heavy metals or pesticides. Notwithstanding, in Latin America, the Cuban, Salvadoran, and Argentine regulations specify that permissible limits for lead and arsenic are a maximum of 2 and 1 ppm, respectively. Both Mexican and Brazilian propolis have been current samples that exceeded the limits of Pb [186,187].

5. Conclusions

The differences in scientific production between the two countries regarding the same product are notorious. Both countries' geographic and ecosystem differences provide propolis with different chemical profiles and well studies in terms of chemical markers (in the case of Brazil). These chemical composition characteristics of propolis help in its commercial valorization. The biological potential was explored differently in terms of forms of extraction and application of the obtained extracts. In terms of antimicrobial activity, the number of species studied was higher with Brazilian propolis (all types and forms of extraction) and its cytotoxic activity, considering the variety of cells tested. The majority of the biological studies in both countries present findings very promising to combat important health problems, however, data in both countries indicate that more research is still needed in order to determine the optimal concentrations of propolis or its components, period of intake or type of extract to use before it can be administered to humans. In terms of contamination, information is still incipient, as propolis is considered an environmental bioindicator, so it needs to be better studied. New studies are needed in all areas to value propolis from a commercial and a scientific point of view.

Supplementary Materials: The following are available online at https://www.mdpi.com/article/10.3390/app112311417/s1, Table S1: Chemical composition of Mexican propolis, Table S2: Chemical composition of Brazilian propolis.

Author Contributions: Conceptualization, M.A.U.-G., A.P.B.-C. and N.P.S.-B.; methodology, M.A.U.-G., D.M.R.R., A.P.B.-C. and N.P.S.-B.; formal analysis, M.A.U.-G., A.P.B.-C. and N.P.S.-B.; original draft preparation and searching for literature, M.A.U.-G., D.M.R.R., A.P.B.-C. and N.P.S.-B.; data curation, M.A.U.-G., D.M.R.R., A.P.B.-C. and N.P.S.-B.; writing—original draft preparation, M.A.U.-G., A.P.B.-C. and N.P.S.-B.; writing—review and editing, M.A.U.-G., A.P.B.-C., T.L.d.P.C., J.C.G.-R. and N.P.S.-B.; visualization, M.A.U.-G., A.P.B.-C. and N.P.S.-B.; supervision, M.A.U.-G., A.P.B.-C. and N.P.S.-B.; project administration, M.A.U.-G., A.P.B.-C. and N.P.S.-B. All authors have read and agreed to the published version of the manuscript.

Funding: M.A.U.-G. is a Technological Development fellow from CNPq (Proc. 304747/2020-3). NPSB used Mixed-Fund Research project (USO313007158) and A.P.B.-C. was supported with Universidad de Sonora internal project funds (USO513007024).

Institutional Review Board Statement: Not applicable.

Informed Consent Statement: Not applicable.

Acknowledgments: The authors would like to thank Sonora University (Mexico) and Federal University of Bahía (Brazil).

Conflicts of Interest: The authors declare no conflict of interest.

References

1. Ghisalberti, L.E. Propolis: A review. *Bee World* **1979**, *60*, 59–84. [CrossRef]
2. Castaldo, S.; Capasso, F. Propolis, an old remedy used in modern medicine. *Fitoterapia* **2002**, *73*, S1–S6. [CrossRef]
3. Crane, E. *Bee Products*; Springer: Boston, MA, USA; Tel Aviv, Israel, 1996; ISBN 9780123741448.
4. Helfenberg, K.D. The analysis of beeswax and propolis. *Chem. Ztg.* **1908**, *2*, 987–988.
5. Sosnowski, Z.M. Method for Extracting Propolis and Water Soluble Dry Propols Powder. U.S. Patent 4,382,886, 10 May 1983.
6. Pereira, D.S.; Iberê, C.; Freitas, A.; Freitas, M.O.; Berg, J.; Agra, R. Histórico e principais usos da própolis apícola. *Agropecuária Clientífica No Semi-Árido* **2015**, *11*, 1–21.
7. Quezada-Euán, J.J.G. The Past, Present, and Future of Meliponiculture in Mexico. *Stingless Bees Mex.* **2018**, 243–269. [CrossRef]
8. Norma Mexicana. Propóleos, Producción y Especificaciones para su Procesamiento. SAGARPA, NOM-003-SAG/GAN-2017. Mexico. 2017. Available online: http://www.dof.gob.mx/normasOficiales/6794/sagarpa11_C/sagarpa11_C.html (accessed on 15 August 2021).
9. Machado, B.A.S.; Santos Cruz, L.; Baptista Nunes, S.; Andres Umsza Guez, M.; Ferreira Padilha, F. Estudo Prospectivo Da Própolis E Tecnologias Correlatas Sob O Enfoque Em Documentos De Patentes Depositados No Brasil. *Rev. Gestão Inovação E Tecnol.* **2012**, *2*, 221–235. [CrossRef]
10. Fraga, É.E.A.; De Oliveira, C.R.; Da Cruz, C.A.B.; De Vasconcelos, C.R.; Almeida Paixão, A.E. Prospecção tecnológica: Um mapeamento de patentes da própolis vermelha. *Cad. De Prospecção* **2017**, *10*, 524. [CrossRef]
11. Araujo Neto, E.; da Morais, L.S.; da Cunha, A.F.S. Mapeamento Tecnológico: Uma prospecção de patentes e trabalhos científicos à própolis verde. *Cad. De Prospecção* **2019**, *13*, 268–279. [CrossRef]
12. Trusheva, B.; Trunkova, D.; Bankova, V. Different extraction methods of biologically active components from propolis: A preliminary study. *Chem. Cent. J.* **2007**, *1*, 1–4. [CrossRef]
13. Katekhaye, S.; Fearnley, H.; Fearnley, J.; Paradkar, A. Gaps in propolis research: Challenges posed to commercialization and the need for an holistic approach. *J. Apic. Res.* **2019**, *58*, 604–616. [CrossRef]
14. Hodel, K.V.S.; MacHado, B.A.S.; Santos, N.R.; Costa, R.G.; Menezes-Filho, J.A.; Umsza-Guez, M.A. Metal Content of Nutritional and Toxic Value in Different Types of Brazilian Propolis. *Sci. World J.* **2020**, *2020*. [CrossRef]
15. Umsza-Guez, M.A.; Silva-Beltrán, N.P.; Machado, B.A.S.; Balderrama-Carmona, A.P. Herbicide determination in Brazilian propolis using high pressure liquid chromatography. *Int. J. Environ. Health Res.* **2021**, *31*, 507–517. [CrossRef]
16. Salatino, A.; Salatino, M.L.F. Scientific note: Often quoted, but not factual data about propolis composition. *Apidologie* **2021**, *52*, 312–314. [CrossRef]
17. Papotti, G.; Bertelli, D.; Bortolotti, L.; Plessi, M. Chemical and Functional Characterization of Italian Propolis Obtained by Different Harvesting Methods. *J. Agric. Food Chem.* **2012**, *60*, 2852–2862. [CrossRef]
18. Funari, C.S.; Ferro, V.O. Análise de própolis. *Food Sci. Technol.* **2006**, *26*, 171–178. [CrossRef]
19. Falcão, S.I.; Lopes, M.; Vilas-Boas, M. A First Approach to the Chemical Composition and Antioxidant Potential of Guinea-Bissau Propolis. *Nat. Prod. Commun.* **2019**, *14*. [CrossRef]
20. Diriba Jobir, M.; Belay, A. Comparative study of different Ethiopian propolis: In vivo wound healing, antioxidant, antibacterial, physicochemical properties and mineral profiles. *J. Apitherapy* **2020**, *7*. [CrossRef]
21. Popova, M.; Lyoussi, B.; Aazza, S.; Antunes, D.; Bankova, V.; Miguel, G. Antioxidant and α-Glucosidase Inhibitory Properties and Chemical Profiles of Moroccan Propolis. *Nat. Prod. Commun.* **2015**, *10*, 1961–1964. [CrossRef]

22. Brasil 2001. Brasil. Instrução Normativa n.3, de 19 de Janeiro de 2001. Ministério da Agricultura, Pecuária e Abastecimento Aprova os Regulamentos Técnicos de Identidade e Qualidade de Apitoxina, cera de Abelha, Geléia Real, Geléia Real Liofilizada, Pólen Apícola, Pró. 2001. Available online: https://www.apacame.org.br/mensagemdoce/115/artigo3.htm (accessed on 19 February 2021).
23. Servicio de Información Agroalimentaria y Pesquera (SIAP). Alemania Demanda el 50% de las Exportaciones Mexicanas de miel. Available online: https://www.gob.mx/siap/articulos/mexico-pais-exportador-de-miel (accessed on 15 August 2021).
24. Martinez, O.A.; Soares, A.E.E. Genetic improvement in the commercial beekeeping in production of propolis. *Rev. Bras. De Saude E Prod. Anim.* **2012**, *13*, 982–990. [CrossRef]
25. Barros, K.B.N.T.; Neto, E.M.R.; de Fonteles, M.M.F. Propolis and its Cosmetic Applications: A Technological Prospection. *J. Young Pharm.* **2019**, *11*, 350–352. [CrossRef]
26. Silva, R.P.D.; Machado, B.A.S.; Costa, S.S.; Barreto, G.A.; Padilha, F.F.; Umsza-Guez, M.A. Application of propolis extract in food products: A prospecting based in patent documents. *Rev. Virtual Quim.* **2016**, *8*, 1251–1261. [CrossRef]
27. Vidal, M.D.F. Potencial da produção de própolis no nordeste. *Escritório Técnico Estud. Econômicos Nordeste-ETENE* **2021**, *153*, 1–9.
28. México. *Atlas Nacional de las Abejas y Derivados Apícolas, Instituto Nacional de Geografía y Estadística*; Instituto Nacional de Estadística y Geografía (INEGI): Aguascalientes, Mexico, 2020. Available online: https://atlasnacionaldelasabejasmx.github.io/atlas/index.html (accessed on 15 August 2021).
29. Vargas Sánchez, R.D.; Martínez Benavidez, E.; Hernández, J.; Torrescano Urrutia, G.R.; Sánchez Escalante, A. Effect of physico-chemical properties and phenolic compounds of bifloral propolis on antioxidant and antimicrobial capacity. *Nova Sci.* **2020**, *12*. [CrossRef]
30. Teixeira, É.W.; Message, D.; Meira, R.M.S.A.; Salantino, A. Indicadores Da Origem Botânica Da Própolis Importância E Perspectivas. *B Indústr.Anim.* **2003**, *60*, 83–106.
31. Park, Y.; Masaharu, I.; Alencar, S.M. de Classification of Brazilian propolis by both physicochemical methods and biological activity. *Mensagem Doce* **2000**, *58*.
32. Daugsch, A.; Moraes, C.S.; Fort, P.; Park, Y.K. Brazilian red propolis-Chemical composition and botanical origin. *Evid.-Based Complementary Altern. Med.* **2008**, *5*, 435–441. [CrossRef] [PubMed]
33. Huang, S.; Zhang, C.-P.; Wang, K.; Li, G.Q.; Hu, F.-L. Recent Advances in the Chemical Composition of Propolis. *Molecules* **2014**, *19*, 19610–19632. [CrossRef]
34. Machado, B.A.S.; Silva, R.P.D.; Barreto, G.D.A.; Costa, S.S.; Da Silva, D.F.; Brandão, H.N.; Da Rocha, J.L.C.; Dellagostin, O.A.; Henriques, J.A.P.; Umsza-Guez, M.A.; et al. Chemical composition and biological activity of extracts obtained by supercritical extraction and ethanolic extraction of brown, green and red propolis derived from different geographic regions in Brazil. *PLoS ONE* **2016**, *11*, e0145954. [CrossRef]
35. Devequi-Nunes, D.; Machado, B.A.S.; De Abreu Barreto, G.; Silva, J.R.; Da Silva, D.F.; Da Rocha, J.L.C.; Brandão, H.N.; Borges, V.M.; Umsza-Guez, M.A. Chemical characterization and biological activity of six different extracts of propolis through conventional methods and supercritical extraction. *PLoS ONE* **2018**, *13*, e0207676. [CrossRef] [PubMed]
36. Rufatto, L.C.; dos Santos, D.A.; Marinho, F.; Henriques, J.A.P.; Roesch Ely, M.; Moura, S. Red propolis: Chemical composition and pharmacological activity. *Asian Pac. J. Trop. Biomed.* **2017**, *7*, 591–598. [CrossRef]
37. do Nascimento, T.G.; dos Santos Arruda, R.E.; da Cruz Almeida, E.T.; dos Santos Oliveira, J.M.; Basílio-Júnior, I.D.; Celerino de Moraes Porto, I.C.; Rodrigues Sabino, A.; Tonholo, J.; Gray, A.; Ebel, R.A.E.; et al. Comprehensive multivariate correlations between climatic effect, metabolite-profile, antioxidant capacity and antibacterial activity of Brazilian red propolis metabolites during seasonal study. *Sci. Rep.* **2019**, *9*, 1–16. [CrossRef] [PubMed]
38. Coelho, J.; Falcão, S.I.; Vale, N.; Almeida-Muradian, L.B.; Vilas-Boas, M. Phenolic composition and antioxidant activity assessment of southeastern and south Brazilian propolis. *J. Apic. Res.* **2017**, *56*, 21–31. [CrossRef]
39. Fausto Rivero-Cruz, J.; Rodríguez de San Miguel, E.; Robles-Obregón, S.; Hernández-Espino, C.C.; Rivero-Cruz, B.E.; Pedraza-Chaverri, J.; Esturau-Escofet, N. Prediction of antimicrobial and antioxidant activities of Mexican propolis by 1H-NMR spectroscopy and chemometrics data analysis. *Molecules* **2017**, *22*, 1184. [CrossRef]
40. Rivero-Cruz, J.F.; Granados-Pineda, J.; Pedraza-Chaverri, J.; Pérez-Rojas, J.M.; Kumar-Passari, A.; Diaz-Ruiz, G.; Rivero-Cruz, B.E. Phytochemical Constituents, Antioxidant, Cytotoxic, and Antimicrobial Activities of the Ethanolic Extract of Mexican Brown Propolis. *Antioxidant (Basel Switz.)* **2020**, *9*, 70. [CrossRef] [PubMed]
41. Rivera-Yañez, N.; Rodriguez-Canales, M.; Nieto-Yañez, O.; Jimenez-Estrada, M.; Ibarra-Barajas, M.; Canales-Martinez, M.M.; Rodriguez-Monroy, M.A. Hypoglycaemic and Antioxidant Effects of Propolis of Chihuahua in a Model of Experimental Diabetes. *Evid.-Based Complementary Altern. Med.* **2018**, *2018*. [CrossRef]
42. Li, F.; He, Y.M.; Awale, S.; Kadota, S.; Tezuka, Y. Two new cytotoxic phenylallylflavanones from mexican propolis. *Chem. Pharm. Bull.* **2011**, *59*, 1194–1196. [CrossRef]
43. Park, Y.K.; Paredes-Guzman, J.F.; Aguiar, C.L.; Alencar, S.M.; Fujiwara, F.Y. Chemical Constituents in Baccharis dracunculifolia as the Main Botanical Origin of Southeastern Brazilian Propolis. *J. Agric. Food Chem.* **2004**, *52*, 1100–1103. [CrossRef]
44. Lugo, E.; Martinez-Robinson, K.; Morales Ramírez, G.; De la Rosa López, R.; Noriega Rodríguez, J.A. Determinación del Origen Botánico de los Propóleos Sonorenses (CABORCA) Mediante HPLC Determination of Botanic Origen of Sonoran Propolis (CABORCA) by HPLC analysis. *Invurnus* **2009**, *4*, 24–30.

45. Riani, L.R.; Silva, L.M.; da Silva, O.O.Z.; Junqueira, L.R.; Nascimento, J.W.L.; da Filho, A.A.S. Development and validation of a uhplc-esi-ms/ms method for the quantification of artepillin c in brazilian green propolis. *Rev. Cienc. Farm. Basica E Appl.* **2021**, *42*. [CrossRef]

46. Guimarães, N.S.S.; Mello, J.C.; Paiva, J.S.; Bueno, P.C.P.; Berretta, A.A.; Torquato, R.J.; Nantes, I.L.; Rodrigues, T. Baccharis dracunculifolia, the main source of green propolis, exhibits potent antioxidant activity and prevents oxidative mitochondrial damage. *Food Chem. Toxicol.* **2012**, *50*, 1091–1097. [CrossRef]

47. Marcucci, M.C.; Bankova, V. Chemical composition, plant origin and biological activity of Brazilian propolis. *Curr. Top. Phytochem.* **1999**, *2*, 115–123.

48. Silva, B.B.; Rosalen, P.L.; Cury, J.A.; Ikegaki, M.; Souza, V.C.; Esteves, A.; Alencar, S.M. Chemical composition and botanical origin of red propolis, a new type of Brazilian propolis. *Evid.-Based Complementary Altern. Med.* **2008**, *5*, 313–316. [CrossRef]

49. de Mendonça, I.C.G.; de Porto, I.C.C.M.; do Nascimento, T.G.; de Souza, N.S.; dos Oliveira, J.M.S.; dos Arruda, R.E.S.; Mousinho, K.C.; dos Santos, A.F.; Basílio-Júnior, I.D.; Parolia, A.; et al. Brazilian red propolis: Phytochemical screening, antioxidant activity and effect against cancer cells. *BMC Complementary Altern. Med.* **2015**, *15*, 1–12. [CrossRef] [PubMed]

50. Xu, X.; Yang, B.; Wang, D.; Zhu, Y.; Miao, X.; Yang, W. The chemical composition of brazilian green propolis and its protective effects on mouse aortic endothelial cells against inflammatory injury. *Molecules* **2020**, *25*, 4612. [CrossRef]

51. Park, Y.; Alencar, S.; Aguiar, S. Botanical origin and chemical composition of Brazilian propolis. *J. Agric. Food Chem.* **2002**, *50*, 2502–2506. [CrossRef] [PubMed]

52. De da Freitas, A.S.; Barth, O.M.; da Luz, C.F.P. Própolis marrom da vertente atlântica do Estado do Rio de Janeiro, Brasil: Uma avaliação palinológica. *Rev. Bras. Botânica* **2010**, *33*, 343–354. [CrossRef]

53. Lotti, C.; Fernandez, M.C.; Piccinelli, A.L.; Cuesta-Rubio, O.; Hernández, I.M.; Rastrelli, L. Chemical Constituents of Red Mexican Propolis. *J. Agric. Food Chem.* **2010**, *58*, 2209–2213. [CrossRef] [PubMed]

54. Boisard, S.; Huynh, T.H.T.; Escalante-Erosa, F.; Hernández-Chavez, L.I.; Peña-Rodríguez, L.M.; Richomme, P. Unusual chemical composition of a Mexican propolis collected in Quintana Roo, Mexico. *J. Apic. Res.* **2016**, *54*, 350–357. [CrossRef]

55. Guzmán-Gutiérrez, S.L.; Nieto-Camacho, A.; Castillo-Arellano, J.I.; Huerta-Salazar, E.; Hernández-Pasteur, G.; Silva-Miranda, M.; ello-Nájera, O.A.; Sepúlveda-Robles, O.; Espitia, C.I.; Reyes-Chilpa, R. Mexican propolis: A source of antioxidants and anti-inflammatory compounds, and isolation of a novel chalcone and ε-caprolactone derivative. *Molecules* **2018**, *23*, 334. [CrossRef]

56. Valencia, D.; Alday, E.; Robles-Zepeda, R.; Garibay-Escobar, A.; Galvez-Ruiz, J.C.; Salas-Reyes, M.; Jiménez-Estrada, M.; Velazquez-Contreras, E.; Hernandez, J.; Velazquez, C. Seasonal effect on chemical composition and biological activities of Sonoran propolis. *Food Chem.* **2012**, *131*, 645–651. [CrossRef]

57. Bankova, V.; Trusheva, B.; Popova, M. Propolis extraction methods: A review. *J. Apic. Res.* **2021**, 1–10. [CrossRef]

58. Oliveira, A.; França, H.; Kuster, R.; Teixeira, L.; Rocha, L. Chemical composition and antibacterial activity of Brazilian propolis essential oil. *J. Venom. Anim. Toxins* **2010**, *16*, 121–130. [CrossRef]

59. Silva, C.S.R.; Villaça, C.L.P.B.; de Peixoto, R.M.; Mota, R.A.; de Ribeiro, M.F.; da Costa, M.M. Antibacterial effect of Brazilian brown propolis in different solvents against staphylococcus spp. isolated from caprine mastitis. *Cienc. Anim. Bras.* **2012**, *13*, 247–251. [CrossRef]

60. Sauri-Duch, E.; Gutiérrez-Canul, C.; Cuevas-Glory, L.F.; Ramón-Canul, L.; Pérez-Pacheco, E.; Moo-Huchin, V.M. Determination of quality characteristics, phenolic compounds and antioxidant activity of propolis from southeastern Mexico. *J. Apic. Sci.* **2021**, *65*, 109. [CrossRef]

61. Machado, B.A.S.; Pereira, C.G.; Nunes, S.B.; Padilha, F.F.; Umsza-Guez, M.A. Supercritical Fluid Extraction Using CO_2: Main Applications and Future Perspectives. *Sep. Sci. Technol.* **2013**, *48*, 2741–2760. [CrossRef]

62. Machado, B.A.S.; De Abreu Barreto, G.; Costa, A.S.; Costa, S.S.; Silva, R.P.D.; Da Silva, D.F.; Brandao, H.N.; Da Rocha, J.L.C.; Nunes, S.B.; Umsza-Guez, M.A.; et al. Determination of parameters for the supercritical extraction of antioxidant compounds from green propolis using carbon dioxide and ethanol as co-solvent. *PLoS ONE* **2015**, *10*, e0134489. [CrossRef]

63. De Oliveira Reis, J.H.; De Abreu Barreto, G.; Cerqueira, J.C.; Dos Anjos, J.P.; Andrade, L.N.; Padilha, F.F.; Druzian, J.I.; MacHado, B.A.S. Evaluation of the antioxidant profile and cytotoxic activity of red propolis extracts from different regions of northeastern Brazil obtained by conventional and ultrasoundassisted extraction. *PLoS ONE* **2019**, *14*, e0219063. [CrossRef]

64. Margeretha, I.; Fatma Suniarti, D.; Herda, E.; Mas'ud, Z.A. Optimization and comparative study of different extraction methods of biologically active components of Indonesian propolis Trigona spp. *J. Nat. Prod.* **2012**, *5*, 233–242.

65. Orsat, V.; Routray, W. Microwave-Assisted Extraction of Flavonoids. *Water Extr. Bioact. Compd. Plants Drug Dev.* **2017**, 221–244. [CrossRef]

66. Saito, É.; Sacoda, P.; Paviani, L.C.; Paula, J.T.; Cabral, F.A. Conventional and supercritical extraction of phenolic compounds from Brazilian red and green propolis. *Sep. Sci. Technol.* **2020**, 1–8. [CrossRef]

67. Xie, M.; Fan, D.; Zhao, Z.; Li, Z.; Li, G.; Chen, Y.; He, X.; Chen, A.; Li, J.; Lin, X.; et al. Nano-curcumin prepared via supercritical: Improved anti-bacterial, anti-oxidant and anti-cancer efficacy. *Int. J. Pharm.* **2015**, *496*, 732–740. [CrossRef] [PubMed]

68. Zhang, X.; Wu, J.Z.; Lin, Z.X.; Yuan, Q.J.; Li, Y.C.; Liang, J.L.; Zhan, J.Y.X.; Xie, Y.L.; Su, Z.R.; Liu, Y.H. Ameliorative effect of supercritical fluid extract of Chrysanthemum indicum Linnén against D-galactose induced brain and liver injury in senescent mice via suppression of oxidative stress, inflammation and apoptosis. *J. Ethnopharmacol.* **2019**, *234*, 44–56. [CrossRef] [PubMed]

69. You, G.S.G.S.; Lin, S.C.; Chen, C.R.; Tsai, W.C.; Chang, C.M.J.; Huang, W.W. Supercritical carbon dioxide extraction enhances flavonoids in water-soluble propolis. *J. Chin. Inst. Chem. Eng.* **2002**, *33*, 233–241.

70. Garcia-Mendoza, M.P.; Paula, J.T.; Paviani, L.C.; Cabral, F.A.; Martinez-Correa, H.A. Extracts from mango peel by-product obtained by supercritical CO_2 and pressurized solvent processes. *LWT-Food Sci. Technol.* **2015**, *62*, 131–137. [CrossRef]

71. Chen, C.R.; Shen, C.T.; Wu, J.J.; Yang, H.L.; Hsu, S.L.; Chang, C.M.J. Precipitation of sub-micron particles of 3,5-diprenyl-4-hydroxycinnamic acid in Brazilian propolis from supercritical carbon dioxide anti-solvent solutions. *J. Supercrit. Fluids* **2009**, *50*, 176–182. [CrossRef]

72. De Zordi, N.; Cortesi, A.; Kikic, I.; Moneghini, M.; Solinas, D.; Innocenti, G.; Portolan, A.; Baratto, G.; Dall'Acqua, S. The super-critical carbon dioxide extraction of polyphenols from Propolis: A central composite design approach. *J. Supercrit. Fluids* **2014**, *95*, 491–498. [CrossRef]

73. Monroy, Y.M.; Rodrigues, R.A.F.; Rodrigues, M.V.N.; Sant'Ana, A.S.; Silva, B.S.; Cabral, F.A. Brazilian green propolis extracts obtained by conventional processes and by processes at high pressure with supercritical carbon dioxide, ethanol and water. *J. Supercrit. Fluids* **2017**, *130*, 189–197. [CrossRef]

74. Bastos, E.M.A.F.; Simone, M.; Jorge, D.M.; Soares, A.E.E.; Spivak, M. In vitro study of the antimicrobial activity of Brazilian propolis against Paenibacillus larvae. *J. Invertebr. Pathol.* **2008**, *97*, 273–281. [CrossRef]

75. Botteon, C.E.A.; Silva, L.B.; Ccana-Ccapatinta, G.V.; Silva, T.S.; Ambrosio, S.R.; Veneziani, R.C.S.; Bastos, J.K.; Marcato, P.D. Biosynthesis and characterization of gold nanoparticles using Brazilian red propolis and evaluation of its antimicrobial and anticancer activities. *Sci. Rep.* **2021**, *11*, 1–16. [CrossRef]

76. Vasconcelos, W.A.; Braga, N.M.A.; Chitarra, V.R.; Santos, V.R.; Andrade, Â.L.; Domingues, R.Z. Bioactive Glass-Green and Red Propolis Association: Antimicrobial Activity Against Oral Pathogen Bacteria. *Nat. Prod. Chem. Res.* **2014**, *2*. [CrossRef]

77. Bueno-Silva, B.; Marsola, A.; Ikegaki, M.; Alencar, S.M.; Rosalen, P.L. The effect of seasons on Brazilian red propolis and its botanical source: Chemical composition and antibacterial activity. *Nat. Prod. Res.* **2017**, *31*, 1318–1324. [CrossRef]

78. Kalil, M.A.; Santos, L.M.; Barral, T.D.; Rodrigues, D.M.; Pereira, N.P.; da Sá, M.C.A.; Umsza-Guez, M.A.; Machado, B.A.S.; Meyer, R.; Portela, R.W. Brazilian Green Propolis as a Therapeutic Agent for the Post-surgical Treatment of Caseous Lymphadenitis in Sheep. *Front. Vet. Sci.* **2019**, *6*. [CrossRef]

79. Martins, M.L.; Monteiro, A.S.N.; Ferreira-Filho, J.C.C.; Vieira, T.I.; de Guimarães, M.B.C.T.; Farah, A.; Romanos, M.T.V.; Maia, L.C.; Cavalcanti, Y.W.; Fonseca-Gonçalves, A. Antibacterial and cytotoxic potential of a brazilian red propolis. *Pesqui. Bras. Em Odontopediatria E Clin. Integr.* **2019**, *19*, 1–9. [CrossRef]

80. Orozco, A.L.; Acevedo, J.G.A.; Martinez, M.M.C.; Delgado, C.T.H.; Serrano, P.R.; Ortiz, C.M.F.; Diaz, A.D.; Carrillo, J.G.P.; Tovar, C.G.G.; Sanchez, T.A.C. Antibacterial comparative study between extracts of mexican propolis and of three plants which use apis mellifera for its production. *J. Anim. Vet. Adv.* **2010**, *9*, 1250–1254. [CrossRef]

81. Carrillo, M.L.; Castillo, L.N.; Mauricio, R. Evaluación de la Actividad Antimicrobiana de Extractos de Propóleos de la Huasteca Potosina (México). *Inf. Tecnol.* **2011**, *22*, 21–28. [CrossRef]

82. Enciso-Díaz, O.J.; Méndez-Gutiérrez, A.; De Jesús, L.H.; Sharma, A.; Villarreal, M.L.; Taketa, A.C. Antibacterial Activity of *Bougainvillea Glabra, Eucalyptus Globulus, Gnaphalium Attenuatum*, and Propolis Collected in Mexico. *Pharmacol. Pharm.* **2012**, *03*, 433–438. [CrossRef]

83. Navarro-Navarro, M.; Lugo-Sepúlveda, R.E.; del García-Moraga, M.C.; de la Rosa-López, R.; Robles-Zepada, R.E.; Ruis-Bustos, E.; Veláquez-Contreras, C.A. Antibacterial and antioxidant activities of propolis methanolic extracts from Magdalena de Kino and Sonoyta, Sonora. *Rev. Cienc. Biológicas Salud* **2012**, *14*, 9–15.

84. Rodríguez Pérez, B.; Canales Martínez, M.M.; Penieres Carrillo, J.G.; Cruz Sánchez, T.A. Composición química, propiedades antioxidantes y actividad antimicrobiana de propóleos mexicanos. *Acta Univ.* **2020**, *30*, 1–30. [CrossRef]

85. Tolosa Cañizares, E.L. Obtención, caracterización y evaluación de la actividad antimicrobiana de extractos de propóleos de Campeche. *ARS Pharm.* **2002**, 187–204.

86. Velazquez, C.; Navarro, M.; Acosta, A.; Angulo, A.; Dominguez, Z.; Robles, R.; Robles-Zepeda, R.; Lugo, E.; Goycoolea, F.M.; Velazquez, E.F.; et al. Antibacterial and free-radical scavenging activities of Sonoran propolis. *J. Appl. Microbiol.* **2007**, *103*, 1747–1756. [CrossRef]

87. Navarro-Navarro, M.; Ruiz-Bustos, P.; Valencia, D.; Robles-Zepeda, R.; Ruiz-Bustos, E.; Virués, C.; Hernandez, J.; Domínguez, Z.; Velazquez, C. Antibacterial activity of sonoran propolis and some of its constituents against clinically significant vibrio species. *Foodborne Pathog. Dis.* **2013**, *10*, 150–158. [CrossRef] [PubMed]

88. Popova, M.; Giannopoulou, E.; Skalicka-Woźniak, K.; Graikou, K.; Widelski, J.; Bankova, V.; Kalofonos, H.; Sivolapenko, G.; Gaweł-Bęben, K.; Antosiewicz, B.; et al. Characterization and Biological Evaluation of Propolis from Poland. *Molecules* **2017**, *22*, 1159. [CrossRef]

89. Herrera-López, M.G.; Rubio-Hernández, E.I.; Leyte-Lugo, M.A.; Schinkovitz, A.; Richomme, P.; Calvo-Irabién, L.M.; Peña-Rodríguez, L.M. Botanical origin of triterpenoids from Yucatecan propolis. *Phytochem. Lett.* **2019**, *29*, 25–29. [CrossRef]

90. Koru, O.; Toksoy, F.; Acikel, C.H.; Tunca, Y.M.; Baysallar, M.; Uskudar Guclu, A.; Akca, E.; Ozkok Tuylu, A.; Sorkun, K.; Tanyuksel, M.; et al. In vitro antimicrobial activity of propolis samples from different geographical origins against certain oral pathogens. *Anaerobe* **2007**, *13*, 140–145. [CrossRef] [PubMed]

91. Vardar-Ünlü, G.; Silici, S.; Ünlü, M. Composition and in vitro antimicrobial activity of Populus buds and poplar-type propolis. *World J. Microbiol. Biotechnol.* **2008**, *24*, 1011–1017. [CrossRef]

92. Kim, Y.H.; Chung, H.J. The effects of Korean propolis against foodborne pathogens and transmission electron microscopic examination. *New Biotechnol.* **2011**, *28*, 713–718. [CrossRef] [PubMed]

93. Silva, J.C.; Rodrigues, S.; Feás, X.; Estevinho, L.M. Antimicrobial activity, phenolic profile and role in the inflammation of propolis. *Food Chem. Toxicol.* **2012**, *50*, 1790–1795. [CrossRef]
94. Silici, S.; Kutluca, S. Chemical composition and antibacterial activity of propolis collected by three different races of honeybees in the same region. *J. Ethnopharmacol.* **2005**, *99*, 69–73. [CrossRef]
95. Mohammadzadeh, S.; Shariatpanahi, M.; Hamedi, M.; Ahmadkhaniha, R.; Samadi, N.; Ostad, S.N. Chemical composition, oral toxicity and antimicrobial activity of Iranian propolis. *Food Chem.* **2007**, *103*, 1097–1103. [CrossRef]
96. Santos, L.M.; Rodrigues, D.M.; Kalil, M.A.; Azevedo, V.; Meyer, R.; Umsza-Guez, M.A.; Machado, B.A.; Seyffert, N.; Portela, R.W. Activity of Ethanolic and Supercritical Propolis Extracts in Corynebacterium pseudotuberculosis and Its Associated Biofilm. *Front. Vet. Sci.* **2021**, 965. [CrossRef] [PubMed]
97. Sawaya, A.C.H.F.; Cunha, I.B.S.; Marcucci, M.C.; De Oliveira Rodrigues, R.F.; Eberlin, M.N. Brazilian Propolis of Tetragonisca angustula and Apis mellifera 1. *Apidologie* **2006**, *37*, 398–407. [CrossRef]
98. Ibitoye, O.B.; Ajiboye, T.O. Ferulic acid potentiates the antibacterial activity of quinolone-based antibiotics against Acinetobacter baumannii. *Microb. Pathog.* **2019**, *126*, 393–398. [CrossRef] [PubMed]
99. Borges, A.; Ferreira, C.; Saavedra, M.J.; Simões, M. Antibacterial activity and mode of action of ferulic and gallic acids against pathogenic bacteria. *Microb. Drug Resist.* **2013**, *19*, 256–265. [CrossRef]
100. Yoshimasu, Y.; Ikeda, T.; Sakai, N.; Yagi, A.; Hirayama, S.; Morinaga, Y.; Furukawa, S.; Nakao, R. Rapid bactericidal action of propolis against Porphyromonas gingivalis. *J. Dent. Res.* **2018**, *97*, 928–936. [CrossRef]
101. Vasconcelos, N.G.; Croda, J.; Simionatto, S. Antibacterial mechanisms of cinnamon and its constituents: A review. *Microb. Pathog.* **2018**, *120*, 198–203. [CrossRef]
102. Wu, M.; Brown, A.C. Applications of catechins in the treatment of bacterial infections. *Pathogens* **2021**, *10*, 546. [CrossRef]
103. Bezerra, C.R.F.; Borges, K.R.A.; de Alves, R.N.S.; Teles, A.M.; Rodrigues, I.V.P.; da Silva, M.A.C.N.; do Nascimento, M.D.S.B.; de Bezerra, G.F.B. Highly efficient antibiofilm and antifungal activity of green propolis against Candida species in dentistry materials. *PLoS ONE* **2020**, *15*, e0228828. [CrossRef] [PubMed]
104. Deegan, K.R.; Fonseca, M.S.; Oliveira, D.C.P.; Santos, L.M.; Fernandez, C.C.; Hanna, S.A.; Machado, B.A.S.; Umsza-Guez, M.A.; Meyer, R.; Portela, R.W. Susceptibility of Malassezia pachydermatis Clinical Isolates to Allopathic Antifungals and Brazilian Red, Green, and Brown Propolis Extracts. *Front. Vet. Sci.* **2019**, *6*. [CrossRef]
105. Londoño Orozco, A.; Guillermo Penieres Carrillo, J.; Gerardo García Tovar, C.; Carrillo, L.M.; Leonor Quintero Mora, M.; Elvira García Vázquez, S.; Antonio Mendoza Saavedra, M.; natiuh Alejandro Cruz Sánchez, T.; Orozco, L.; Carrillo, P.; et al. Palabras clave Estudio de la actividad antifúngica de un extracto de propóleo de la abeja Apis mellifera proveniente del estado de México. *Rev. Tecnol. Marcha* **2008**, *21*, ág-49.
106. Ramón-Sierra, J.; Peraza-López, E.; Rodríguez-Borges, R.; Yam-Puc, A.; Madera-Santana, T.; Ortiz-Vázquez, E. Partial characterization of ethanolic extract of Melipona beecheii propolis and in vitro evaluation of its antifungal activity. *Rev. Bras. Farmacogn.* **2019**, *29*, 319–324. [CrossRef]
107. Rodríguez Pérez, B.; Penieres Carrillo, J.G.; Canales Martínez, M.M.; Luna Mora, R.A.; Cruz Sánchez, T.A. Potencialización de la actividad antifúngica de propóleos mexicanos con la adición de chalconas sintetizadas. *Biotecnia* **2019**, *21*, 76–85. [CrossRef]
108. Pascoal, A.; Feás, X.; Dias, T.; Dias, L.G.; Estevinho, L.M. The Role of Honey and Propolis in the Treatment of Infected Wounds. *Microbiol. Surg. Infect. Diagn. Progn. Treat.* **2014**, 221–234. [CrossRef]
109. Kurek-Górecka, A.; Górecki, M.; Rzepecka-Stojko, A.; Balwierz, R.; Stojko, J. Bee Products in Dermatology and Skin Care. *Molecules* **2020**, *25*, 556. [CrossRef]
110. González-Búrquez, M.D.J.; González-Díaz, F.R.; García-Tovar, C.G.; Carrillo-Miranda, L.; Soto-Zárate, C.I.; Canales-Martínez, M.M.; Penieres-Carrillo, J.G.; Crúz-Sánchez, T.A.; Fonseca-Coronado, S. Comparison between in Vitro Antiviral Effect of Mexican Propolis and Three Commercial Flavonoids against Canine Distemper Virus. *Evid.-Based Complementary Altern. Med.* **2018**, *2018*. [CrossRef] [PubMed]
111. Amoros, M.; Simõs, C.M.O.; Girre, L.; Sauvager, F.; Cormier, M. Synergistic Effect of Flavones and Flavonols Against Herpes Simplex Virus Type 1 in Cell Culture. Comparison with the Antiviral Activity of Propolis. *J. Nat. Prod.* **2004**, *55*, 1732–1740. [CrossRef]
112. Anjum, S.I.; Ullah, A.; Khan, K.A.; Attaullah, M.; Khan, H.; Ali, H.; Bashir, M.A.; Tahir, M.; Ansari, M.J.; Ghramh, H.A.; et al. Composition and functional properties of propolis (bee glue): A review. *Saudi J. Biol. Sci.* **2019**, *26*, 1695–1703. [CrossRef]
113. Bachevski, D.; Damevska, K.; Simeonovski, V.; Dimova, M. Back to the basics: Propolis and COVID-19. *Dermatol. Ther.* **2020**, *33*, e13780. [CrossRef] [PubMed]
114. Silva-Beltrán, N.P.; Balderrama-Carmona, A.P.; Umsza-Guez, M.A.; Souza Machado, B.A. Antiviral effects of Brazilian green and red propolis extracts on Enterovirus surrogates. *Environ. Sci. Pollut. Res.* **2020**, *27*, 28510–28517. [CrossRef]
115. Silveira, M.A.D.; De Jong, D.; Berretta, A.A.; dos Galvão, E.B.S.; Ribeiro, J.C.; Cerqueira-Silva, T.; Amorim, T.C.; da Conceição, L.F.M.R.; Gomes, M.M.D.; Teixeira, M.B.; et al. Efficacy of Brazilian green propolis (EPP-AF®) as an adjunct treatment for hospitalized COVID-19 patients: A randomized, controlled clinical trial. *Biomed. Pharmacother.* **2021**, *138*, 111526. [CrossRef]
116. Lobo-Galo, N.; Gálvez-Ruíz, J.C.; Balderrama-Carmona, A.P.; Silva-Beltrán, N.P.; Ruiz-Bustos, E. Recent biotechnological advances as potential intervention strategies against COVID-19. *3 Biotech* **2021**, *11*. [CrossRef]
117. Rebouças-Silva, J.; Celes, F.S.; Lima, J.B.; Barud, H.S.; De Oliveira, C.I.; Berretta, A.A.; Borges, V.M. Parasite Killing of Leishmania (V) braziliensis by Standardized Propolis Extracts. *Evid.-Based Complementary Altern. Med.* **2017**, *2017*. [CrossRef]

118. Forma, E.; Bryś, M. Anticancer Activity of Propolis and Its Compounds. *Nutrients* **2021**, *13*, 2594. [CrossRef] [PubMed]
119. Marcucci, M.C.; Rodriguez, J.; Ferreres, F.; Bankova, V.; Groto, R.; Popov, S. Chemical composition of Brazilian propolis from Sao Paulo State. *Z. Nat.-Sect. C J. Biosci.* **1998**, *53*, 117–119. [CrossRef]
120. Marcucci, M.C.; Ferreres, F.; García-Viguera, C.; Bankova, V.S.; De Castro, S.L.; Dantas, A.P.; Valente, P.H.; Paulino, N. Phenolic compounds from Brazilian propolis with pharmacological activities. *J. Ethnopharmacol.* **2001**, *74*, 105–112. [CrossRef]
121. Franchi, G.C.; Moraes, C.S.; Toreti, V.C.; Daugsch, A.; Nowill, A.E.; Park, Y.K. Comparison of effects of the ethanolic extracts of brazilian propolis on human leukemic cells as assessed with the MTT assay. *Evid.-Based Complementary Altern. Med.* **2012**, *2012*. [CrossRef] [PubMed]
122. de Carvalho, F.M.A.; Schneider, J.K.; de Jesus, C.V.F.; de Andrade, L.N.; Amaral, R.G.; David, J.M.; Krause, L.C.; Severino, P.; Soares, C.M.F.; Bastos, E.C.; et al. Brazilian red propolis: Extracts production, physicochemical characterization, and cytotoxicity profile for antitumor activity. *Biomolecules* **2020**, *10*, 726. [CrossRef] [PubMed]
123. Dantas Silva, R.P.; Machado, B.A.S.; de Barreto, G.A.; Costa, S.S.; Andrade, L.N.; Amaral, R.G.; Carvalho, A.A.; Padilha, F.F.; Barbosa, J.D.V.; Umsza-Guez, M.A. Antioxidant, antimicrobial, antiparasitic, and cytotoxic properties of various Brazilian propolis extracts. *PLoS ONE* **2017**, *12*, e0172585. [CrossRef]
124. Silva, F.R.G.; Matias, T.M.S.; Souza, L.I.O.; Matos-Rocha, T.J.; Fonseca, S.A.; Mousinho, K.C.; Santos, A.F. Phytochemical screening and in vitro antibacterial, antifungal, antioxidant and antitumor activities of the red propolis Alagoas. *Braz. J. Biol.* **2018**, *79*, 452–459. [CrossRef]
125. Awale, S.; Li, F.; Onozuka, H.; Esumi, H.; Tezuka, Y.; Kadota, S. Constituents of Brazilian red propolis and their preferential cytotoxic activity against human pancreatic PANC-1 cancer cell line in nutrient-deprived condition. *Bioorg. Med. Chem.* **2008**, *16*, 181–189. [CrossRef] [PubMed]
126. Ishiai, S.; Tahara, W.; Yamamoto, E.; Yamamoto, R.; Nagai, K. Histone deacetylase inhibitory effect of Brazilian propolis and its association with the antitumor effect in Neuro2a cells. *Food Sci. Nutr.* **2014**, *2*, 565–570. [CrossRef]
127. Bonamigo, T.; Campos, J.F.; Oliveira, A.S.; Torquato, H.F.V.; Balestieri, J.B.P.; Cardoso, C.A.L.; Paredes-Gamero, E.J.; de Souza, K.P.; Santos, E.L. dos Antioxidant and cytotoxic activity of propolis of Plebeia droryana and Apis mellifera (Hymenoptera, Apidae) from the Brazilian Cerrado biome. *PLoS ONE* **2017**, *12*, e0183983. [CrossRef] [PubMed]
128. Messerli, S.M.; Ahn, M.-R.; Kunimasa, K.; Yanagihara, M.; Tatefuji, T.; Hashimoto, K.; Mautner, V.; Uto, Y.; Hori, H.; Kumazawa, S.; et al. Artepillin C (ARC) in Brazilian green propolis selectively blocks oncogenic PAK1 signaling and suppresses the growth of NF tumors in mice. *Phytother. Res.* **2009**, *23*, 423–427. [CrossRef] [PubMed]
129. Yasukawa, K.; Yu, S.Y.; Tsutsumi, S.; Kurokawa, M.; Park, Y.K. Inhibitory effects of Brazilian propolis on tumor promotion in two-stage mouse skin carcinogenesis. *J. Pharm. Nutr. Sci.* **2012**, *2*, 71–76. [CrossRef]
130. Conti, B.J.; Santiago, K.B.; Búfalo, M.C.; Herrera, Y.F.; Alday, E.; Velazquez, C.; Hernandez, J.; Sforcin, J.M. Modulatory effects of propolis samples from Latin America (Brazil, Cuba and Mexico) on cytokine production by human monocytes. *J. Pharm. Pharmacol.* **2015**, *67*, 1431–1438. [CrossRef]
131. Silveira, M.A.D.; Teles, F.; Berretta, A.A.; Sanches, T.R.; Rodrigues, C.E.; Seguro, A.C.; Andrade, L. Effects of Brazilian green propolis on proteinuria and renal function in patients with chronic kidney disease: A randomized, double-blind, placebo-controlled trial. *BMC Nephrol.* **2019**, *20*, 1–12. [CrossRef]
132. Begnini, K.R.; Moura De Leon, P.M.; Thurow, H.; Schultze, E.; Campos, V.F.; Martins Rodrigues, F.; Borsuk, S.; Dellagostin, O.A.; Savegnago, L.; Roesch-Ely, M.; et al. Brazilian red propolis induces apoptosis-like cell death and decreases migration potential in bladder cancer cells. *Evid.-Based Complementary Altern. Med.* **2014**, *2014*. [CrossRef]
133. da Silva Frozza, C.O.; Garcia, C.S.C.; Gambato, G.; de Souza, M.D.O.; Salvador, M.; Moura, S.; Padilha, F.F.; Seixas, F.K.; Collares, T.; Borsuk, S.; et al. Chemical characterization, antioxidant and cytotoxic activities of Brazilian red propolis. *Food Chem. Toxicol.* **2013**, *52*, 137–142. [CrossRef] [PubMed]
134. da Frozza, C.O.S.; Santos, D.A.; Rufatto, L.C.; Minetto, L.; Scariot, F.J.; Echeverrigaray, S.; Pich, C.T.; Moura, S.; Padilha, F.F.; Borsuk, S.; et al. Antitumor activity of Brazilian red propolis fractions against Hep-2 cancer cell line. *Biomed. Pharmacother.* **2017**, *91*, 951–963. [CrossRef]
135. Costa, A.G.; Yoshida, N.C.; Garcez, W.S.; Perdomo, R.T.; de Matos, M.F.C.; Garcez, F.R. Metabolomics Approach Expands the Classification of Propolis Samples from Midwest Brazil. *J. Nat. Prod.* **2020**, *83*, 333–343. [CrossRef]
136. Hernandez, J.; Goycoolea, F.M.; Quintero, J.; Acosta, A.; Castañeda, M.; Dominguez, Z.; Robles, R.; Vazquez-Moreno, L.; Velazquez, E.F.; Astiazaran, H.; et al. Sonoran Propolis: Chemical Composition and Antiproliferative Activity on Cancer Cell Lines. *Planta Med.* **2007**, *73*, 1469–1474. [CrossRef] [PubMed]
137. Alday, E.; Valencia, D.; Garibay-Escobar, A.; Domínguez-Esquivel, Z.; Piccinelli, A.L.; Rastrelli, L.; Monribot-Villanueva, J.; Guerrero-Analco, J.A.; Robles-Zepeda, R.E.; Hernandez, J.; et al. Plant origin authentication of Sonoran Desert propolis: An antiproliferative propolis from a semi-arid region. *Sci. Nat.* **2019**, *106*. [CrossRef]
138. Mendez-Pfeiffer, P.; Alday, E.; Carreño, A.L.; Hernández-Tánori, J.; Montaño-Leyva, B.; Ortega-García, J.; Valdez, J.; Garibay-Escobar, A.; Hernandez, J.; Valencia, D.; et al. Seasonality modulates the cellular antioxidant activity and antiproliferative effect of sonoran desert propolis. *Antioxidants* **2020**, *9*, 1294. [CrossRef]
139. Li, F.; Awale, S.; Tezuka, Y.; Esumi, H.; Kadota, S. Study on the constituents of mexican propolis and their cytotoxic activity against PANC-1 human pancreatic cancer cells. *J. Nat. Prod.* **2010**, *73*, 623–627. [CrossRef] [PubMed]

140. Li, F.; Awale, S.; Tezuka, Y.; Kadota, S. Cytotoxic constituents from Brazilian red propolis and their structure-activity relationship. *Bioorg. Med. Chem.* **2008**, *16*, 5434–5440. [CrossRef]

141. Alday-Provencio, S.; Diaz, G.; Rascon, L.; Quintero, J.; Alday, E.; Robles-Zepeda, R.; Garibay-Escobar, A.; Astiazaran, H.; Hernandez, J.; Velazquez, C. Sonoran propolis and some of its chemical constituents inhibit in vitro growth of Giardia lamblia trophozoites. *Planta Med.* **2015**, *81*, 742–747. [CrossRef] [PubMed]

142. Kimoto, T.; Arai, S.; Kohguchi, M.; Aga, M.; Nomura, Y.; Micallef, M.; Kurimoto, M.; Mito, K. Apoptosis and suppression of tumor growth by artepillin C extracted from Brazilian propolis. *Cancer Detect. Prev.* **1998**, *22*, 506–515. [CrossRef]

143. Matsumoto, K.; Akao, Y.; Kobayashi, E.; Ito, T.; Ohguchi, K.; Tanaka, T.; Iinuma, M.; Nozawa, Y. Cytotoxic benzophenone derivatives from Garcinia species display a strong apoptosis-inducing effect against human leukemia cell lines. *Biol. Pharm. Bull.* **2003**, *26*, 569–571. [CrossRef]

144. Chen, W.; Sun, Y.; Lu, C.; Chao, C. Thermal cycling as a novel thermal therapy to synergistically enhance the anticancer effect of propolis on PANC-1 cells. *Int. J. Oncol.* **2019**, *55*, 617–628. [CrossRef]

145. Santos, L.M.; Fonseca, M.S.; Sokolonski, A.R.; Deegan, K.R.; Araújo, R.P.C.; Umsza-Guez, M.A.; Barbosa, J.D.V.; Portela, R.D.; Machado, B.A.S. Propolis: Types, composition, biological activities, and veterinary product patent prospecting. *J. Sci. Food Agric.* **2020**, *100*, 1369–1382. [CrossRef]

146. Estrada, S.T.A.; López, G.P.A.; Autran, Z.C.I.; Pérez, M.M.; Londoño, V. V Use of Propolis for Topical Treatment of Dermatophytosis in Dog. *Open J. Vet. Med.* **2014**, *4*, 239–245. [CrossRef]

147. del Rodríguez, I.S.F.; Monteagudo, M.M.; Orozco, A.L.; Sánchez, T.A.C. Use of Mexican Propolis for the Topical Treatment of Dermatomycosis in Horses. *Open J. Vet. Med.* **2016**, *06*, 1–8. [CrossRef]

148. Aguilar-Ayala, F.J.; Rejón-Peraza, M.E.; Cauich-Rodríguez, J.V.; Borges-Argáez, R.; Pinzón-Te, A.L.; González-Alam, C.J.; Aguilar-Perez, F.J. Biophysicochemical study of propolis and its clinical and radiographic assessment in dental pulpectomy. *Drug Invent. Today* **2019**, *12*, 2928–2933.

149. Walkwork-Baber, M.; Ferenbaugh, R.; Glandney, E. The use of honey bees as monitor of environment pollution. *Am. Bee J.* **1982**, *112*, 770–772.

150. Celli, G.; Maccagnani, B. Honey bees as bioindicators of environmental pollution. *Bull. Insectology* **2003**, *56*, 137–139. [CrossRef]

151. Borg, D.; Attard, E. Honeybees and theis products as bioindicators for heavy metal pollution in Malta. *Acta Bras.* **2020**, *4*, 60–69. [CrossRef]

152. Johnson, R.M. Honey bee toxicology. *Annu. Rev. Entomol.* **2015**, *60*, 415–434. [CrossRef]

153. Mizrahi, A.; Lensky, Y. *Bee Products*; Springer: Berlin/Heidelberg, Germany, 1997.

154. Navntoft, S.; Strandberg, B.; Nimgaard, R.; Esbjerg, P.; Axelsen, J. *Effects of Herbicide-Free Field Margins on Bumblebee and Butterfly Diversity in and along Hedgerows*; Environmental Protection Agency: Copenhagen, Denmark, 2011.

155. Moreira, J.C.; Peres, F.; Simões, A.C.; Pignati, W.A.; de Dores, E.C.; Vieira, S.N.; Strüssmann, C.; Mott, T. Contaminação de águas superficiais e de chuva por agrotóxicos em uma região do estado do Mato Grosso. *Ciência Saúde Coletiva* **2012**, *17*, 1557–1568. [CrossRef]

156. Kogan, W.R.; Kogam, M.; Parada, A.M. Phytotoxic activity of root absorbed glyphosate in corn seedlings (*Zea mays* L.). *Weed Biol. Manag.* **2003**, *3*, 228–232. [CrossRef]

157. Nandula, V.K.; Vencill, W.K. Herbicide Absorption and Translocation in Plants using Radioisotopes. *Weed Sci.* **2015**, *63*, 140–151. [CrossRef]

158. Brian, P.W. *How Herbicides Work*; Alberta Agriculture and Rural Development: Edmonton, AB, Canada, 1999; Volume 246, ISBN 0773261311.

159. Medina-Dzul, K.; Muñoz-Rodríguez, D.; Moguel-Ordoñez, Y.; Carrera-Figueiras, C. Application of mixed solvents for elution of organophosphate pesticides extracted from raw propolis by matrix solid-phase dispersion and analysis by GC-MS. *Chem. Pap.* **2014**, *68*, 1474–1481. [CrossRef]

160. González-Martín, M.I.; Revilla, I.; Vivar-Quintana, A.M.; Betances Salcedo, E.V. Pesticide residues in propolis from Spain and Chile. An approach using near infrared spectroscopy. *Talanta* **2017**, *165*, 533–539. [CrossRef] [PubMed]

161. González-Martín, M.I.; Revilla, I.; Betances-Salcedo, E.V.; Vivar-Quintana, A.M. Pesticide residues and heavy metals in commercially processed propolis. *Microchem. J.* **2018**, *143*, 423–429. [CrossRef]

162. Gérez, N.; Pérez-Parada, A.; Cesio, M.V.; Heinzen, H. Occurrence of pesticide residues in candies containing bee products. *Food Control* **2017**, *72*, 293–299. [CrossRef]

163. Acosta-Tejada, G.M.; Medina-Peralta, S.; Moguel-Ordóñez, Y.B.; Muñoz-Rodríguez, D. Matrix solid-phase dispersion extraction of organophosphoruspesticides from propolis extracts and recovery evaluationby GCM. *Anal. Bioanal. Chem.* **2011**, *400*, 885–891. [CrossRef]

164. de Orsi, R.O.; Barreto, L.M.R.C.; Gomes, S.M.A.; Kadri, S.M. Pesticidas na própolis do Estado de São Paulo, Brasil. *Acta Sci.-Anim. Sci.* **2012**, *34*, 433–436. [CrossRef]

165. Valdovinos-Flores, C.; Alcantar-Rosales, V.M.; Gaspar-Ramírez, O.; Saldaña-Loza, L.M.; Dorantes-Ugalde, J.A. Agricultural pesticide residues in honey and wax combs from Southeastern, Central and Northeastern Mexico. *J. Apic. Res.* **2017**, *56*, 667–679. [CrossRef]

166. Tette, P.A.S.; Da Silva Oliveira, F.A.; Pereira, E.N.C.; Silva, G.; De Abreu Glória, M.B.; Fernandes, C. Multiclass method for pesticides quantification in honey by means of modified QuEChERS and UHPLC-MS/MS. *Food Chem.* **2016**, *211*, 130–139. [CrossRef]

167. Orso, D.; Floriano, L.; Ribeiro, L.C.; Bandeira, N.M.G.; Prestes, O.D.; Zanella, R. Simultaneous Determination of Multiclass Pesticides and Antibiotics in Honey Samples Based on Ultra-High Performance Liquid Chromatography-Tandem Mass Spectrometry. *Food Anal. Methods* **2016**, *9*, 1638–1653. [CrossRef]

168. Woodcock, B.A.; Isaac, N.J.B.; Bullock, J.M.; Roy, D.B.; Garthwaite, D.G.; Crowe, A.; Pywell, R.F. Impacts of neonicotinoid use on long-term population changes in wild bees in England. *Nat. Commun.* **2016**. [CrossRef]

169. Sánchez-Bayo, F.; Wyckhuys, K.A.G. Worldwide decline of the entomofauna: A review of its drivers. *Biol. Conserv.* **2019**, *232*, 8–27. [CrossRef]

170. Grab, H.; Branstetter, M.G.; Amon, N.; Urban-Mead, K.R.; Park, M.G.; Gibbs, J.; Blitzer, E.J.; Poveda, K.; Loeb, G.; Danforth, B.N. Agriculturally dominated landscapes reduce bee phylogenetic diversity and pollination services. *Science* **2019**, *363*, 282–284. [CrossRef] [PubMed]

171. Martinello, M.; Manzinello, C.; Borin, A.; Avram, L.E.; Dainese, N.; Giuliato, I.; Gallina, A.; Mutinelli, F. A survey from 2015 to 2019 to investigate the occurrence of pesticide residues in dead honeybees and other matrices related to honeybee mortality incidents in Italy. *Diversity* **2020**, *12*, 15. [CrossRef]

172. Banks, J.E.; Banks, H.T.; Myers, N.; Laubmeier, A.N.; Bommarco, R. Lethal and sublethal effects of toxicants on bumble bee populations: A modelling approach. *Ecotoxicology* **2020**, *29*, 237–245. [CrossRef]

173. Naiara Gomes, I.; Ingred Castelan Vieira, K.; Moreira Gontijo, L.; Canto Resende, H. Honeybee survival and flight capacity are compromised by insecticides used for controlling melon pests in Brazil. *Ecotoxicology* **2020**, *29*, 97–107. [CrossRef] [PubMed]

174. Tomé, H.V.V.; Schmehl, D.R.; Wedde, A.E.; Godoy, R.S.M.; Ravaiano, S.V.; Guedes, R.N.C.; Martins, G.F.; Ellis, J.D. Frequently encountered pesticides can cause multiple disorders in developing worker honey bees. *Environ. Pollut.* **2020**, *256*. [CrossRef]

175. Sgolastra, F.; Medrzycki, P.; Bortolotti, L.; Maini, S.; Porrini, C.; Simon-Delso, N.; Bosch, J. Bees and pesticide regulation: Lessons from the neonicotinoid experience. *Biol. Conserv.* **2020**, *241*. [CrossRef]

176. Orsi, R.O.; Barros, D.C.B.; Silva, R.C.M.; Queiroz, J.V.; Araújo, W.L.P.; Shinohara, A.J. Toxic metals in the crude propolis and its transfer rate to the ethanolic extract. *Sociobiology* **2018**, *65*, 640–644. [CrossRef]

177. Schmidt, E.M.; Stock, D.; Chada, F.J.; Finger, D.; Christine Helena Frankland Sawaya, A.; Eberlin, M.N.; Felsner, M.L.; Quináia, S.P.; Monteiro, M.C.; Torres, Y.R. A comparison between characterization and biological properties of brazilian fresh and aged propolis. *BioMed Res. Int.* **2014**, *2014*. [CrossRef]

178. Finger, D.; Filho, I.K.; Torres, Y.R.; Quináia, S.P. Propolis as an indicator of environmental contamination by metals. *Bull. Environ. Contam. Toxicol.* **2014**, *92*, 259–264. [CrossRef]

179. Divan, A.M.; de Oliveira, P.L.; Perry, C.T.; Atz, V.L.; Azzarini-Rostirola, L.N.; Raya-Rodriguez, M.T. Using wild plant species as indicators for the accumulation of emissions from a thermal power plant, Candiota, South Brazil. *Ecol. Indic.* **2009**, *9*, 1156–1162. [CrossRef]

180. Zoffoli, H.J.O.; do Amaral-Sobrinho, N.M.B.; Zonta, E.; Luisi, M.V.; Marcon, G.; Tolón-Becerra, A. Inputs of heavy metals due to agrochemical use in tobacco fields in Brazil's Southern Region. *Environ. Monit. Assess.* **2012**, *185*, 2423–2437. [CrossRef]

181. Sattler, J.A.G.; Melo, A.A.M.; do Nacimento, K.S.; de Melo, I.L.P.; Mancini-Filho, J.; Sattler, A.; de Almeida-Muradian, L.B. Essential minerals and inorganic contaminants (barium, cadmium, lithium, lead and vanadium) in dried bee pollen produced in Rio Grande do Sul State, Brazil. *Food Sci. Technol.* **2016**, *36*, 505–509. [CrossRef]

182. Pierini, G.D.; Pistonesi, M.F.; Di Nezio, M.S.; Centurión, M.E. A pencil-lead bismuth film electrode and chemometric tools for simultaneous determination of heavy metals in propolis samples. *Microchem. J.* **2016**, *125*, 266–272. [CrossRef]

183. Brasil RESOLUÇÃO-RDC No 42, DE 29 DE AGOSTO DE. 2013. Available online: https://bvsms.saude.gov.br/bvs/saudelegis/anvisa/2013/rdc0042_29_08_2013.html (accessed on 15 August 2021).

184. Montiel, J.; Marmolejo, Y.; Castellanos, I.; Pérez, F.; Prieto, F.; Gaytán, J.; Fonseca, M. Niveles de cadmio, cromo y plomo en abejas (Apis mellifera) y sus productos en Hidalgo, México. *Rev. Iberoam. Cienc.* **2020**, *7*, 57–58.

185. Cuba 1994. Norma Cubana (1994) Propóleos Materia Prima. Especificaciones, NRAG-1135-94. Ministerio de Agricultura. 1994. Available online: https://www.minag.gob.cu/ (accessed on 9 October 2020).

186. El Salvador. Norma Salvadoreña (2003) Calidad del Propólero Crudo, Norma NSO 65.19.02:03. Diario Oficial.Tomo 360. San Salvador. 2003. Available online: https://www.diariooficial.gob.sv/ (accessed on 28 October 2020).

187. Argentina 2008. Instituto Argentino de Normalización y Certificación (2008) Productos del NOA (Noroeste Argentino). Propóleos. Parte 1-Propóleos en Bruto. Norma IRAM-INTA 15935-1. Available online: https://catalogo.iram.org.ar/#/normas/detalles/9314 (accessed on 19 February 2021).